T0258138

Encyclopedia of Ionizing Radiation Research

Volume II

Encyclopedia of Ionizing Radiation Research
Volume II

Edited by **Peggy Sparks**

New York

Published by NY Research Press,
23 West, 55th Street, Suite 816,
New York, NY 10019, USA
www.nyresearchpress.com

Encyclopedia of Ionizing Radiation Research
Volume II
Edited by Peggy Sparks

International Standard Book Number: 978-1-63238-142-2 (Hardback)

Contents

Preface

The world is advancing at a fast pace like never before. Therefore, the need is to keep up with the latest developments. This book was an idea that came to fruition when the specialists in the area realized the need to coordinate together and document essential themes in the subject. That's when I was requested to be the editor. Editing this book has been an honour as it brings together diverse authors researching on different streams of the field. The book collates essential materials contributed by veterans in the area which can be utilized by students and researchers alike.

X-rays were discovered by Roentgen in 1895, and ever since, the ionizing radiation has been largely used in various industrial as well as medical applications. However, its harmful characteristics are gradually recognized through accidental uses. The experience of nuclear power plant mishaps in Fukushima and Chernobyl, has unveiled the risk of ionizing radiation and its prevalence in our contemporary society. Therefore, it is extremely necessary that more engineers, students and scientists become known to ionizing radiation research regardless of the research field they are engaged in. On the basis of this idea, this book was formulated to analyze the current achievements in this field inclusive of medical uses, principles of radiation measurement, and its impact on health.

Each chapter is a sole-standing publication that reflects each author's interpretation. Thus, the book displays a multi-facetted picture of our current understanding of application, resources and aspects of the field. I would like to thank the contributors of this book and my family for their endless support.

Editor

Part 1

Health Effects

1

Ionizing Radiation Carcinogenesis

Otto G. Raabe
University of California Davis,
USA

1. Introduction

1.1 Ionizing radiation

Human exposure to ionizing radiation is a natural part of life on earth. These exposures occur every day from radiation associated with naturally occurring radionuclides in soil, air, and food, and also from cosmic rays. In addition, many people are exposed to dental and medical diagnostic procedures and therapy involving: x rays, gamma rays, charged particles, radionuclides, or other ionizing radiation sources. Others may be exposed in their workplace such as in laboratories, hospitals, underground mines, or nuclear power plants. Excessive exposures may lead to the development of cancer by promotion of ongoing carcinogenic biological processes or by independent cancer induction.

Ionization converts a neutral atom to a charged atom. The γ rays and x rays are ionizing electromagnetic radiation with energies above ultraviolet in the electromagnetic spectrum. The difference between x rays and γ rays is a matter of nomenclature. The γ rays are emitted from the nucleus of a radioactive atom while x rays originate in the orbital electrons of a energized atom. The α^{2+} particle has been identified as a positively charged helium nucleus with 2 protons and two neutrons emitted from the nucleus of certain radionuclides. The β-radiation was identified as negatively charged electrons emitted from the nucleus of certain radionuclides. There are other types of ionizing radiation including most prominently uncharged neutrons emitted from the nucleus of atoms that ionize atoms via nuclear interactions, and positrons and that are positive beta particles, β^+. The measure of absorbed radiation dose in matter or tissue is the gray (Gy) equal to one joule of energy per kilogram of matter or tissue. To account for different types of ionizing radiation, such as gamma, alpha, and beta radiations, the dose from ionizing radiation is corrected for the theoretical relative biological effectiveness of the different radiations in causing cellular damage using a radiation weighting factor, w_R (Sv/Gy), and this is reported as the equivalent dose in sieverts (Sv). Hence, the measure of biological dose is the sievert (Sv), equal to the absorbed dose in Gy times the radiation weighting factor. The γ rays, x rays, and β radiation typically have a radiation weighting factor of 1, while α radiation has a radiation weighting factor of 20 and neutrons may have a radiation weighting factor from 5 to 20 depending on energy. The quantity of radioactivity is measured in becquerel (Bq), equal to one radioactive atomic event or nuclear disintegration per second.

Radiation induced cancer is a complex and not completely understood process involving multiple events including but not limited to cellular DNA damage, up and down regulation of genes, intercellular communication, tissue and organ responses, clonal expansion of altered cell lines, and possibly eventual malignancy. The current understanding of radiation carcinogenesis is informed by epidemiological studies of human populations exposed to elevated levels of ionizing radiation and controlled studies utilizing laboratory animals. The two major human epidemiological studies have led to sharply contrasting results. Studies of the atomic bomb survivors indicate a linear no-threshold dose-response relationship. Studies of the radium dial painters have led to a sharp threshold relationship. These contrasting findings occur because quite different mechanisms of ionizing radiation carcinogenesis are involved in brief acute exposures at relatively high dose rates and in protracted exposures. Brief acute high dose-rate exposures act by promoting ongoing biological processes that lead to cancer in later life and this promotion effect is proportional to the dose. Protracted exposures induce cancer with a latency that depended on the lifetime average dose-rate of exposure to the target organ. At low dose rates the latent period may exceed the natural lifespan leading to a lifetime virtual threshold for cancer induction. Another observation at lower dose rates is the amelioration of an ongoing pre-malignant process. These findings have important implications with respect to ionizing radiation safety standards.

1.2 Naturally occurring ionizing radiation and radioactivity

There are naturally occurring sources of ionizing radiation that have existed on the planet since its formation (Eisenbud & Gesell, 1997). Naturally occurring radiation sources include cosmic rays, cosmogenic radionuclides, and primordial radionuclides and their decay products. Cosmic rays consist primarily of extremely high energy (mean energy ~ 10 billion electron volts) particulate radiation (primarily protons) and high energy gamma rays. When the particulate radiations collide with the earth's atmosphere a shower of "secondary" radiations are produced, which include high energy electrons and photons. The average person's dose from cosmic radiation is 0.27 mSv per year or ~ 7% of natural background. Exposure to cosmic radiation increases with altitude as there is less atmosphere to absorb the radiation, so populations at high elevations receive higher cosmic doses. For example, people living in Leadville, Colorado, at 3,200 meters above sea level, receive ~ 1.25 mSv y^{-1} or five times the average exposure at sea level. A fraction of the secondary particulate cosmic radiation collides with stable atmospheric nuclei making them radioactive. These cosmogenic radionuclides contribute very little (~ 0.004 mSv y^{-1} or less than 1%) to natural background radiation. The majority of this component of natural background is from the formation of carbon-14 and tritium (hydrogen-3).

Terrestrial radioactive materials that have been present on earth since its formation are called primordial radionuclides. Population exposure from primordial radionuclides comes from external exposure, inhalation, and incorporation of radionuclides into the body. The decay chains of ^{238}U, whose half-life is 4.5 billion years (uranium series), and ^{232}Th, whose half-life is 14 billion years (thorium series), produce several dozen radionuclides that together with natural potassium-40, whose half life is 1.3 billion years, are responsible for the majority of the external terrestrial average equivalent dose rate of 0.28 mSv per year or ~9% of natural background. Traces of these primordial isotopes of uranium and thorium are found in all the soil and rock on the face of the earth. Some regions of the world with high concentrations of primordial radionuclides produce local equivalent dose rates of environmental gamma radiation as high as 25 mSv per year.

Radon, ^{222}Rn, and its decay products, which are constituents of the ^{238}U decay series (Fig. 1), are the most significant source of natural background radiation exposure. Once inhaled, the majority of the dose is deposited in the tracheobronchial region by its short-lived daughters rather than by ^{222}Rn itself. Radon concentrations in the environment vary widely due to differences in ^{238}U concentration in the soil and differences in ventilation and construction of buildings. All other factors being equal, buildings with less ventilation will tend to have higher radon concentrations and thus, higher level of background radiation exposure. Outdoor concentrations of radon and its decay products are normally low because of the rapid mixing and dilution of radon gas with ambient air.

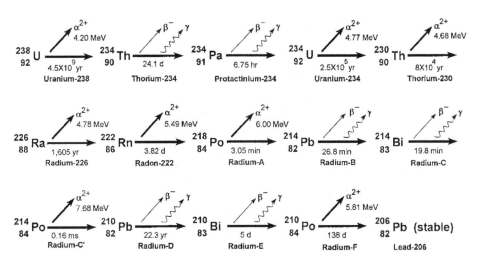

Fig. 1. Schematic illustration of the natural ^{238}U decay series as found in all soil and rock on the surface of planet Earth.

The second largest source of natural background radiation exposure comes from the ingestion of food and water that contain primordial radionuclides (and their decay products) of which ^{40}K is the most important. Altogether this pathway is responsible for an average effective dose rate of 0.4 mSv y^{-1}, or ~ 13% of natural background.

1.3 Typical human exposures to ionizing radiation

Every person who has ever lived has been exposed to ionizing radiation every minute of every day. Among the natural exposures are radiation emitted by nuclides naturally found on the earth (3% terrestrial background), internal exposure to natural radionuclides in food and water (5% internal background), cosmic radiation (5% space background), and radon and thoron gases released from the earth by the decay of naturally occurring radium isotopes (37% radon isotopes alpha irradiation of the lung). A major change has occurred here in the 21st Century from increased exposures that may occur through medical diagnostic use of x rays and radionuclides.

The average per capita "effective" radiation dose in the United States from naturally occurring (or "background") sources is about 3 mSv) per year. The "effective" dose is calculated using the International Commission on Radiological Protection (ICRP) methodology in which a tissue weighting factor is assigned to each organ or tissue and multiplied by the actual equivalent dose to the organ or tissue (ICRP 26, 1977). In the United States diagnostic medical radiology and nuclear medicine now add about 3 mSv of effective dose to ionizing radiation exposure per year per average person. The percent average contribution of various sources of ionizing radiation exposures of typical residents of the United States population in 2006 is illustrated in a pie-chart (Fig.2) prepared by the National Council for Radiation Protection and Measurement.

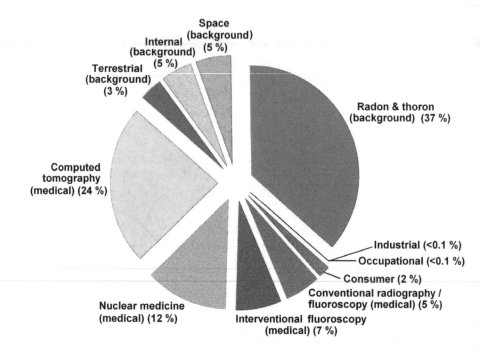

Fig. 2. Percent contribution of various sources of exposure to ionizing radiation to the average annual effective dose (6.2 mSv) per person in the United States population (about 300 million people) in 2006 (NCRP 160, 2009).

There are many places on the earth where natural background doses are much higher because of the natural presence of high concentrations of naturally occurring radionuclides in soil, rock, and water. Studies of these areas of naturally elevated ionizing radiation have not yielded any major deleterious effects associated with these natural exposures. For example, people in some areas of Ramsar, Iran, receive an annual ionizing radiation biological dose from natural background that is up to 260 mSv per year (over 40 times the USA national average). Chromosomal, hematological, and immunological studies and life span and cancer data show no untoward effects or responses (Ghiassi et al., 2002).

1.4 Naturally occurring radium isotopes and decay products

There are many naturally occurring radioactive nuclides that are continually being produced from decay chains in which the parent has a very long physical half life. For example, ^{226}Ra that is present in all the soil and rock on the earth at concentrations of about

25 Bq kg^{-1} is a product of the decay series headed by ^{238}U whose half life is comparable to the age of the earth (4.5 billion years). Consequently, it is also naturally found in ground water and the human body, usually in trace amounts (about 1 Bq per person). The ^{238}U decay series is shown in Figure 1. As shown in the figure, various types of ionizing radiation are emitted by the radionuclides in the chain including alpha radiation (α^{2+}), beta radiation (β^-) and gamma radiation (γ). In addition, similar or smaller amounts of natural ^{228}Ra, a decay product of natural ^{232}Th, are also found in soil and water. A similar ^{232}Th decay chain yields ^{228}Ra, ^{228}Ac, ^{228}Th, ^{224}Ra, ^{220}Rn, ^{216}Po, ^{212}Pb, ^{212}Bi, ^{212}Po, and stable ^{208}Pb (Eisenbud & Gesell, 1997).

As an analogue of calcium, radium that enters the body tends to incorporate into mineral bone where it and its progeny irradiate portions of the skeleton, primarily by alpha radiation emissions. Thorium isotopes tend to deposit on bone surfaces and irradiate surface cells, primarily by alpha radiation emissions. The radionuclide undergoing decay is referred to as the parent nuclide and the transformed nuclide is called the daughter or decay product. The daughter may be no more stable than its parent and may also be radioactive. Successive transformations will occur in a so-called decay chain, possibly yielding several radioactive progeny, until a stable nuclide is reached (Fig. 1).

In this context for ^{226}Ra, the word dose is used specifically to refer to the mean alpha radiation dose absorbed by the irradiated skeleton measured in units of energy deposited per unit mass of skeletal tissue (Gy). Dosage refers to the amount of material that enters the body or is administered (Bq). Burden refers to the amount retained in the body or skeleton with time post intake (Bq). Intake can refer to systemic intake into the blood or other organs of the body, as for the U.S. radium cases (Rowland, 1994), or it can refer only to that material that passes into the mouth or nose during ingestion or inhalation as in radiation protection practice (ICRP-30, 1979). Note that there is a really big difference between ingestion or inhalation intake and systemic intake, sometimes a factor of thousands. Cumulative dose, D, to an organ of the body is derived from the time integral of concentration (or dose rate) in that organ and requires a specified time limit to be meaningful.

2. Internally deposited radionuclides

2.1 Radium in people

Early in the last century people were exposed to high concentrations of ^{226}Ra and ^{228}Ra and their decay products in the luminous dial industry, in laboratory work, and in medical or private therapeutic use of radium. In particular, luminous-dial painters, who were mostly young women, were taught to tip their paint brushes on the tongue to make a sharp brush point; this procedure resulted in the ingestion of considerable radium leading to systemic uptakes of some of the ingested radium. These massive intakes of emulsions of pure radium salts resulted in life-time absorbed doses to the irradiated skeleton of some dial painters as high as a few thousand Sv, with consequent cases of severe injury to the skeleton and/or bone sarcoma as well as carcinomas of the head associated with retained ^{222}Rn and daughters in the nasal sinuses and mastoid spaces.

Detailed evaluations of these human cases have been described by Evans (1943), Evans et al. (1972), Evans (1974) and further documented in recent years (Rowland, 1994). Studies of individuals who had accidental intakes of massive amounts of radium have shown that the principal risk of exposure to large quantities of radium is direct damage to bone, bone cancer (but not leukemia), and cancers of the head caused by trapped decay of radon gas produced by decay of ^{226}Ra in the skeleton. It is interesting to note that no person among the

U.S. radium dial painters whose skeletal dose was less than 10 Gy (200 Sv for ionizing radiation weighting factor of 20 for alpha radiation) developed bone cancer because of exposure to [226]Ra or [228]Ra. An effective or virtual threshold dose, below which there is no radiation-induced cancer from exposure to radium, has been observed (Evans, 1974; Raabe et al., 1980; Raabe, et al., 1990; Raabe, 2010). In addition, serious non-neoplastic (non-cancer) radiation bone injury, such a bone osteodystrophy and fractures, occurred at even higher doses than did cancer.

Since deposition in and protracted irradiation of the skeleton follows systemic uptake, even a single brief intake initiates a period of chronic irradiation from radium and its retained decay products that depends in duration on skeletal retention or radioactive decay of the deposited radium. About 30% of radon-222 is retained from decay of radium-226, and almost all radon-220 is retained following the decay of radium-224 that forms from the decay products of radium-228. In the case of inhalation of radium (or thorium), irradiation of the lung is also an important aspect of the exposure. However, radium (or thorium) deposited in the lung is cleared by the lymphatic flow, coughed up and swallowed or expectorated, or enters the blood. Upon entry into the blood from the lung, inhaled radium is mostly excreted in the urine, while about 20% is deposited and tenaciously retained in the skeleton. The retention of radium in the human body, particularly the skeleton, has been the subject of considerable study and has been described in a mathematical form that allows the ionizing radiation dose to body tissues to be estimated from time of systemic intake until any time after intake (ICRP-20, 1973).

Fig. 3. Early 20th Century photo of young women painting clock dials with luminous paint containing radium. They tipped their brushes on their tongues and swallowed considerable quantities of [226]Ra or [228]Ra.

Lifetime studies were conducted in the 20th Century of the American radium dial painters (Fig. 3) and others with internally deposited radium isotopes (^{226}Ra and ^{228}Ra). Internally deposited radium chemically behaves somewhat like calcium and deposits somewhat uniformly in bone mineral in the skeleton. Radium is cleared from the body slowly so that a few percent of the intake can still be detected decades after intake. Alpha radiation from radium and its decay products lead to bone sarcoma in some exposed people at the highest doses. For ^{226}Ra (half life 1,600 years), the first decay product is ^{222}Rn gas much of which migrates in the blood to other parts of the body and is exhaled via the lung. Some ^{222}Rn gas sequesters in the sinus and mastoid regions of the head where alpha radiation from radon and its decay products lead to head carcinoma in some of the most highly exposed people. The most extensive studies of people exposed to radium were conducted at Massachusetts Institute of Technology (MIT) and Argonne National Laboratory (ANL). Results have been reported for over 2,000 people including about 1,700 who had significant intakes of radium, mostly to ^{226}Ra (Rowland, 1994). The nuclear physicist, Robley Evans, author of The Atomic Nucleus (1955) lead the early studies. Evans observed that only those people who received skeletal absorbed alpha radiation doses exceeding 1,000 cGy, developed bone sarcoma (Fig. 4), which he called a "practical threshold". He showed that the absence of sarcoma cases below 1,000 cGy was a statistically significant finding (Evans et al, 1972). This finding did not deter others from reorganizing the data into selective groups of cases and advancing a linear no-threshold model for radium induced bone sarcoma (Fig. 5).

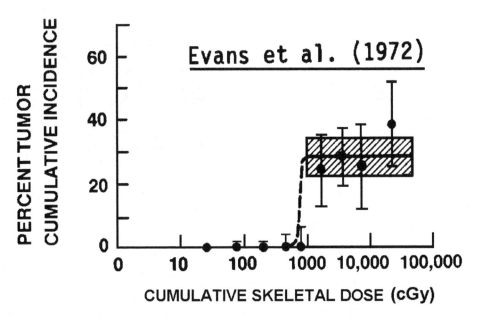

Fig. 4. Bone sarcoma incidence in people exposed to ^{226}Ra as a function of cumulative dose to the skeleton as reported by Evans et al. (1972).

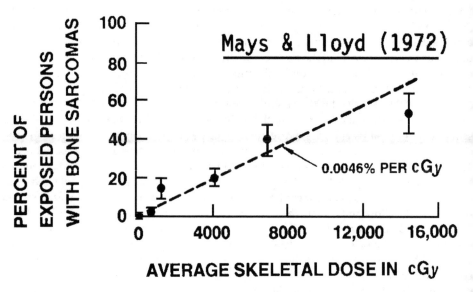

Fig. 5. Misleading linear dose-response model fit to bone sarcoma incidence in people who were exposed to [226]Ra as a function of cumulative dose by Mays and Lloyd (1972).

2.2 Radium-226 in beagle dogs

Since human data for exposure to internally deposited radionuclides are usually difficult to evaluate and readily confounded, in the late 1950's, the Division of Biology and Medicine of the United States Atomic Energy Commission conceived a series of carefully controlled scientific studies utilizing purpose-bred dog colonies of pedigreed beagles. These studies were conducted primarily at the University of Utah, the University of California, Davis (UC Davis), Hanford Laboratory, Richland, Washington, and Lovelace Foundation in Albuquerque, New Mexico.

The largest lifetime study of beagles exposed to [226]Ra was conducted at UC Davis from 1961 to 1989. Dogs were bred in a manner to maintain and randomize the gene pool and entered into the study over several years in a randomized-block design so that and all dosage levels and controls were represented contemporaneously in each treatment block. Exposure to [226]Ra, temporally simulating the dial painter exposures, were by eight intravenous injections two weeks apart in young adults from 435 to 540 days of age (Raabe 1989).

The distribution of deaths and causes are show in Fig. 6 (Raabe, 2010). The bone sarcoma deaths followed a tight lognormal power function of average radiation absorbed dose rate to skeleton with geometric standard deviation of only 1.22. Although the induced cancer response is a tight power function of dose rate, it is not a meaningful function of cumulative dose. For example, at an alpha radiation dose rate of about 0.1 Gy d^{-1} the median time to bone cancer death is about 1,200 days with a total cumulative dose of 120 Gy. In sharp contrast, at a alpha radiation dose rate of 0.003 Gy d^{-1} the median time to bone cancer death is about 4,000 days with a total cumulative dose of only 12 Gy. It would seem that the radiation is ten times more effective at the lower dose rate than at the higher dose rate, but that observation is misleading since it is the dose rate that controls the cancer induction and latent period. It is the lifetime average dose rate that controls the risk of radiation-induced cancer. Since the extrapolated bone cancer latent period exceeds the natural life span of the beagles at low dose rates, there is a virtual threshold for bone cancer from radium. None of the dogs in the lowest dose group developed bone cancer because they died first from causes associated with natural aging.

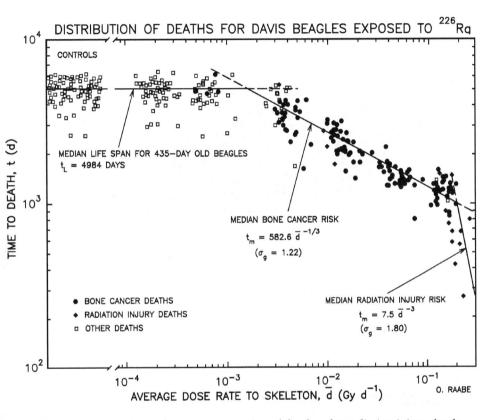

Fig. 6. Two-dimensional logarithmic representation of the data for radiation injury deaths, bone sarcoma deaths, and other deaths in beagles injected with [226]Ra at UC Davis, showing time from initial intake at age 435 days to death for each individual dog versus average dose rate to skeleton and fitted lognormal risk functions (Raabe, 2010).

The data shown in Fig. 6 is a two-dimensional representation of a three dimensional phenomenon. The three dimensions are lifetime average dose rate to the skeleton, time to death, and the frequency distribution of bone cancer deaths (Fig. 7). This three-dimensional phenomenon can be expressed in terms of the risk of death from various causes during the beagle lifetime (Fig. 8). The occurrence of beagle bone cancer deaths from [226]Ra-induced cancer is displayed as a mound rising out of a Euclidian plane. At low dose rates there are no cases of bone cancer observed because all of the beagles have died of causes associated with natural aging. At high dose rates there are radiation injury deaths.

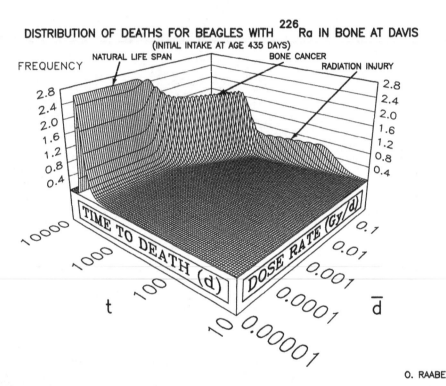

TIME AFTER INTAKE & AVERAGE ALPHA DOSE RATE TO SKELETON (LOG SCALES)

Fig. 7. Three-dimensional logarithmic representation of average-dose-rate/time/response relationships of Fig. 6, for beagles injected with 226Ra at UC Davis, shown as the probability density distribution frequency of the combined risk of dying from causes associated with natural life span, radiation-induced bone sarcoma, and radiation injury, as a function of average dose rate to skeleton and elapsed time after intake at age 435 days (Raabe, 2010).

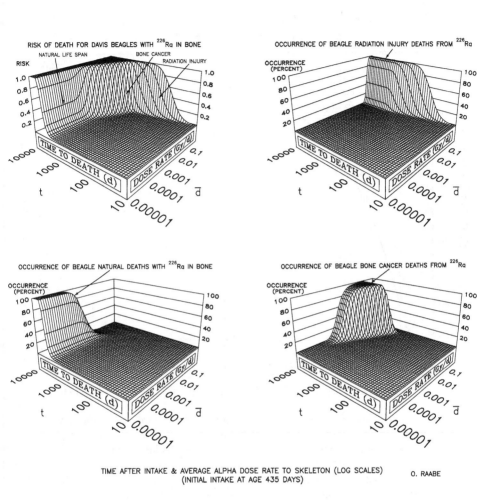

TIME AFTER INTAKE & AVERAGE ALPHA DOSE RATE TO SKELETON (LOG SCALES)
(INITIAL INTAKE AT AGE 435 DAYS)   O. RAABE

Fig. 8. Three dimensional representation from Fig. 6 of average-dose-rate/time/response relationships for beagles injected with 226Ra at UC Davis shown in the top left panel as the combined risk of dying, and in the successive panels as the occurrence of deaths from radiation injury, natural life span, and radiation-induced bone sarcoma (Raabe, 2010).

The common but inappropriate practice of assuming that the cancer induction risk from exposure to ionizing radiation is proportional to cumulative radiation dose leads to misunderstandings about the true nature of the dose-response relationship. For example, the data from the 226Ra beagle study can be plotted in the traditional fashion (Fig. 9). The common practice of fitting a linear no-threshold line starting at the zero-zero coordinates leads to an imprecise model that completely obscures the virtual threshold at low doses shown by the precise relationship to dose rate (Fig. 6).

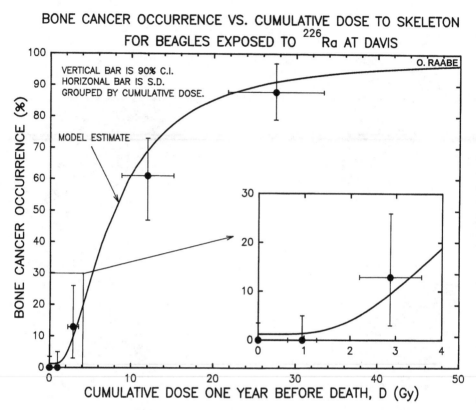

Fig. 9. Two-dimensional representation of the data for radiation-induced bone cancer deaths, in beagles injected with [226]Ra at age 435 days of age at UC Davis, showing lifetime bone sarcoma occurrence as percent of population versus calculated cumulative dose to skeleton one year before death. Note the life-span virtual threshold below 1 Gy as shown in the inset that would be obscured by a linear no-threshold (LNT) model of these data that starts at the origin with the only zero or near-zero risk at zero dose (Raabe, 2010).

2.3 Human cancer risk from radium-226

When a three-dimensional analysis is used for an interspecies comparison for radium-induced cancer, the different species display a similar relationship displaced in time based on natural life span (Fig. 10) (Raabe et al., 1980). The greater scatter of data for mice and people is the result of uncertainty and inaccuracy for the dose data compared to the precise beagle study. When the time scale is normalized with respect to natural lifespan for the species, all three sets of data overlay (Fig. 11). These results provided a basis for scaling cancer induction dose-response relationships among difference species. For example, the precise beagle data can be used to predict radium-induced bone cancer deaths in people by life-span normalization. The resulting prediction of median bone cancer risk for people falls almost perfectly at the median of the observed risk although the human data are quite scattered because of the dosimetric limitations (Fig. 12).

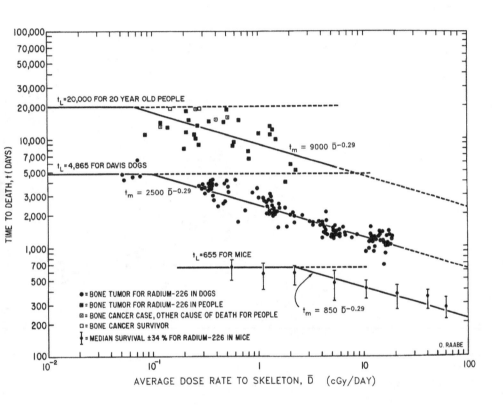

Fig. 10. Primary bone sarcoma deaths from [226]Ra in people, beagles, and female mice showing similar dose-response functions indicated by parallel median lines with negative slope of about one-third and t_L are the typical life spans for the three species (Raabe et al., 1980).

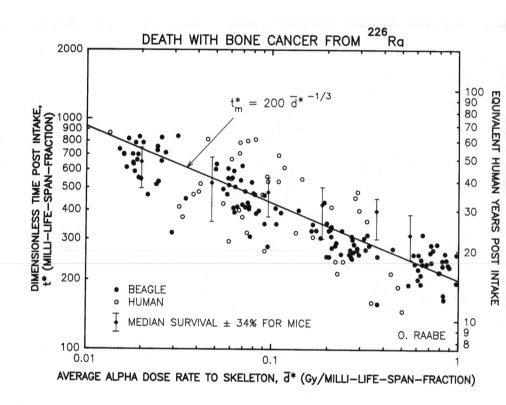

Fig. 11. Life span normalization with respect to dimensionless time and average alpha radiation absorbed dose rate to skeleton for fatal bone sarcoma in beagles, mice and people (Fig. 10) showing a single logarithmic regression line (determined for beagles) represents the median risk for all three species (Raabe, 2010).

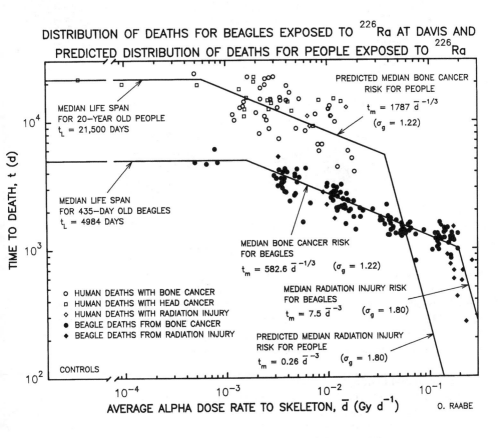

Fig. 12. Distribution of deaths for beagles and predicted distribution of deaths for people exposed to [226]Ra along with human data from the U.S. radium studies (Raabe, 2010).

Another form of alpha radiation-induced cancer occurs in people but not in laboratory animals. The radioactive decay of [226]Ra yields [222]Rn gas, much of which enters the blood stream and some of which sequesters in the nasal sinus and mastoid region of the head which leads to alpha irradiation of the internal tissues in these regions of the head from radon and its decay products. These deaths are somewhat delayed in time. The resulting three-dimensional frequency distribution of causes of death in people with internally deposited [226]Ra is shown in Fig. 13. Both head carcinoma and bone sarcoma display virtual threshold cancer risk relationships at low doses and dose rates.

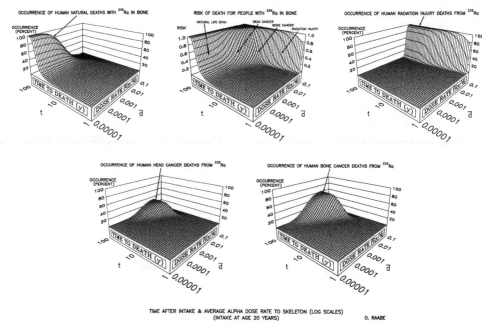

Fig. 13. Three-dimensional representation of the average-dose-rate/time/response relationships for people after intake of 226Ra shown in the top middle panel as the cumulative risk of dying, in the top left panel as the occurrence of deaths associated with natural life span, in the top right panel as the occurrence of deaths from radiation injury to the skeleton, in the lower panels have the occurrence percent of deaths from radiation-induced head carcinoma and bone sarcoma, respectively (Raabe, 2010).

2.4 Strontium-90 in people

Radioactive 90Sr (half life 29 years), a product of nuclear fission of uranium and plutonium, has been released into the atmosphere of the earth by nuclear weapons in Japan in 1945, by numerous atmospheric nuclear weapons tests up until 1966, and by the Chernobyl nuclear reactor accident in Russia in 1986. Strontium is chemically similar to calcium and deposits somewhat uniformly in bone mineral in the skeleton. Released to the environment, strontium is adsorbed and retained by clay in soil, and is found in small amounts in agricultural food products. The 90Sr, a beta particle emitter, decays to 90Y, a short-lived beta particle emitter that decays to stable 90Zr. Everyone born in the Northern hemisphere after about 1950 probably has measurable trace amounts 90Sr in bones and teeth.

2.5 Strontium-90 in beagle dogs

Because of the widespread exposure of people to 90Sr from nuclear weapons testing in the 20th Century, the U. S. Atomic Energy Commission initiated controlled studies of ingested 90Sr in purebred beagles. The largest lifetime study of beagles exposed to 90Sr by ingestion was conducted at UC Davis from 1961 to 1989 in parallel with the study of injected 226Ra.Dogs were bred in a manner to maintain and randomize the gene pool and entered into the study over several years in a randomized-block design so that and all dosage levels and controls were represented contemporaneously in each treatment block. Exposed beagles received measured amounts of ⁹⁰Sr in food via the mother during mid-gestation and via their food through adulthood at 540 days of age. The exposed young adult beagles in the study had bones and teeth that were uniformly labeled with ⁹⁰Sr. The controls were fed food with a high-dose mass equivalent of non-radioactive stable natural strontium, mostly ⁸⁸Sr (Raabe & Parks, 1993).

Unlike the short-range alpha radiation from ²²⁶Ra and decay products in bone that irradiated primarily the bone mineral of the skeleton, the longer-range beta radiation from ⁹⁰Sr-Y irradiated the bone mineral, the bone marrow, tissue and tissues adjacent to bone and teeth. Induced cancer deaths were associated with four different types of cancer: (a) bone sarcoma, (b) periodontal carcinoma, (c) oral carcinoma, and (d) leukemia. Each of these displayed a virtual threshold relationship (Fig. 14). Although there were 2 cases of bone sarcoma deaths among the 80 controls, there were no cases among the 183 exposed beagles

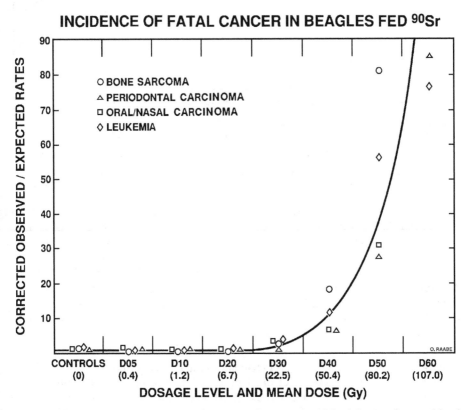

Fig. 14. Statistical evaluation by survival analysis (Peto et al. 1980) of the incidence of fatal leukemia, bone sarcoma, and carcinoma in beagles fed ⁹⁰Sr from before birth to adulthood at UC Davis demonstrating a life-span virtual threshold with all radiation-induced cancers occurring at calculated cumulative skeletal beta radiation doses above 20 Gy (20 Sv). The absence of bone sarcoma in the lowest three dosage groups is significantly less than those found in the controls (p<0.047)(Raabe, 2010).

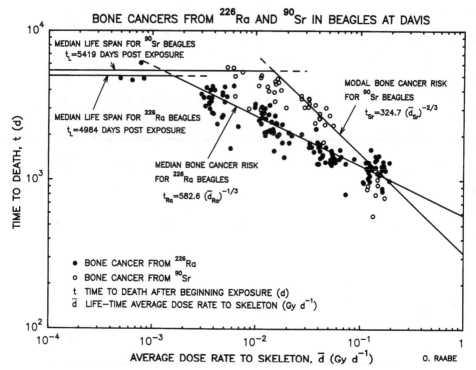

Fig. 15. Comparison of bone sarcoma in beagles at UC Davis from injected ²²⁶Ra and ingested ⁹⁰Sr (Raabe, 2010).

in the three lowest dose groups. The chance of this is p=0.092 (Fisher's Exact Test). Among all 162 controls in the ²²⁶Ra and ⁹⁰Sr studies there were 2 fatal bone sarcoma cases and two with bone sarcoma that died of other causes. With 4 bone sarcoma cases among 162 controls, the absence of bone sarcoma cases in the three lowest ⁹⁰Sr dose groups is a statistically significant reduction in tumor cancer incidence with p=0.047 (Fisher's Exact Test). The shape of the dose response relationship for ⁹⁰Sr-induced bone sarcoma is similar to that for ²²⁶Ra except it follows a Weibull distribution and the power function slope is twice as steep as for ²²⁶Ra indicating that an average of two beta particles are needed to equal the effect of one alpha particle for induction of bone cancer (Fig. 15).

2.6 Inhaled, ingested, and injected radionuclide studies in beagles
An analysis of 25 lifetime studies of inhaled, ingested, and injected internally-deposited radionuclides in beagles show remarkable consistency (Raabe, 2010). Three-dimensional models have been fit to selected data from life-time studies of internally deposited radionuclides in young adult beagles at four laboratories: University of California, Davis (Raabe and Abell, 1990), Lovelace Inhalation Toxicology Research Institute (ITRI) and University of Utah (Mauderly and Daynes, 1994), and Battelle Pacific Northwest Laboratory, PNL (Park et al., 1993). Dose response data were evaluated for beagles with skeletal burdens of ⁹⁰Sr after exposure by ingestion at Davis, by injection at Utah, and by inhalation at ITRI,

for lung burdens of inhaled ¹⁴⁴Ce, ⁹¹Y, and ⁹⁰Sr in fused aluminosilicate particles (FAP) at ITRI, inhaled ²³⁹Pu dioxide at PNL, skeletal burdens of injected ²²⁶Ra at Davis and Utah and inhaled ²³⁸Pu at ITRI, and skeletal burdens of injected ²²⁸Ra and ²⁴¹Am at Utah. Analyses were based on the mean organ absorbed doses to to the target tissues from parent and corresponding doses from decay products in their appropriate proportion, where all x ray and gamma emissions were ignored because of their minor contribution, and where beta emissions are also ignored in the cases where the primary exposures were from alpha radiation. Overall, there were separate injection studies of ²²⁶Ra, ²²⁸Ra, ²²⁴Ra, ²²⁸Th, ²³⁹Pu, ²⁴⁹Cf, ²⁵²Cf, ²⁴¹Am and ⁹⁰Sr. There were separate inhalation studies of ²³⁹Pu, ²³⁸Pu, ⁹⁰Y, ⁹¹Y, ⁹⁰Sr, and ¹⁴⁴Ce (Raabe, 2010).

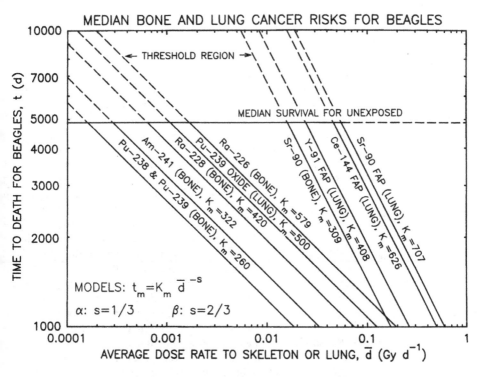

Fig. 16. Summary of bone sarcoma and lung carcinoma risk functions for beagles demonstrating similar target organ dose-rate/time/response patterns illustrating the life-span virtual thresholds at low dose rates. The positions of the lines vary because of inherent differences in irradiation of the target cells by the different radionuclides. Data were selected from life-time laboratory studies of internally deposited radionuclides in young adult beagles including skeletal deposits of ⁹⁰Sr after exposure by ingestion at Davis, by injection at Utah, lung deposits of inhaled ¹⁴⁴Ce, ⁹¹Y, and ⁹⁰Sr in fused aluminosilicate particles (FAP) at ITRI, lung deposits of inhaled ²³⁹PuO₂ at PNL, skeletal deposits of injected ²²⁶Ra at Davis, inhaled ²³⁸PuO₂ at ITRI, and skeletal deposits of injected ²²⁸Ra and ²⁴¹Am at Utah (Raabe, 2010).

A selection of these results are shown in Fig. 16. All of the alpha radiation studies followed lognormal distributions with power function slope of negative one-third. All of the beta radiation studies followed Weibull distributions with power function slope of negative two-thirds. The different positions of these functions were shown to be associated with the efficiency of irradiation of the target sensitive tissue in each case. At low dose rates and associated low life-time cumulative doses there was a virtual threshold for ionizing radiation induced cancer. For young adult beagles, bone sarcoma induction from alpha-emitting or beta-emitting radionuclides was unlikely for cumulative doses below about 20 Sv delivered specifically to the sensitive tissues at bone surfaces. For inhaled alpha-emitting radionuclides lung carcinoma was unlikely for cumulative lung doses less than 10 Sv and for inhaled beta-emitting radionuclides lung carcinoma was unlikely for cumulative lung doses less than 5 Sv (Raabe 2010).

3. Radon exposures

3.1 Environmental radon

All of the earth's atmosphere contains low concentrations of radioactive airborne particles and gases associated with various naturally occurring radionuclides. Of particular interest are radioactive isotopes of the inert gas, radon. Three radioactive isotopes of the radon gas ^{219}Rn from the natural ^{235}U decay series, ^{220}Rn from the natural ^{232}Th decay series and ^{222}Rn from the natural ^{238}U decay series are produced in earth's crust and released by diffusion into the outdoor air. The ^{222}Rn with a decay half-life of 3.82 days is therefore the most important. In some cases ^{220}Rn, commonly called thoron, can be important also. The ^{238}U decay series that yields ^{222}Rn and its decay products is shown in Figure 1. The ^{222}Rn is the decay product of the naturally occurring radium isotope ^{226}Ra that occurs naturally in all the soils and rock on the surface of the earth at concentrations of about 25 Bq kg^{-1}. Consequently, ^{226}Ra is also naturally found in ground water and the human body, usually in trace amounts (about 1 Bq per person). When ^{226}Ra undergoes radioactive decay (half-life 1,600 y) it forms gaseous ^{222}Rn, which can percolate through and diffuse out of the soil or rock and into the air. This process has definite temporal limits since the half-life of ^{222}Rn is only about 3.8 d. However, enough radon reaches to the earth's atmosphere to provide an average outdoor concentration of about 10 Bq m^{-3} in outdoor air in the most populated part of the world. Much lower concentrations occur over the oceans and in cold polar locations. The ^{222}Rn decays in air to form radioactive, solid decay products that form radioactive aerosols. In addition to ^{226}Ra, similar or smaller amounts of ^{228}Ra, a decay product of ^{232}Th, are also found on the earth leading to the atmospheric release of the radon gas isotope, thoron, ^{220}Rn. However, the concentration of thoron and its decay products is usually negligible in comparison the ^{222}Rn and its decay products. Very high air concentrations of radon can be found in some mines and elevated levels can be found in most buildings.

When air containing radon gas is inhaled by a person, a small amount of this inert gas dissolves in body fluids. During its radioactive decay emitting an alpha particle and the following decay of its decay products some cells of the body are irradiated, but the level of whole body irradiation is quite small. Much more important is the irradiation of the lungs associated with the deposition in the bronchial airways of the radon decay products which are solid metal atoms rather than gases. While radon gas is inhaled and exhaled, the unattached decay products and those attached to small inhalable airborne particles can deposit in the bronchial airways of the lungs during normal breathing. For example, in the case of ^{222}Rn , the short-lived alpha-particle emitting isotopes of polonium, ^{218}Po and ^{214}Po, that deposit in the lungs can result in relatively high alpha-particle irradation of bronchial airways. The comparatively small dose associated with the beta-particle emitting decay products is insignificant compared to the alpha irradiation which is about 20 times more biologically effective than beta radiation per unit of energy. The equivalent dose to the bronchial airways from radon and thoron decay products in ambient air in homes for the average person in the USA has been estimated to be about 28 mSv per year (NCRP 160, 2009). In some areas of the USA, the annual dose to the bronchial airways from radon and thoron may be double this average value.

Given sufficient time and other favorable conditions, the decay products of ^{222}Rn will come into radioactivity equilibrium with the radon with each of the short-lived decay products having the same activity concentration in the air as the ^{222}Rn. However, this ideal equilibrium is rarely attained in air containing elevated levels of radon so that measurement of the radon gas concentration does not precisely indicate the concentration of decay products unless that state of disequilibrium is known. Since essentially all of the biologically important radiation dose to the respiratory epithelium is derived from the alpha-emitting radon decay products, their concentration is sometimes described in special units called the working level, WL. The WL unit is defined as any combination of the short-lived radon progeny in one liter of air that will result in the emission of 130,000 MeV of alpha particle energy. Air having a ^{222}Rn concentration of 3.7 kBq m^{-3} with the progeny in secular equilibrium would represent 1 WL. The exposure associated with a typical work month in a uranium mine for 170 h at 1 WL is called an exposure of 1 working level month, WLM. Dosimetric models indicate that the nominal dose to the bronchial epithelium associated with inhalation of radon decay product aerosols by a uranium miner is about 6 mGy/WLM. Assuming an alpha radiation weighting factor of 20, this yields about 120 mSv equivalent dose per WLM to the bronchial region of the lung.

Upon inhalation, the airborne particles containing radon decay products may deposit upon contact onto the surfaces of the respiratory airways. Because of their diffusivity, the very small molecular clusters may efficiently deposit in the head airways or in the trachea and bronchial airways of the lung. Other somewhat larger particles may reach the alveolar region of the lung, as well. Because of their short radioactive half-lives, they usually decay prior to being cleared from the respiratory tract and irradiate the respiratory epithelium. Of primary concern in this regard is the irradiation of the bronchial epithelium by the highly ionizing alpha radiation emitted by radium-A (^{218}Po) and radium-C' (^{214}Po). Since radon itself is an inert gas, it does not readily deposit in the respiratory airways during inhalation and is mostly exhaled. Occasionally an atom of radon gas may decay and emit alpha radiation in air present in the lung irradiating the epithelium, but the fraction of the dose contributed by the radon itself is small compared to that associated with the deposited decay product particles. Because naturally occurring radon decay products are in ambient air outdoors and within a building, the lung is the most highly irradiated organ of the body of a typical person from background radiation sources.

3.2 Radon and radon decay product dosimetry

The dosimetry methodology associated with estimating the alpha radiation dose to the lungs of people from exposure to atmospheres containing radon and thoron is quite complicated and the dosimetry estimates are subject to quite large uncertainties. Consider the case of radon gas entering an air space as it diffuses from the earth. Initially there are no decay products since they are produced as the radon atoms decay and convert from gas atoms to metallic atoms. This in-growth of 218Po (commonly called Ra-A), 214Pb (commonly called Ra-B), and 214Bi (commonly called Ra-C) is shown graphically in Figure 17 for a concentration of radon equal to 3.7 kBq m−3 which at decay-product equilibrium is the level commonly called one working level (WL) associated with underground mining. While 218Po comes to equilibrium with radon in about 20 minutes, it takes more than 2 hours for 214Pb and 214Bi to approach equilibrium. In that time more radon is entering the airspace, ventilation is changing the relative concentrations, and the metallic decay products are being surrounded by water molecules, are undergoing Brownian diffusion and are attaching to larger airborne particles and onto the surfaces of walls, floor, ceiling, furniture, and people. Measurements that are made of radon gas concentrations in a living space or underground mine do not describe the concentration of the airborne solid decay products that deliver more than 90% of the dose the bronchial airways of people in that room or underground mine.

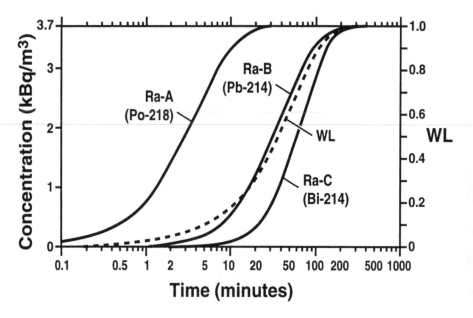

Fig. 17. Ingrowth of radon decay products as a function of time starting with 3.7 kBq m−3 of 222Rn yielding one working level, WL.

As each radon gas atom in air decays by emitting alpha radiation it forms an atom of radioactive polonium, 218Po (called radium-A, half-life about 3 minutes). This metallic atom quickly oxidizes and forms the center of a particulate molecular cluster of about 8 Å in diameter. When it decays emitting alpha radiation, a series of short-lived decay products are formed in air. These also form particulate clusters. This radon decay process is shown schematically in Fig 1. The long-lived lead isotope, 210Pb (half-life 22.3 y) provides negligible radioactivity to the atmospheric aerosols and is, in effect, virtually non-radioactive compared to its short-lived progenitors. These radon decay product clusters represent the smallest airborne particles normally found in ambient air. In addition, these clusters attach upon contact to other, larger airborne particles to an extent that depends upon their concentration. Typically, more than 90% of radon decay product aerosols are associated with particles smaller than 0.5 μm in aerodynamic diameter.

Actually, the dosimetry is even more complicated. The deposition of airborne particles in the bronchial airways during inhalation is a function of the aerodynamic and diffusive properties of the airborne particles to which the radon decay products attach (ICRP 66, 1994). This depends on both the size distribution and concentration of the the airborne particles in the air. When the concentration of airborne particles is quite small, many of the decay products may not be attached to a larger particle (the so called unattached fraction). The unattached fraction efficiently attaches in all parts of the airways because of Brownian diffusion properties of very small particles causing somewhat efficient deposition in the upper airways of the nose and throat as well as the small bronchial airways. The irradiation of the living cells occurs primarily after the airborne particles deposit on the surface of the airways. If the radon decay products are associated with larger airborne parrticles, the deposition in the lung may be less because some of the inhaled particles may be exhaled. If the airborne particles are relatively large, they may deposit in the nose and throat and not readily reach the bronchial airways. Both the concentration and the aerodynamic particle size distribution of the airborne particles will affect the bronchial dose. Smoke in the air space can have a large effect on the bronchial radiation dose from radon decay products. Clearly, a simple measurement of the average concentration of radon in the air is insufficient to provide an accurate estimate of the lifetime average radiation dose rate to the bronchial epithelium of a person. Many reported studies only involve average radon measurements. Many reported studies only reported cumulative exposures to radon decay products in units such as working level months (WLM).

3.3 Lung cancer from exposure to radon and its decay products

Uranium mining in the 20th Century led to the clear realization that protracted and repeated exposures to high concentrations of radon can cause lung cancer. For example, miners working uranium mines in the Colorado Plateau region of the USA were found to have a high incidence of lung cancer depending on the level of exposure. People chronically exposed over extended period of employment to high levels of airborne radon decay products such as are found in uranium mines have developed bronchogenic carcinoma at incidence rates that significantly exceeded the expected rates in either smokers or non-smokers. However, almost all the lung cancer cases in the western U.S. mines occurred in smokers (Saccomanno et al., 1988). The reported excess relative risks of lung cancer as a function of cumulative dose expressed in working level months of exposure are shown in Figure 18. The miners did not have radon dosimeters and it could never be clear how long anyone worked in any of the areas within the mines, hence the working level month (WLM) values are estimates that are intended to represent cumulative radiation dose. As with all cumulative dose plots it is common to fit or draw a linear no-threshold function that begins at the origin obscuring the apparent virtual threshold. This is facilitated by the large dosimetric uncertainties (Lubin et al., 1995).

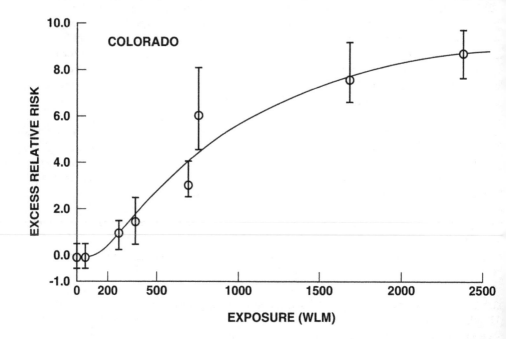

Fig. 18. Historical data of observed lung cancer risk as a function of exposure in working level months (WLM) among uranium miners from the Colorado Plateau in the United States (BEIR IV, 1988). Uranium miner data were routinely expressed in estimated cumulative doses in WLM so that the virtual threshold was badly obscured or ignored and linear no-threshold (LNT) models were assumed to apply in every case (Figure 19).

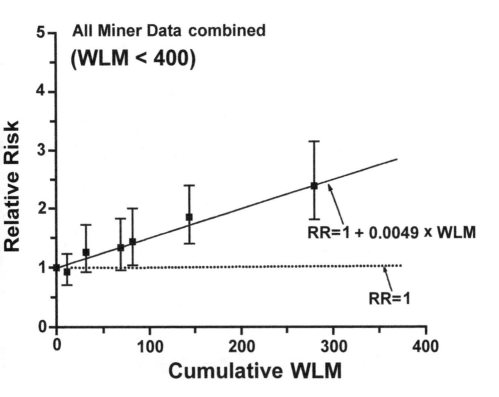

Fig. 19. Observed relative risk (RR) of lung cancer in underground miners versus cumulative airborne radon decay product exposure in working level months (WLM) for WLM less than 400 with a linear no-threshold risk model fit to the data (Lubin et al., 1994).

The reported epidemiological studies of radon in homes are not conclusive and are usually based on average radon concentration measurements rather than on decay product levels and the results were displayed using the standard linear no-threshold assumption. Also, the dosimetric uncertainties are usually not displayed. In Figure 20 the authors created a mathematical relative risk of 1 at a radon concentration exposure of zero but there is no such point in the real world. The major cause of lung cancer in the United States is known to be cigarette smoking (up to 95% incidence) so that cigarette dosimetry is actually more important than radon dosimetry in these types of studies.

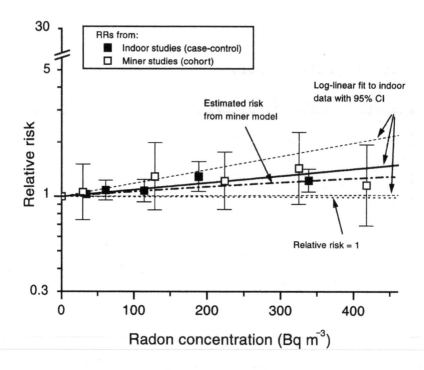

Fig. 20. Summary of calculated relative lung cancer risk mathematical meta-analysis of indoor radon studies along with miner studies (BEIR VI, 1999).

A careful ecological study was conducted of the relationship of lung cancer mortality as a function of measured air concentrations in homes in over 1,600 counties in the United States (Cohen, 1995). The remarkable results showed that the counties with homes having the higher radon concentrations tended to have the lowest lung cancer mortality rates (Figure 21). This relation was robust and was demonstrated for widely separated counties in various parts of the United States. His finding were in conflict with the prevailing linear no-threshold theory promulgated by the ICRP and United States Environmental Protection Agency (EPA). A scientific committee formed to review the radon risk data concluded that the results were faulty no matter how robust and reproducible because they were based on an ecological study that could not directly match cases and radon exposures (BEIR VII-2, 2006). Cohen argued to no avail that under the LNT hypothesis the distribution of the doses was irrelevant and his data proved the linear no-threshold model did not apply to exposure to radon in homes. Although it is mathematically possible that Cohen's ecological study is misleading, it seems highly unlikely that such an anomaly would uniformly appear everywhere in the United States. Cohen's result is substantiated by a simple comparison of the County by County radon zones and lung cancer mortality rates (Figure 22).

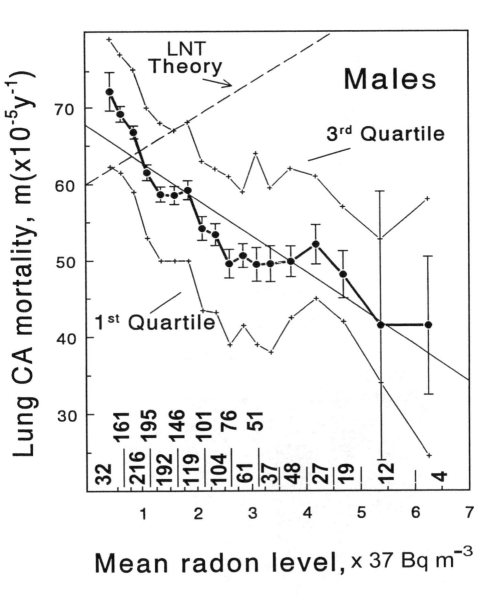

Fig. 21. Lung cancer mortality rates in men versus mean radon level for 1,601 counties spread throughout the continental United States as a function of mean radon level with the abscissa in units of 37 Bq m⁻³ (Cohen, 1995).

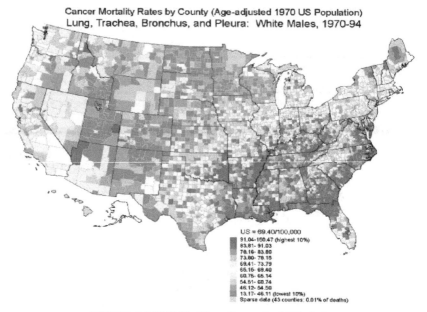

Cancer Mortality Rates by County (Age-adjusted 1970 US Population)
Lung, Trachea, Bronchus, and Pleura: White Males, 1970-94

US = 69.40/100,000
91.04-156.47 (highest 10%)
83.81- 91.03
78.16- 83.80
73.80- 78.15
69.41- 73.79
65.16- 69.40
60.75- 65.14
54.51- 60.74
46.12- 54.50
13.17- 46.11 (lowest 10%)
Sparse data (43 counties: 0.01% of deaths)

LUNG CANCER: Blue Low, Red High ↑

↓ RADON: Red HIGH, Yellow LOW

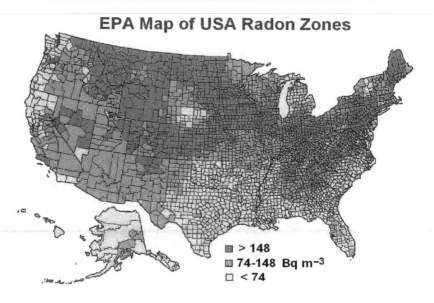

EPA Map of USA Radon Zones

■ > 148
□ 74-148 Bq m⁻³
□ < 74

Fig. 22. Comparison of lung cancer mortality rates and radon concentration in homes among the counties of the United States of America.

A conditional logistic regression case-control study of lung cancer risk from residential radon was conducted in the State of Massachusetts in the United States by Thompson et al. (2008). Although their radon dosimetry was limited to average air concentration, they also evaluated cigarette smoking dose in a precise manner by stratifying the data with nine categories of smoking history. This ingenious approach led to results that agreed well with Cohen's ecological study (Cohen 1995). They found that lung cancer rates were the highest in those people with the smallest radon expose levels (<25 Bq m^{-3}). As radon levels were increased the lung cancer rate went down in a consistent manner which was statistically significant in one dose group (p<0.05 at 59-75 Bq m^{-3}) and statistically strong in three others (p<0.1). Radon induced lung cancer appeared to be present only for average radon air concentrations greater than 250 Bq m^{-3}.

Fornalski and Dobrzynski (2011) reanalyzed the data from 28 published studies of the relationship of lung cancer and radon exposure using Bayesian statistical methods. They concluded that for people exposed to radon concentrations less than 838 Bq m^{-3} in those 28 studies there is no evidence that there is an association between radon exposure and lung cancer incidence.

These studies suggest that radon exposures at low dose levels can interfere with ongoing neoplastic processes in the lung as may be associated with cigarette smoking. Mechanisms such as stimulation of DNA repair may be invoked by low level exposures to ionizing radiation helping to inhibit the damage associated with chemical carcinogens such as cigarette smoke exposures. Similarly, up-regulating and down-regulating of genes at lower ionizing radiation doses can interfere with ongoing carcinogenic biological processes and lead to reduced cancer development.

4. External ionizing radiation epidemiology

4.1 Typical study problems

Several external ionizing radiation epidemiology studies have been conducted. However, some of these studies involve technical problems that affect the reliability of the findings. A distortion of the possible carcinogenic response is associated with the use of cumulative dose as the radiation exposure measure rather than the more appropriate lifetime average dose rate. Several confounders such as exposure to toxic and carcinogenic chemicals are quite difficult to properly evaluate and are often ignored. In years past two carcinogenic agents, benzene and trichloroethylene, were commonly used as cleaning agents in laboratories including nuclear research laboratories. There were few if any records of these exposures or chemical doses. The most important confounder among the predominantly male radiation workers is cigarette smoking. Confounding occurs when the workers who are most likely to receive the highest exposures to ionizing radiation are also the most likely to have been smokers. It is possible that those workers at nuclear facilities (mostly men) who worked most directly in the radiation areas were more predominantly cigarette smokers than workers who had managerial, supervisory, or clerical assignments. This confounding from smoking is most likely important prior to 1960 before the health risks of smoking were well understood. This possibility is most prominent in studies of workers at Oak Ridge National Laboratory in the United States State of Tennessee where nuclear operations began in 1943. Some epidemiological studies of Oak Ridge workers have attributed lung cancer and other smoking-related illnesses to radiation exposure (Wing et al., 1991).

4.2 Atomic worker studies

Two major series and the most important are the ionizing radiation epidemiological studies of multinational radiation works in several countries of the world that have been conducted by Cardis and her numerous co-workers (Cardis et al. 1995, Cardis et al. 2005). The two studies were similar in approach, but the second was much larger and included subjects who were in the first study. In these studies emphasis was on workers who had available ionizing radiation dosimetry records of external photon exposure (x rays and gamma rays) but who did not have major internal exposure or neutron exposure. In the second study internally deposited radionuclide dose was specifically limited to ten percent of the total. Two cause of death risk types were separately evaluated: (a) leukemia excluding chronic lymphocytic leukemia and (b) all cancers excluding leukemia. Simple linear dose-response models of risk that was assumed to be proportional to cumulative external ionizing radiation dose were statistically evaluated. Radiation doses were lagged by 10 years for cancer cases and 2 years for leukemia. Hence, they assumed the cancer risk was proportional to the cumulative dose received prior to the lag period just before death.

If a person developed lung cancer, the radiation dose received in the prior 10 years was assumed to have not been a contributor to the risk. Typical lifetime doses in these studies were well below 1 Sv, so that the risk of radiation-induced cancer would be expected to be very near zero, based on the internal emitter studies discussed above.

In the first study (Cardis et al., 1994; Cardis et al., 1995) 95,673 nuclear industry workers (85.4% men) in the United States, United Kingdom, and Canada were monitored for external exposure to ionizing radiation prior to 1988. A linear cumulative dose response model with Poisson regression was used to estimate excess relative risk per Sv. Eleven dose categories were used. Also, it was assumed that radiation might increase cancer risk but it would not decrease cancer risk, so a one-tailed statistical test was used with significance at the 10% level instead of the usual 5% level. In this study about 99% of the deaths involved lagged doses less than 0.4 Sv. No evidence of association was found between radiation dose and mortality from all causes or from all cancers. However, mortality from leukemia, other than chronic lymphocytic leukemia, had a statistically significant association in a trend test with cumulative external radiation dose (estimated p=0.046, 119 deaths among 15,825 deaths). This estimated trend test p=0.046 is primarily based on 6 cases among 238 deaths above 400 mSv which by itself is not significant. The excess relative risk for leukemia excluding chronic lymphocytic leukemia was reported to be 2.18 per Sv (90% confidence interval 0.1 to 0.99). The excess relative risk for all cancer types excluding leukemia was minus 0.07 per Sv (90% confidence interval minus 0.4 to 0.3). These results may be somewhat affected by the use cumulative dose risk mathematical models rather than the more appropriate lifetime average dose rate models. Exposures to leukemia-causing organic chemicals may have occurred in the workplace.

In the second enlarged study (Cardis et al. 2005) 407,391 nuclear industry workers (90% men) in 15 countries who had available external exposure ionizing radiation dosimetry data were evaluated by a joint analysis group with 50 contributors. These 15 countries included Australia, Belgium, Canada, Finland, France, Hungary, Japan, South Korea, Lithuania, Slovak Republic, Spain, Sweden, Switzerland, United Kingdom, and the United States. Organ doses were estimated by dividing the recorded doses by organ dose bias factors. The colon and bone marrow doses were used for both all cancers excluding leukemia and for leukemia. Eleven dose groups were created. Analyses used a simple linear relative risk ɔoisson regression model of the form where relative risk equals 1+βZ where Z is the ːumulative equivalent dose in Sv; 95% likelihood based confidence intervals were ːalculated. Estimates of excess relative risk were stratified for sex, age, and calendar period, ːacility, duration of employment, and socio-economic status. A total of 24,158 (5.9%) died ɹuring the study period including 6,519 from cancer other than leukemia and 196 from ːeukemia other than chronic lymphocytic leukemia. The leukemia incidence was statistically ːimited and the risk per Sv was not significantly different from zero. ːhe key results of this second study for cancer other than leukemia (Cardis et al. 2005) are ːummarized in Fig. 23. None of the cohorts were found to have excess relative risk of cancer ɔther than those in Canada. The overall combined excess relative risk per sievert for all ːancer excluding leukemia was 0.97 (0.14 to 1.97 ninety-five percent confidence range). The ɐxcess relative risk per sievert for all cohorts for non-leukemia cancers excluding lung and ρleural cancers, which may be related to smoking, was no longer significantly different from ːzero [0.59 (-0.29 to 1.70 ninety-five percent confidence range)]. Of the fifteen countries, the ːalculated excess relative risk of cancer that was observed in Canada appear to be ɐnomalously high for some unknown reason. When the study analyzed only the other 14 ːountries the excess relative risk of cancer (other than leukemia) per sievert was no longer ːsignificantly different from zero [0.58 (-0.22 to 1.55 ninety-five percent confidence range)]. ːWith this abridgement, there would be no significant cancer risks observed in this study.

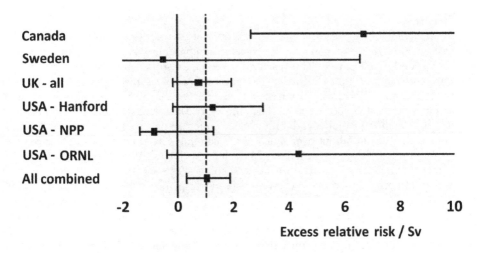

Excess relative risk per Sv for all cancer excluding leukemia in cohorts with more than 100 deaths.

(NPP = nuclear power plants)

(ORNL = Oak Ridge National Laboratory)

Fig. 23. Excess relative risk per Sv for all cancer types except leukemia in cohorts with more then 100 deaths in radiation worker study of 15 countries; NPP = nuclear power plants, ːORNL = Oak Ridge National Laboratory in Tennessee, and Hanford = the Hanford Laboratory in Richland, Washington (Cardis et al., 2005).

5. Nuclear power plant accidents

5.1 Three Mile Island

The accident at the Three Mile Island in Dauphin County, Pennsylvania, USA, on March 28, 1979, drew attention to the possible exposures of people to radionuclides that might be released from a nuclear power during an accident involving the overheating of the nuclear fuel. Because of the design issue and instrument confusion, the plant operator of the Unit 2 reactor was led to think that the cooling water level was high when in fact it was low. Eventually, feed water pumps failed and the reactor overheated. Steam and water were vented through a relief valve and the reactor control rods scrammed in and the reactor shut down. Even though the fission chain-reaction process had been terminated, the heat generated by the highly radioactive fission decay products overheated and melted part of fuel elements and their containment tubes. Volatile radionuclides such as ^{131}I and ^{137}Cs were released from the reactor containment but because of high efficiency particle filters and activated charcoal filters, very little was released to the outdoors. The extent of the risk to people living near the reactor was not clear as the accident progressed. Fortunately, no member of the public received a serious exposure. The maximum total radiation exposure to the members of the public was estimated to be only a few μSv from small releases of radioiodine, ^{131}I. The Unit 2 nuclear reactor unit was damaged beyond repair. The twin Three Mile Island Unit 1 nuclear reactor is still operational.

5.2 Chernobyl

An extremely serious and catastrophic Chernobyl nuclear reactor accident in the Ukraine occurred on 26 April 1986 when a nuclear reactor number four went super-critical during a systems test of the control rods. This reactor did not have a strong containment building and used graphite carbon for a neutron moderator. The resulting super-critical reaction destroyed the fuel elements, ruptured the reactor vessel, and exposed the graphite moderator which, with exposure to air, began to burn. The seriously damaged containment allowed the resulting smoke and effluent of radionuclides to mix with the outdoor atmosphere and spread fallout over much of the Ukraine, Soviet Union, and other parts of Europe. Not only did the most volatile radionuclides such and ^{131}I and ^{137}Cs escape but also many other fission product radionuclides such as ^{90}Sr and ^{144}Ce were part of this widespread airborne contamination. Twenty-six power plant workers died of acute whole body radiation overdoses.

The large population exposure to long-term protracted ionizing radiation associated with the 1986 Chernobyl reactor accident in the Ukraine was predicted with the linear no-threshold model to result in a virtual epidemic of long term radiation-induced cancer (Anspaugh et al. 1988). Instead, there was no apparent major health effects of this widespread protracted exposure to ionizing radiation except for thyroid disease associated with very high acute radiation doses from short-lived (half life 8 days) radioiodine (^{131}I) contamination in cow's milk and dairy products (WHO 2006, Jaworowski 2010).

5.3 Fukushima Daiichi

Three of the six boiling water nuclear power plants at Okuma, Fukushima, Japan, underwent destructive meltdown of the nuclear fuel along with widespread releases of radioactive fission products to the environment as a result of massive flooding associated with the Tohoku earthquake and tsunami on March 11, 2011. Although plants safely shut during the earthquake, the tsunami that followed completely flooded the plants and shut down all sources of local and nearby electricity. Without electricity it was not possible to operate water pumps needed to cool the reactor fuel to prevent overheating associated with the natural radioactive decay of fission product radionuclides produced during normal operations.

When the quake occurred units 1, 2, and 3 were operational, but unit 4 had previously been defueled and units 5 and 6 had been shut down for maintenance. Ultimately the fuel elements in units 1,2, and 3, overheated and ruptured, The special zircaloy cladding reacted with steam to produce large quantities hydrogen gas which vented into the containment buildings and subsequently exploded injuring workers and damaging the building walls and roofs.

Ultimately, there were widespread releases to the outdoor air of smoke and volatile fission products from the damaged nuclear fuel. These releases lofted into the atmosphere and were diluted and carried for long distances. The most important human exposures were probably those associated with contamination of water and food. Two most prominent fission products released in this accident were short-lived (half life 8 days) radioiodine (^{131}I) and long-lived (half life 30 years) radiocesium (^{137}Cs). Both of these radionuclides are beta radiation emitters that can efficiently irradiate body organs if inhaled or ingested. Associated gamma rays contribute a more diffuse and less important portion of organ doses when these radionuclides are internally deposited in the body of a person.

Ingested ^{131}I primarily follows the grass to cow to milk route of human exposure. Ingested iodine tends to concentrate in the small thyroid gland where it can efficiently irradiate the thyroid tissue. The very efficient release of iodine from damaged nuclear fuel and the short radioactive half life of ^{131}I means that the risk of exposure is relatively short-lived after an accidental release but the acute doses to the thyroid gland can be quite large. From studies of children who ingested ^{131}I from the Chernobyl nuclear reactor accident, it has be observed that the risk of thyroid cancer is primarily associated with thyroid cumulative radiation doses larger than about 10 Sv.

Ingestion or inhalation of ^{137}Cs leads to very uniformly widespread irradiation of body because cesium is quite soluble in most of its chemical forms and behaves somewhat like potassium in the human body. Although ^{137}Cs has a long physical half-life, it has a relatively short biological half life on the order of about one month. For an acute single intake, very little remains in the human body after one year. In the environment, cesium tends to strongly bind by ion-exchange with clay in soils, somewhat limiting its uptake into plants.

In order to evaluate the biological risk, including carcinogenesis, associated with the intake of ^{137}Cs, fifty-four beagles were injected with graded dosages of ^{137}Cs as cesium chloride from 32.5 to 148 MBq per kg body weight and held for lifetime study at the Inhalation Toxicology Research Institute (Mauderly and Daynes, 1989)(Fig. 24). Because of the relatively short biological retention half-time of about 30 days (Boecker, 1969), about half of the whole body absorbed dose from ^{137}Cs beta rays and ^{137m}Ba daughter gamma rays was delivered in the first 30 days after the injection and nearly all of the dose was delivered within one year. Those deaths that occurred prior to one-half year after injection were associated with direct injury to the blood forming organs by beagles receiving the highest dosages of ^{137}Cs. There were only 12 unexposed controls assigned to this study, but the fate of 52 other controls assigned to contemporaneous studies was used for comparison to the exposed dogs. Those beagles that survived the acute injury phase in the very highest dose groups lived about as long as unexposed controls and had insufficient cases of radiation induced cancer to develop a meaningful three-dimensional model.

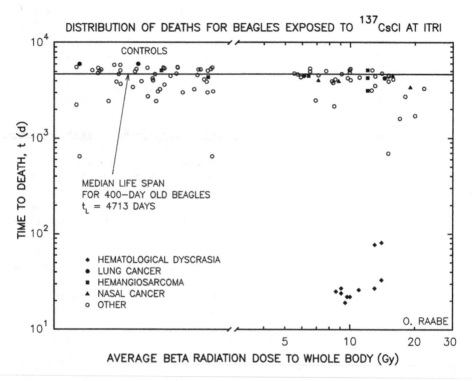

Fig. 24. Distribution of deaths in lifetime studies of beagles injected with [137]CsCl at ITRI (Raabe,1994).

A Peto trend test (Peto et al, 1980) of the cancer cases showed that two cancer types had a statistically significant occurrence: (a) nasal carcinomas reported in 4 exposed dogs and not observed in controls, and (b) hemangiosarcoma in different organs (4 cases) with one case in controls. The lowest lifetime cumulative whole body radiation dose associated with a death with nasal carcinoma was 7.7 Gy; this compares with the lowest skeletal beta radiation dose of 18.4 Gy associated with the occurrence of nasal carcinoma in beagles exposed to ^{90}Sr at Davis. There was no statistically significant incidence of lung, liver, or bone cancer, and no reported cases of leukemia. This result is consistent with the expected risk of radiation-induced cancer from protracted exposure as discussed above.

6. Summary of protracted and repeated exposure cancer induction risk

As presented in the previous sections, the studies of repeated and protracted exposures to ionizing radiation have shown that radiation induced cancer follows a sharply non-linear upward temporal pattern with a virtual threshold at lower life-time average dose rates (Raabe, 2010). The precision and time-delay of the cancer induction phenomenon indicate an underlying gradual biological process involving many altered cells associated with cellular deoxyribonucleic acid (DNA) mutations, clonal development depending on cell division cycles, cellular maturation, and average ionizing radiation dose rate over long latency periods. Studies described above of protracted exposures to ionizing radiation from internally deposited radionuclides in people and laboratory animals have demonstrated that cancer induction risk is actually a somewhat precise function of average dose-rate and that at lower average dose rates cancer latency can exceed natural life-span.

Because of the long latency at low dose rate that may exceed the natural life-span, the radiation-induced cancer risk associated with protracted exposures to ionizing radiation involves a life-span virtual threshold when the lifetime average dose rate is low and the cumulative dose to sensitive tissues is below about 10 Sv.

Life-span virtual thresholds for radiation-induced cancer risk should exist for other types of protracted and fractionated exposures including radon inhalation and external exposures associated with background levels of ionizing radiation from environmental radionuclides. The organs, tissues, and cells of the body cannot distinguish between ionizing radiation emitted by internally deposited radionuclides or entering the body from external sources. Low linear energy transfer (LET) gamma rays from external sources yield ionizing energetic electrons in the body and beta particles from internally deposited radionuclides are also ionizing energetic electrons. High LET neutrons from external sources yield ionizing energetic protons whose radiation quality weighting factors are similar to high LET ionizing alpha particles from internally deposited radionuclides (Raabe, 2011).

Another important finding is that two beta particles are about equal on the average to one alpha particle in the radiation induction process (Raabe 2010). This finding suggests that double strand damage to cellular DNA is involved in the cancer induction process. While one alpha particle hitting DNA may produce the necessary double-strand break, two beta particles almost simultaneously hitting the DNA are apparently needed to produce similar double-stand breaks. Also, the shape of the cancer-induction dose-response curve is not linear but rather sharply increasing as a function of cumulative dose delivered from virtually zero risk at low doses to high risk at high doses as a function of dose rate.

7. Single acute exposures to ionizing radiation

7.1 Atomic bomb survivor studies

A quite different phenomenon has been observed in studies of the Japanese survivors of atomic bombs detonated over Hiroshima, August 6, 1945, and over Nagasaki, August 9, 1945 (Manhattan Engineer District 1946). These Japanese survivors were exposed instantaneously to a large acute pulse of external gamma radiation (and some neutrons) delivered in about one minute. Many of these acute absorbed doses were several hundred times the normal annual radiation exposure from background radiation. All of the cells of the body were exposed to this large ionizing radiation pulse and those cells and body tissues were all affected in some way. Years later some of the highly exposed survivors developed cancer of the same types as occurred in the control population but somewhat earlier and at somewhat higher rates. In addition, the increases in these cancer cases were proportional to that one-minute whole-body absorbed radiation dose. This linear relationship is in sharp contrast with cancer induction observed in the protracted and repeated exposures described above. Except for some early cases of myeloid leukemia caused by acute radiation damage to the blood forming tissues of the body from the higher exposures, there were no apparent radiation induced cancer cases among the atomic bomb survivors (Raabe, 2011).

There has been a long-standing scientific disagreement concerning cancer induction associated with protracted and repeated exposures to ionizing radiation and the carcinogenic effects of the atomic bomb acute instantaneous exposures. The reason for this conflict is readily apparent from the data themselves. It is explained by the fact that the increased risk of cancer observed in the atomic bomb survivor studies was primarily the result of acute high dose-rate promotion of ongoing biological processes that lead to cancer in the Japanese people rather than actual de-novo cancer induction.

The Japanese survivor studies are very important since they have been extremely well done and represent a significant body of human data about the possible effects of acute exposure to ionizing radiation (RERF, 2008). Also, the currently accepted methodology for estimation of solid cancer risk associated with human exposure to ionizing radiation is based primarily on the detailed studies of the Japanese atomic-bomb survivors (ICRP-26, 1977; ICRP-60, 1991; BEIR VII Phase 2, 2006; NCRP 136, 2001; ICRP 103, 2007). The prevailing models of human ionizing radiation exposure risk and relative organ sensitivities (ICRP tissue weighting factors, W_T) are based primarily on those studies. Specifically, the concept that cancer risk as a linear function of cumulative dose and follows a linear no-threshold cancer risk model is based primarily on the acute high dose rate exposures received by the Japanese atomic bomb survivors.

7.2 Radiation promoted cancer from acute exposures

The Radiation Effects Research Foundation (RERF) and its predecessor Atomic Bomb Casualty Commission (ABCC) studied the development of solid malignant tumors in about 80,000 of the Japanese atomic bomb survivors (Pierce and Preston, 2000; Preston et al., 2003; Preston et al., 2004; Preston et al., 2007). The main study considered 79,972 survivors for which there were available calculated radiation doses. Of these, 44,636 survivors had calculated doses that exceeded 5 mSv. Those with less than 5 mSv doses were used as the control group (Preston et al., 2007). These about one-minute acute exposures involved very high-energy gamma radiation and some neutrons. Myeloid leukemia from high dose bone marrow exposures followed a different response course and is usually considered separately from the solid tumor incidence.

The traditional approach is to assume that the solid malignant tumors are the result of simple stochastic (isolated and random) initiating events in individual cells that occurred during that about one-minute exposure. This simplistic stochastic model of ionizing-radiation-induced cancer is based on the idea that a single cell is randomly altered by a unique ionizing radiation event causing a unique pre-malignant mutation in that cellular deoxyribonucleic acid [DNA] (Moolgavkar et al. 1988; Heidenreich et al. 1997). This single random event then ultimately leads to a clone of similar pre-malignant cells. Later, usually much later, a second random DNA alteration occurs in one of the clonal pre-malignant cells that produces a malignant cell that develops into a monoclonal malignant tumor. This process began with the single random (stochastic) cellular event. This process can be advanced by promoter agents including other ionizing radiation that presumably affect the clonal development, quantity, and maturation of the pre-malignant clonal cells (Heidenreich et al., 2007) . A cancer promoter is anything that advances the development of a malignancy other than a directly carcinogenic agent or an intrinsic component of the carcinogenesis process (Casarett, 1968). Mathematical models are readily constructed with unknown parameters and hypothetical modifying factors and process lag times designed to fit the data associated with acute ionizing radiation exposures (Heidenreich et al., 1997).

The internal emitter studies discussed above strongly suggest that double strand DNA damage or a related phenomenon is involved in the cancer induction process associated with ionizing radiation. In particular, two low LET beta particles were found to be required to match the radiation induction process associated with each alpha particle (Raabe 2010). Since the exposure of the Japanese atomic bomb survivors was primarily associated with low LET gamma radiation and associated energetic electrons, two hits at the same region of DNA in a target cell would be expected to be required for the induction of cancer. The resulting increase in cancer by induction in this two-hit process would follow a sharply increasing curvilinear power function of increasing cumulative absorbed dose. In fact, the increase in cancer among the atomic bomb survivors tended to follow a remarkably linear pattern as a function of absorbed ionizing radiation dose. Deterministic cancer promotion of those types found in the exposed population rather than simple isolated stochastic cancer induction better explains the increase of solid cancers in the atomic bomb survivor studies.

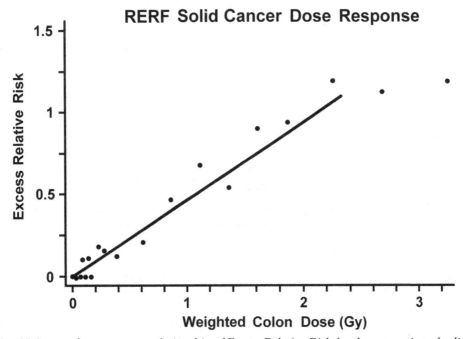

Fig. 25. Linear dose-response relationship of Excess Relative Risk for the promotion of solid cancer in Japanese survivors of the atomic bomb detonations at Hiroshima and Nagasaki in 1945 with respect to survivors who received low radiation exposures as reported by the Radiation Effects Research Foundation [RERF](Preston et al. 2007).

The studies of the atomic bomb survivors demonstrate a linear dose-response promotional effect related to the natural or existing biological processes that may eventually lead to cancer in the exposed population (Fig. 25). These processes involve years of cellular division, clonal expansion, and cellular maturation. The exposure to a sudden high dose of ionizing radiation delivered in about one minute at the time of the nuclear detonations may have advanced or stimulated the cellular changes that eventually lead to various typical types of cancer. Hence, some cancers may have appeared at an earlier time than otherwise would have occurred based on the existing underlying cellular and tissue processes (Fig. 26). This promotional effect was observed to advance cancer rates not only relatively soon after exposure but throughout the lives of the exposed individuals (Fig. 27). This behavior is proportional to the instantaneous dose just as would be expected for any phenomenon that involves augmentation of existing processes rather than a few random or "stochastic" changes in a few select cells.

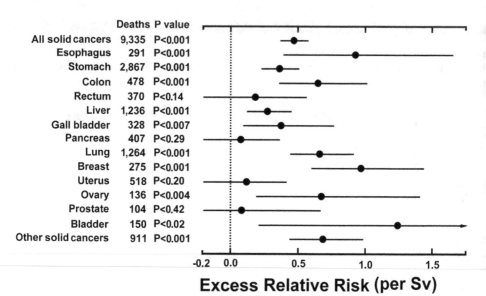

Excess Relative Risk (per Sv)

Fig. 26. Observed Excess Relative Risk by body organ for promoted solid cancer observed for one Sv of equivalent ionizing radiation dose in highly irradiated Japanese atomic bomb survivors compared to survivors who received low radiation exposure as reported by the Radiation Effects Research Foundation [RERF] (Preston et al. 2003).

Concerning the solid tumor incidence in the atomic bomb survivor studies, Pierce and Mendelsohn (1999) pose the question, "How could it be that the excess cancer rate might depend only on age and not on time since exposure or age at exposure?" Fig. 27 shows that the increase in malignant solid tumors in the atomic bomb survivors associated with their radiation exposure follows the same lifetime pattern irrespective of the age at exposure. The simple answer is that the normal progression of cancer incidence in the population was somewhat promoted by the radiation exposure without the actual independent induction of cancer. This promotion is not a single event stochastic process but rather the result of the almost instantaneous delivery of ionizing charged electrons produced in all the tissues by ionizing radiation from the atomic bombs. The tissues response is complex and unfocused. The cells of the tissue communicate among themselves as a result of the exposure in a process called bystander effect with various responses evoked even among the cells in these tissues that were not directly impacted (Hall 2003). Tissues respond to sudden ionizing radiation exposure with up-regulation of various genes and down-regulation of others (Snyder and Morgan, 2004; Coleman et al. 2005). This imprint is apparently not lost and carries forward throughout life promoting the existing biological processes that may lead to cancer in the exposed atomic bomb survivor population. This complex process suggests a systems biology phenomenon rather than a unique stochastic response (Barcellos-Hoff, 2008).

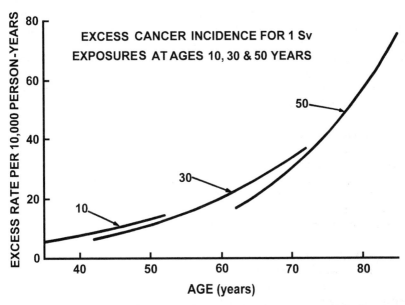

Fig. 27. Observed gender-averaged age-specific excess incidence rates at 1 Sv for most major solid cancers over the 1958-1987 follow-up period for ages at exposure 10 y, 30 y, and 50 for the Japanese atomic bomb survivors (Pierce and Mendelsohn, 1999). The excess rates appear to depend only on age and not on time since exposure or age at exposure as might be expected for radiation-induced promotion of the cancer types normally found in this population.

The Japanese atomic bomb survivor data are unique because they do not in any way predict the observed carcinogenesis associated with protracted exposures as occur in the case of internal emitters. In fact, the acute gamma ray exposures clearly represent a completely different mechanism of carcinogenesis from that which occurs with protracted exposure as with long-lived internal emitters (Raabe 2011). The resulting cancer promotion phenomena in the atomic bomb survivors should not be expected to describe the effects of similar exposures delivered uniformly or fractionated over a relatively long period of time. Since promotion is a relative process rather than an absolute process, it is not meaningful to try to create absolute risk estimates from relative response information.

The tissue-weighting factors (w_T) developed by the ICRP to calculate the so-called effective dose are actually a reflection of the convolution of the underlying incidences of different types of cancer in the control population and the relative promotional effect of the whole body exposure to gamma rays and some neutrons in the Japanese atomic bomb life-span studies (ICRP 26, 1977; ICRP 30, 1979; ICRP 60, 1991; ICRP 103, 2007). Cancers that were somewhat rare in this Japanese population, such as bone cancer, were assigned relatively low tissue weighting factors relative to the whole body cancer or assumed genetic risk (such as w_T = 0.01 for bone surfaces). Cancers that were somewhat common in this Japanese population, such as lung cancer, were assigned relatively high ratio values relative to the whole body cancer or genetic risk (such as w_T = 0.12 for whole lung) . These tissue-weighting factors (w_T) can be related to observed cancer promotion, but are unrelated to the cancer induction associated with protracted or fractionated exposure to ionizing radiation. The use of tissue-weighting factors (w_T) recommended by the ICRP based on the atomic bomb survivor studies is inappropriate for protracted or fractionated exposures.

The elaborate Radiation Effects Research Foundation (RERF) studies of the atomic bomb survivors have investigated in rigorous detail the effect of whole body irradiation by high-energy gamma rays (and some neutrons) delivered in about one minute. The A-bomb RERF life-span study clearly describes a meaningful linear dose model of promotion of ongoing biological processes that lead to increased cancer rates for brief high dose rate exposure to ionizing radiation. The relative risk values might be applicable to other brief high dose-rate ionizing radiation exposures as may occur in occupational exposures or in medical diagnosis and treatment (Hall and Brenner, 2008). However, there is still considerable uncertainty for acute doses less than about 0.05 Sv. Small acute doses may by beneficial as they may promote or stimulate DNA repair or other defensive cellular phenomena that reduce promotional cancer risks associated with ongoing cellular processes that might otherwise lead to cancer (Feinendegen 2005).

7.3 Acute medical exposures
7.3.1 Medical exposure concerns
Among the most common instantaneous exposures of people to ionizing radiation are those associated with diagnostic medical radiographs (X-rays). Modern computed tomography (CT) scans using x rays with higher radiation doses can provide precise three-dimensional images and are of great medical value. Fluoroscopic images that use nearly continuous irradiation are also valuable in certain medical procedures. The question as to whether these acute diagnostic medical exposures result in an increased cancer risk has been considered and even assumed in various studies. The promotion of cancer observed among the Japanese atomic bomb survivors is a reasonable basis for considering the possible cancer promotion associated with medical exposures to ionizing radiation, but protracted exposures do not seem to involve cancer risk at small doses and sometimes appear beneficial.

7.3.2 Neonatal medical exposures
Among the earliest studies of possible effects of medical exposures of patients associated with diagnostic X-rays are those associated with the Oxford Survey of Childhood Cancer (OSCC) of deaths in England, Wales, and Scotland from childhood cancer. These case-control studies were initiated and advanced by Alice Stewart, a British physician (Stewart et al., 1956; Stewart et al., 1958). The basic hypothesis of these studies is that childhood cancer was caused in part by fetal exposures from medical X-rays associated with diagnostic radiology of the pregnant mothers. Healthy control children were matched by sex, by birth location and, as closely as readily possible, by birth date. A comprehensive summary and statistical evaluation has been published (Mole 1990). The overall study considered a majority of the cancer deaths in of children in Britian under the age of 16 primarily spanning

the years 1953 to 1978. Most of the major findings were based on children who were born between 1940 and 1969. The initial studies inquired from the mothers about X-ray films that might have been made during pregnancy. Later studies involved verification by review of medical records, but not all cases could be verified. There was no direct ionizing radiation dosimetry associated with these studies. Overall, there was a strong trend to a higher fraction of diagnostic radiology among the mothers of children who died of cancer compared to the healthy controls. Typical overall raw values of these ratios were 1.33 for children dying under age 6, 1.38 for children age 6 or older dying under age 10, and 1.27 for children age 10 or older dying under age 16. For births that occurred prior to 1957 the odds ratios were significantly above unity, while for later births they were not. It may have been most difficult to verify the X-ray recollections of the mothers for those earlier years. The odds ratios for X-raying in Britian for the four birth years 1958 to 1961 were 1.27. 1.36 and`1.02 for cancer deaths at ages 0 to 5, 6 to 9, and 10 to15, respectively. The overall odds ratio for all ages 0 to 15 years was 1.23 with a 95% confidence interval from 1.04 to 1.48 and the associated excess lethal tumor rate from in utero x ray exposure was 0.00028 with a confidence interval from 0.48 to 0.00058 (Mole. 1990).

The findings from the Oxford Survey of Childhood Cancer tend to show an important trend, but they may not convincingly prove that in utero X-rays were the cause of childhood cancer. Much of the earlier data were based on patient-reported X-ray exposures. A mother whose child died of cancer is more likely to remember or readily believe that she received X-ray exposures than a mother of a healthy child. Early medical records may have been incomplete or difficult to locate.

In many cases the X-rays may have been associated with some medical complication or abnormality associated with the pregnancy especially in many cases where the mother was X-rayed more than once. Those medical abnormalities, if they existed, may have directly contributed to later development of cancer in the child. Another possibility is that the in utero exposure promoted a pre-cancer process that led to earlier cancer development than would have otherwise occurred, but the low radiation doses involved in these studies was too low to correlate with the atomic bomb survivor studies promotional effects (Preston et al., 2007). Totter and McPherson (1981) point out that differences in population groups in the Oxford Survey and other factors can explain for the observed relationships.

Although no dosimetry was available to accurately estimate the maternal or in utero ionizing radiation doses associated with the X-ray exposures in these studies, Mole (1990) attempted to estimate the risk of childhood cancer per unit of absorbed ionizing radiation to the fetus using estimates made in 1966 by the Adrian Committee of the British Ministry of Health. He assumed that average dose from all obstetric radiography was 0.61 cGy. From his calculated excess lethal tumor rate he calculated and reported a risk coefficient for irradiation in the third trimester for childhood cancer deaths at ages up to 15 years equal to 0.00046 per cGy with a confidence interval from 0.8 to 0.00095 per cGy. This estimate is not valid because the Oxford Survey of Childhood Cancer does not provide any dose-response information for anyone in the study nor for the whole population at risk. More than 95% of the cases of childhood cancer are not explained by neonatal radiography. In 1958 there were 840,196 births in Britian and it was likely that at least 100,000 babies were irradiated in utero based on the estimates (Mole 1990). The number of deaths associated with cancer ages 0-14 born in 1958 was 977 and the data show that about 15% of those or about 150 had been irradiated in utero. Therefore the fraction of those irradiated that died of cancer in childhood was about 0.15%.

It might be noted, however, that the promotion of cancer observed in the atomic bomb survivor studies is a mechanism that could explain the observed results, but there are no statistically significant radiation promoted cancer risks for acute doses below about 10 cSv among the atomic bomb survivor data. (Preston et al. 2007). Totter and McPherson (1981) point out that differences in population groups can account for the observed misleading relationships since the sub-population of cancer cases does not have the same probability of being X-rayed as the sub-population of controls.

7.3.3 Medical diagnostic X-ray risks

Medical diagnostic radiology has become one of the highest sources of exposure to ionizing radiation in many countries of the world. This is largely associated with the increasing use of computed tomography (CT) images using x rays which involve much higher radiation doses than conventional X-ray images. Since these doses are delivered almost instantaneously, they can be expected to present similar cancer promotion risks as observed in the atomic bomb survivor studies discussed above. However, the radiation doses involved in the use of CT examination are typically well below the approximately 10 cSv dose above which the cancer promotion risk is statistically significant in the Japanese atomic bomb survivor studies (Preston et al 2007). Hence, attempt to extrapolate the cancer promotion risk associated with these small doses involves considerable imprecision and uncertainty. No one can be sure that the same linear no-threshold model applies precisely to individual organ exposures in radiology as applies to whole body exposures as occurred for the Japanese atomic bomb survivors. In addition, biological processes that affect the behavior of cells and tissues at lower doses may alter or cancel the promotional phenomena. For example, low acute doses may stimulate cellular DNA repair mechanisms.

Some investigators have created somewhat precise models of what may be occurring at low doses to predict lifetime cancer risks from the acute x ray exposures that may be associated with diagnostic radiology using linear no-threshold models based on the atomic bomb survivor studies of cancer promotion. These studies of diagnostic medical exposure to ionizing radiation usually refer to their studies as estimates of risk of radiation induced cancer, but in this chapter the acute exposure dose phenomenon has been be more correctly shown to be cancer promotion. In addition, some of these studies report their results as cancer risks, but they are not measured risks but calculated estimates subject to the theoretical use of the linear model and the approximate parameters of that model obtained from the atomic bomb survivor studies.

Brenner and associates (Brenner et al., 2000; Brenner, 2002; Brenner and Hall, 2007; Hall and Brenner, 2008) have published a series of papers providing estimates of lifetime fatal cancer as a function of dose from CT scans of various organs of the body as a function of exposure age utilizing data from the atomic bomb survivor studies and the risk models developed from those data by the United States National Research Council (BEIR V, 1990) and the International Commission on Radiological Protection (ICRP-60, 1991).

The calculated cancer risk depends on the organ irradiated and the sex and age of the irradiated person. The calculated overall lifetime attributable risk tends to be higher for children than for adults since children have many more years of life during which cancer might develop. Representative values for lifetime attributable risk in these studies are from

).14% for babies to 0.02% for adults. These are typical calculated values using linear no-threshold models following from the separate organ exposure results at high doses from the atomic bomb survivor studies.

Based on BEIR V (1990), Brenner et al. (2000) summarizes observed relative increased risk of cancer for an exposure of Japanese atomic bomb survivors exposed to one Gy for males as a function of age at exposure (Fig. 28). This representation clearly shows cancer promotion. The total risk per year is about the same for everyone. Given an 85 year life span in this figure, the risk per year of life for a five year old boy is about 13% over 80 years = 0.16% per year. For a 25 year old man the risk per year of life if about 9.5% over 60 years = 0.16% per year. For a 45 year old man the risk per year of life if about 6.5% over 40 years = 0.16% per year. The total cancer risk per year is almost independent of age at exposure. This is not explained by any simple stochastic cancer induction model, but is explained by some sort of deterministic cellular reprogramming that promotes the cancer process in a somewhat linear way.

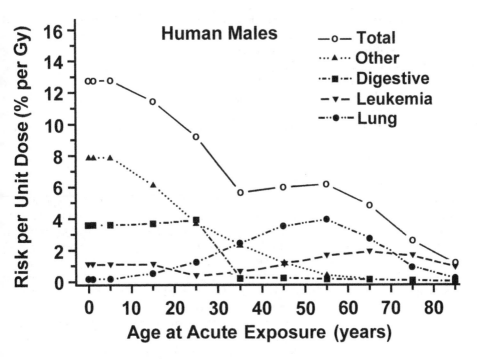

Fig. 28. Lifetime attributable cancer mortality risks for males by cancer type as a function of age at a single acute exposure of one Gy of γ radiation based on BEIR V (Brenner et al., 2000).

Addition evidence for promotion is seen in the lung cancer rates shown in Figure 28. Lung cancer in these males may be related to smoking history. For young boys up to age 15 there is no apparent promotion of lung cancer (total lifetime risk about 0.5%). For these non-smokers there was no ongoing lung cancer development to promote. But lung cancer risk per year climbs for older males many of whom are likely to have been smokers with a peak risk at age 55.

Berrington de Gonzalez and Darby (2004) estimated the hypothetical overall population cancer risks from diagnostic X-rays in the UK and 14 other countries based on frequency of diagnostic X-ray use, doses to different body organs, and linear no-threshold risk models based on the Japanese atomic bomb survival data. Their results led them to report that the cumulative risk of cancer for people up to age 75 from diagnostic X-ray exposures was indicated to be in the range from 0.6% to 1.8%.

In studies of 64,172 tuberculosis patients of whom 39% were exposed externally to highly fractionated x ray chest fluoroscopies, lung cancer deaths showed no evidence of cancer risk associated with the x ray exposures with the relative risk at a cumulative doses of 1 Sv being 1.00 [95% confidence interval 0.94-1.07] (Howe 1995). Also, studies of people exposed to unusually high levels of protracted external ionizing radiation associated with natural background (up to 260 mSv y^{-1}) have not detected increased cancer risks (Ghiass et al., 2002).

Overall, the attributable risk values estimated in these reports from individual X-rays and CT scans are hypothetical and unproved since the doses involved are much smaller than those for which statistically significant elevations in cancer rates have been observed in the atomic bomb survivors. The calculations depend on the unproved assumption that the promotion risk for acute exposures is perfectly linear down to a dose near zero, the linear no-threshold hypothesis. Various biological processes could compete with or nullify the ionizing radiation promotion process for these relatively small individual acute exposures. There seems to be considerable evidence there is no significant risk at very low acute doses, and some possible beneficial effects.

8. Discussion

An important observation is that the dose-response relationships for protracted exposures to internally deposited radionuclides are quite different from the observed effects of a single high dose rate exposure as experienced by the atomic bomb survivors in temporal manifestation and preciseness of the effect. In the internal emitter studies the time to cancer development depends on the lifetime average dose rate and the cases are not randomly distributed with respect to time but rather are tightly grouped over a predictable and narrow range of times. At low lifetime average dose rate the time required to develop cancer may exceed the natural life span yielding a virtual threshold for cancer induction.

Bone cancer and lung cancer are equally sensitive in the internal emitter studies with beagles, but there is no apparent induction of bone cancer in the atomic bomb survivors while there is a considerable increase in lung cancer. The main data are the relative increase that suggests radiation promotion was more effective than radiation induction.

Based on the linear relative risk relationship observed for acute exposure in the RERF studies of the atomic bomb survivors, the ICRP developed an absolute risk model for radiation-induced cancer as a function of absorbed cumulative ionizing radiation dose that was believed to apply to all forms of exposures to ionizing radiation. It is assumed that the linear model can predict cancer induction risk but needs a hypothetical dose and dose-rate effectiveness correction factor (DDREF) for protracted, fractionated, or low dose exposures.

In recognition that protracted or fractionated exposures have been observed to be less carcinogenic than the RERF data directly suggest, the ICRP reduced the calculated risk using a DDREFof 2, but still assumed lifetime LNT linearity of cancer induction was maintained (ICRP 103, 2007). This approach has led to simplistic linear cancer induction risk factors such as " 5% per Sv" (ICRP 60, 1991; ICRP 103, 2007).

The idea that such a risk factor is valid rests on the assumption that each ionizing-radiation-induced malignancy is the result of a simple isolated random or "stochastic" initial event that is equally probable in any exposed person. Although each malignant tumor is monoclonal, the biological processes that lead to that first tumor (and successive tumors) are apparently deterministic systemic processes.

Others have also adopted similar linear cancer induction risk models as a function of person-sievert based on the RERF studies. However, as discussed above, these cancer inductions models do not apply to non-acute exposures that depend on lifetime average dose rate rather than simply on cumulative dose.

These cancer induction risk models are not valid because of the confusion of cancer promotion associated with brief high dose-rate exposures and cancer induction associated with protracted exposures. Unfortunately, these misleading cancer induction risk values may be used to implement expensive cleanup standards for environmental radioactivity.

The LNT model does not predict the observed effects or lack thereof for protracted exposures to ionizing radiation exposure (Jaworowski 2010). The LNT model does not readily predict the results of a cumulative set of fractionated medical diagnostic x ray exposures of the lung where there was no indication of increased lung cancer with the observed relative risk at a cumulative doses of 1 Sv being 1.00 [95% confidence interval 0.94-1.07] contrasting with the high expected relative risk based on the atomic bomb survivors being 1.60 [95% confidence interval 1.27-1.99] (Howe 1995). Also, the large population exposures to long-term protracted ionizing radiation associated with the 1986 Chernobyl reactor accident in the Ukraine was predicted with LNT to result in a virtual epidemic of long term radiation-induced cancer (Anspaugh et al. 1988). Instead, there was no apparent major effect of this widespread protracted exposure to ionizing radiation except for thyroid disease associated with very high acute radiation doses from short-lived ^{131}I in milk (WHO, 2006; Jaworowski, 2010).

Failure to realize the fundamental differences between cancer promotion and cancer induction has been the source of scientific misunderstandings and risk estimate disagreements. A logical barrier has stood between the linear no-threshold (LNT) model of cancer promotion in the acute exposures associated with the Japanese survivors and the virtual threshold associated with induction of cancer associated with protracted exposures as received from long-lived internally deposited radionuclides in humans or animals. Further, it has led to a systematic overestimation of cancer induction risk from protracted exposures to ionizing radiation.

The underlying assumption in these current recommendations is that risk of radiation-induced cancer is proportional to cumulative dose without threshold. This assumption obviously conflicts with the induction of cancer observed in the cases of lifetime protracted and repeated exposures to ionizing radiation. An understanding of the source of this conflict of data interpretation is essential for sound estimates of cancer risk associated with exposures to ionizing radiation.

Radiation cancer induction is not a simple stochastic process in the sense that it is not the result of a random radiation event in a single cell that converts it into some kind of pre-malignant neoplastic cell. In reality, both radiation promoted and radiation-induced cancers are the result of complex biological processes involving multiple cellular events. These events include but are not limited to intercellular communication in irradiated tissues, up-regulation and down-regulation of genes, DNA mutations, cell division rates, clonal expansion of altered cells, and various responses to numerous specific radiation events such as DNA double-strand breaks. Induced cancer risk from protracted radiation exposures of body organs depends on the lifetime average dose rate to the irradiated organs rather than on some function of cumulative dose. The middle Twentieth Century radiation protection standards promulgated by the ICRP were properly focused on minimizing dose rate (ICRP 2, 1959).

9. Conclusion

Clearly the development of a radiation-induced malignant tumor from either protracted ionizing radiation exposures or acute exposures is not the result of a single random interaction of the ionizing radiation with an isolated cell. Hence, the term stochastic as used by the ICRP is not appropriate. The following conclusions indicate that major revisions of methodology and standards are needed and other currently accepted ionizing radiation risk models should be improved to provide more meaningful and realistic estimates of ionizing radiation cancer risk:

(1) Cancer induction risk associated with protracted or fractionated ionizing radiation exposure is a non-linear function of lifetime average dose rate to the affected tissues and exhibits a virtual threshold at low lifetime average dose rates. (2) Cumulative radiation dose is neither an accurate nor an appropriate measure of cancer induction risk for protracted or fractionated ionizing radiation exposure except for describing the virtual threshold for various exposures. (3) Cancer promotion risk for ongoing lifetime biological processes is a relative process as seen in the RERF studies of the Japanese atomic bomb survivors for brief high dose-rate exposures to ionizing radiation. It cannot be used to estimate cancer induction risk from protracted or fractionated ionizing radiation exposures over long times and at low dose rates.

10. References

Anspaugh, L.R.; Catlin R.J. & Goldman M. (1988). The Global Impact of the Chernobyl Reactor Accident. *Science* 242:1513–1519.

Barcellos-Hoff, M.H. (2008). Cancer as an Emergent Phenomenon in Systems Radiation Biology. *Radiat. Environ. Biophys.* 47: 33–38.

BEIR IV. (1988). *Health Risks of Radon and Other Internally Deposited Alpha-Emitters.* National Research Council, Washington DC: The National Academy Press.

BEIR V. (1990). *Health Effects of Exposure to Low Levels of Ionizing Radiation.* National Research Council, Washington DC: The National Academy Press.

BEIR VI (1999). *Health Effects of Exposure to Radon.* National Research Council, Washington DC: The National Academies Press.

BEIR VII Phase 2. (2006). *Health Risks From Exposure to Low Levels of Ionizing Radiation.* National Research Council, Washington DC: The National Academies Press.

Berrington de Gonzalez A. & Darby S. (2004). Risk of Cancer from Diagnostic X-rays: Estimates for the UK and 14 other countries. *Lancet* 363:345-51.

Boecker, B.B. (1969). Comparison of ^{137}Cs Metabolism in the Beagle Dog Following Inhalation and Intravenous Injection. *Health Phys.* 16: 785-788.

Brenner, D.J. (2002). Estimating Cancer Risks From Pediatric CT: Going From the Qualitative to the Quantitative. *Pediatr. Radiol.* 32:228-231.

Brenner, D.J.; Elliston. C.D.; Hall, E.J. & Berdon, W.E. (2000). Estimated Risks of Radiation-Induced Fatal Cancer From Pediatric CT. *American J. Radiology* 176:289-296.

Brenner, D.J. & Hall, E.J. (2007). Computed Tomography - An Increasing Source of Radiation Exposure. *N. Engl. J. Med.* 357:2277-2284.

Cardis, E. & International Agency for Research on Cancer Study Group on Cancer Risk Among Nuclear Industry Workers. (1994). Direct Estimates of Cancer Mortality Due to Low Doses of Ionising Radiation: An International Study. *Lancet* 344:1039-1043.

Cardis, E.; Gilbert, E.S.; Carpenter, L.; Howe, G.; Kato, I.; Armstrong, B.K.; Beral, V.; Cowper G.; Douglas, A.; Fix, J.; Kaldor, J.; Lave, C.; Salmon, L.; Smith, P.G.; Voelz, G.L. & Wiggs, L.D. (1995). Effects of Low Doses and Low Dose Rates on External Ionizing Radiation: Cancer Mortality Among Nuclear Industry Workers in Three Countries. *Radiat. Res.* 142:117-132.

Cardis, E.; Vrijheid, M.; Blettner, M.; Gilbert, E.; Hakama, M.; Hill, C,; Howe, G.; Kaldor, J.; Muirhead, C.R.; Schubauer-Berigan, M.;Yoshimura, T. & International Study Group.(2005). Risk of Cancer After Low Doses of Ionizing Radiation: Retrospective Cohort Study in 15 Countries. *British Med. J.* 331:77-80.

Casarett, A.P. (1968). *Radiation Biology.* Prentice-Hall, Englewood Cliffs, New Jersey.

Cohen, B.L. (1995). Test of the Linear-no Threshold Theory of Radiation Carcinogenesis for Inhaled Radon Decay Products. *Health Phys.* 68:157-174.

Coleman, M.A.; Yin, E.; Peterson, L.E.; Nelson,D.; Sorensen, K.;Tucker, J.D.; Wyrobek, A.J. Low-dose irradiation alters the transcript profiles of human lymphoblastoid cells including genes associated with cytogenetic radioadaptive response. (2005). *Radiat. Res.* 164:369-382.

Eisenbud, M. & Gesell, T.F. (1997) *Environmental Radioactivity From Natural, Industrial, and Military Sources,* Fourth Edition, Academic Press, San Diego, California.

Evans, R.D. (1955). *The Atomic Nucleus.* McGraw-Hill Book Company, New York.

Evans, R.D. (1943). Protection of Radium Dial Workers and Radiologists from Injury by Radium. *Journal of Industrial Hygiene and Toxicology* 25: 253-269.

Evans, R.D.; Keane, A.T. & Shanahan, M.M. (1972). Radiogenic Effects In Man of Long-term Skeletal Alpha-irradiation. In: Stover BJ, Jee WSS, eds. *Radiobiology of Plutonium.* Salt Lake City, UT: The J.W. Press; 431-468.

Evans, R.D. (1974). Radium In Man. *Health Phys.* 27: 497-510.

Finkel, M.P.; Biskis, B.P. & Jinkins. P.B. (1969) Toxicity of radium-226 in mice. In: *Radium Induced Cancer.* Vienna: International Atomic Energy Agency [IAEA-SM-118/11]; 369-391.

Feinendegen, L.E. (2005). Evidence For Beneficial Low Level Radiation Effects and Radiation Hormesis. *Br. J. Radiology* 78:3-7.

Fornalski, K.W. & Dobrzynski, L. (2011). Pooled Bayesian Analysis of Twenty-eight Studies on Radon Induced Lung Cancers. *Health Phys.* 101:265-273.

Ghiassi, M.; Mortazavi, S.M.J.; Cameron, J.R.; Niroomand-rad, A. & Karam, P.A. (2002). Very High Background Radiation Areas of Ramsar, Iran: Preliminary Biological Studies. *Health Phys.* 81:87-93.

Hall, E.J. (2003). The Bystander Effect. *Health Phys.* 85:31-35.

Hall, E.J. & Brenner, D.J. (2008). Cancer risks from diagnostic radiology. *Br. J. Radiology* 81:362-378.

Heidenreich, W.F.; Luebeck, E.G.; Moolgavkar, S.H. (1997). Some properties of the Hazard Function of the Two-Mutation Clonal Expansion Model. *Risk Anal.* 17:391-399.

Heidenreich, W.F.; Cullings, H.M.; Funamoto, S. & Paretske, H.G. (2007). Promoting action of radiation in the atomic bomb survivor Carcinogenesis Data? *Radiat. Res.* 168: 750-756.

Howe, G.R. (1995) Lung Cancer Mortality between 1950 and 1987 after Exposure to Fractionated Moderate-Dose-Rate Ionizing Radiation in the Canadian Fluoroscopy Cohort Study and a Comparison with Lung Cancer Mortality in the Atomic Bomb Survivors Study. *Radiat. Res.* 142: 295-304.

ICRP 2. (1959) *Report of Committee II on Permissible Dose for Internal Radiation.* The International Commission on Radiological Protection Publication 2. Pergamon Press, Headington Hill Hall, Oxford, UK.

ICRP 20. (1973) *Alkaline earth metabolism in adult man.* International Commission on Radiological Protection Publication 20. Pergamon Press, Elmsford, NY.

ICRP 26. (1977). *Recommendations of the International Commission on Radiological Protection.* International Commission on Radiological Protection Publication 26. *Annals of the ICRP* 1:3. Pergamon Press, Oxford.

ICRP 30. (1979). *Limits for Intakes of Radionuclides by Workers.* International Commission on Radiological Protection Publication 30 Part 1. *Annals of the ICRP* 2: 3-4. Pergamon Press, Oxford, UK.

ICRP 60. (1991). *1990 Recommendations of the International Commission on Radiological Protection.* International Commission on Radiological Protection Publication 60. *Annals of the ICRP* 21: 1-3. Pergamon Press, Elmsford, NY.

ICRP 66. (1994). *Human Respiratory Tract Model for Radiological Protection.* International Commission on Radiological Protection Pub. 66. Elsevier Science Ltd., Oxford.

ICRP 103. (2007). *The 2007 Recommendations of the International Commission on Radiological Protection.* Exeter, International Commission on Radiological Protection. Publication 103. Annals of the ICRP 37: 2-4. Elsevier, UK.

Jaworowski, Z. (2010). Observations on the Chernobyl Disaster and LNT. *Dose Response* 8: 148-171.

Lloyd, R.D.; Miller, S.C.; Taylor, G.N.; Bruenger, F.W.; Jee, W.S.S. & Angus, W. (1994). Relative Effectiveness of ^{239}Pu and Some Other Internal Emitters for Bone Cancer Induction in Beagles. *Health Phys.* 67: 346-353.

Lubin, J.H.; Boice, J.D.; Edling, C.; Horning, R.W.; Howe, G.; Kunz, E.; Kusiak, R.A.; Morrison, H.I.; Radford, E.P.; Samet, J.M.; Tirmarche, M.; Woodward, A.; Yao, S.X. & Pierce, D.A. (1994). *Radon and Lung Cancer Risk: A Joint Analysis of 11 Underground Miner Studies*, National Institutes of Health Publication No. 94-3644, United States Department of Health and Human Services, Washington, DC.

Lubin, J.H.; Boice, J.D.; Edling, C.; Horning, R.W.; Howe, G.R.; Kunz, E.; Kusiak, R.A.; Morrison, H.I.; Radford, E.P.; Samet, J.M.; Tirmarche, M.; Woodward, A.; Yao, S.X. & Pierce, D.A. (1995). Lung Cancer In Radon-Exposed Miners and Estimation of Risk From Indoor Exposure. *J. Natl. Cancer Inst.* 87: 817-827.

Manhattan Engineering District. (1946). *The Atomic Bombings of Hiroshima and Nagasaki.* Manhatten Engineer District, United States Army, Washington, DC.

Mauderly, J.L. & Daynes, R.A. (1994). *Biennial Report on Long-term Dose-response Studies of Inhaled or Injected Radionuclides 1991-93.* Inhalation Toxicology Research Institute, Albuquerque, New Mexico. Springfield, VA: National Technical Information Service; ITRI-139.

Mays, C.W. & Lloyd, R.D. (1972). Bone Sarcoma Incidence Versus. Alpha Particle Dose. In: Stover BJ, Jee WSS, eds. *Radiobiology of Plutonium.* Salt Lake City, UT: The J.W. Press, 409-430.

Ministry of Health. (1966). *Radiological Hazards to Patients.* Final Report. HMSO: London, UK.

Mole, R.H. (1990). Childhood Cancer After Prenatal Exposure To Diagnostic X-ray Examinations in Britian. *Br. J. Cancer* 62: 152-168.

Moolgavkar, S.H. Dewanji, A.; & Venzon, D.J. (1988). A Stochastic Two-Stage Model for Cancer Risk Asssessment. I. The Hazard Function and the Probability of Tumor. *Risk Anal.* 8:383-392.

NCRP 136. (2001). *Evaluation of the Linear-Nonthreshold Dose-Response Model for Ionizing Radiation.* National Council on Radiation Protection and Measurements Report No. 136. Bethesda MD.

NCRP 160. (2009). *Ionizing Radiation Exposure of the Population of the United States.* National Council on Radiation Protection and Measurements Report No. 160. Bethesda, MD.

Park, J.F. (1993). *Pacific Northwest Laboratory Annual Report for 1992 to the DOE Office of Energy Research: Part 1: Biomedical Sciences.* Pacific Northwest Laboratory, Richland, Washington. Springfield, VA: National Technical Information Service.

Peto, R.; Pike, M.C.; Day, N.E.; Gray, R.G.; Lee, P.N.; Parish, S.; Peto, J.; Richards, S. & Wahredorf, J. (1980). *Guidelines For Simple, Sensitive Significance Tests for Carcinogenic Effects in Long-Term Animal Experiments, Supplement 2 In: Long-Term and Short-Term Screening Assays for Carcinogens: A Critical Appraisal.* IARC Monographs. Lyon, France: World Health Organization International Agency for Research on Cancer; Sup 2, Annex 311-426.

Pierce, D.A. & Mendelsohn, M.L.(1999). A Model For Radiation-Related Cancer Suggested by Atomic Bomb Survivor Data. *Radiat. Res.* 152: 642-654.

Pierce, D.A. & Preston, D.L. (2000). Radiation-Related Cancer Risks at Low Doses Among Atomic Bomb Survivors. *Radiat. Res.* 154: 178-186.

Preston, D.L.; Shimizu, Y.; Pierce, D.A.; Suyama, A. & Mabuchi, K. (2003). Studies of Mortality of Atomic Bomb Survivors. Report 13: Solid Cancer and Noncancer Disease Mortality: 1950-1997. *Radiat. Res.* 160:381-407.

Preston, D.L.; Pierce, D.A.; Shimizu, Y.; Cullings, H.M.; Fujita, S.; Funamoto, S. & Kodama, K. (2004). Effect of Recent Changes in Atomic Bomb Sutvivor Dosimetry on Cancer Mortality Risk Estimates. *Radiat. Res.* 162:377-389.

Preston, D.L.; Ron, E.; Tokuoka, S.; Funamoto, S.; Nishi, N.; Soda. M.; Mabuchi, K. & Kodama, K. (2007). Solid Cancer Incidence in Atomic Bomb Survivors: 1958-1998. *Radiat. Res.* 168:1-64;

Raabe, O.G. (1987). Three-dimensional Dose-Response Models of Competing Risks and Natural Life Span. *Fundam. Appl. Toxicology* 8:465-473.

Raabe, O.G. (1989). Extrapolation and Scaling of Animal Data to Humans: Scaling of Fatal Cancer Risks From Laboratory Animals to Man. *Health Phys.* 57, Sup. 1:419-432.

Raabe, O.G. (1994). Cancer and Injury Risks From Internally Deposited Radionuclides. In: *Actualités sur le Césium,* Comité de Radioprotection, Electricité de France, Publication Numéro 8: 39-48.

Raabe, O.G.. (2010). Concerning the Health Effects of Internally Deposited Radionuclides. *Health Phys.* 98: 515-536.

Raabe, O.G. (2011). Toward Improved Ionizing Radiation Safety Standards. *Health Phys.* 101: 84-93.

Raabe, O.G.; Book, S.A. & Parks, N.J. (1980). Bone cancer from radium: Canine dose response explains data for mice and humans. *Science* 208:61-64.

Raabe, O.G. & Abell, D.L. (1990). *Laboratory for Energy-Related Health Research Final Annual Report, Fiscal Year 1989,* UCD 472-13. University of California, Davis, CA. Springfield VA: National Technical Information Service.

Raabe, O.G.; Rosenblatt, L.S. & Schlenker, R.A. (1990). Interspecies Scaling of Risk for Radiation-induced Bone Cancer. *Int. J. Radiat. Biol.* 57: 1047-1061.

Raabe, O.G. & Parks, N.J. (1993). Skeletal uptake and lifetime retention of ^{90}Sr and ^{226}Ra in beagles. *Radiat. Res.* 133:204-218.

RERF. (2008). 60-year Anniversary of ABCC/RERF, Annual Report 1 April 2007-31 March 2008 Radiation Effects Research Foundation Hiroshima, Japan.

Rowland, R.E. (1994). *Radium in Humans, A Review of U.S. Studies.* ANL/ER-3, Argonne, IL: Argonne National Laboratory.

Saccomanno, G; Huth, G.C.; Auerbach, O. & Kuschner, M. (1988). Relationship of Radioactive Radon Daughters and Cigarette Smoking in the Genesis of Lung Cancer in Uranium Miners. *Cancer* 62:1402-1408.

Snyder, A.R. & Morgan, W.F. (2004). Gene Expression Profiling After Irradiation: Clues To Understanding Acute and Persistent Responses? *Cancer Metastasis Reviews* 23:259-268.

Stewart, A., Webb, J., Giles, D. & Hewitt, D. (1956). Malignant Disease in Childhood and Diagnostic Irradiation In Utero, *Lancet* ii:447-448.

Stewart, A. , Webb, J., Hewitt, D. (1958). A Survey of Childhood Malignancies. *Br. Med. J.* 1:1495-1508.

Totter, J.R. & MacPherson, H.G. (1981). Do Childhood Cancers Result From Prenatal X-rays. *Health Phys.* 40:511-524.

Wing, S.; Shy, C.M.; Wood, J.L.; Wolf, S.; Cragle, D.L. & Frome, E.L. (1991). Mortality Among Workers at Oak Ridge National Laboratory. *J. American Med. Assoc.* 265:1397-2402.

WHO. (2006) *Health Effects of the Chernobyl Accident and Special Health Care Programmes.* World Health Organizion, Geneva, Switzerland.

2

Ionizing Radiation in Medical Imaging and Efforts in Dose Optimization

Varut Vardhanabhuti and Carl A. Roobottom
*Derriford Hospital and
Peninsula College of Medicine and Dentistry, Plymouth,
United Kingdom*

1. Introduction

Medical-related radiation is the largest source of controllable radiation exposure to humans and it accounts for more than 95% of radiation exposure from man-made sources. Its direct benefits in modern day medical practices are beyond doubt but risks-benefits ratios need to be constantly monitored as the use of ionizing radiation is increasing rapidly. From 1980 to 2006, the per-capita effective dose from diagnostic and interventional medical procedures in the United States increased almost six fold, from 0.5 to 3.0mSv, while contributions from other sources remained static (NCRP report no 160, 2009).

This chapter will review radiation exposure from medical imaging initially starting from a historical viewpoint as well as discussing innovative technologies on the horizon. The challenges for the medical community in addressing the increasing trend of radiation usage will be discussed as well as the latest research in dose justification and optimization.

2. Sources and trends

Medical radiation is by far the largest artificial source of population exposure to ionising radiation, accounting for 90% of all doses from artificial sources (United Nations, 2010). This radiation burden is increasing. In the US, this increase has been primarily due to the increasing use of computed tomography (CT). Despite the fact that CT only accounts for 11% of the examinations it contributes 68% of collective dose. In comparison, conventional radiography constitutes 90% of the examinations but only 19% of collective dose (Mettler et al., 2009). From 1980s to 2006, there is an increase of approximately 6-fold in cumulative effective dose per individual in the US from 0.5 to 3mSV (NCRP report no 160., 2009).

As an overall exposure to all kinds of radiation to humans, medical-related radiation exposure now accounts for 48% (an increase from 15%) with background radiation remaining relative static at 50%. The remaining 2% of total radiation exposure came from consumer-related products and activities. These include cigarette smoking, building materials, commercial air travel, mining and agriculture.

Exposure Category	Collective Effective Dose (person-sievert)		Effective Dose per Individual in the U.S. Population (mSv)	
	Early 1980s	2006	Early 1980s	2006
Ubiquitous natural background	690,000	933,000	3	3.11
Medical Procedures	123,000	899,000	0.53	3
Consumer products	12,000 - 29,000	39,000	0.05 - 0.13	0.13
Industrial, security, medical, educational, and research exposures	200	1,000	0.001	0.003
Occupational exposures	2,000	1,400	0.009	0.005
Overall	835,000	1,870,000	3.6	6.2

Data from NCRP report no 160 (2006 data) and no 93 (1980s data).

Table 1. Changes in collective effective dose and effective dose per individual in the US population between early 1980s and 2006.

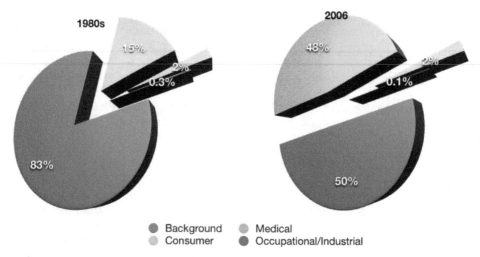

Fig. 1. Exposure of the US population in early 1980s and in 2006 according to NCRP report no 93 and 160.

From data in 2006, in radiology and nuclear medicine procedures, the biggest contribution as previously mentioned is from computed tomography (49%), followed by nuclear medicine studies (26%), then interventional procedures (14%). The remaining 11% came from radiographic and fluoroscopic studies despite accounting for more than 74% of total number of procedures. Not surprisingly, the per capita dose for CT is highest at 1.47 mSv. Data is summarised in the following table.

Type of Procedure	No. of Procedures (in millions)	Percentage of Total No. of Procedures	Collective Effective Dose (person-sievert)	Percentage of Collective Dose from Procedures	Per-Capita Dose (mSv)
Diagnostic radiographic and fluoroscopic studies	293	74%	100,000	11%	0.33
Interventional procedures	17	4%	128,000	14%	0.43
CT scanning	67	17%	440,000	49%	1.47
Nuclear medicine studies	18	5%	231,000	26%	0.77
Total	395	100%	899,000	100%	3.01

Data taken from NRCP report no 160

Table 2. Estimated number and collective effective doses from radiologic and nuclear medicine procedures in the US for 2006.

The overall trend is similar in the UK. A recent Health Protection Agency publication (Hart et al., 2010) showed a 28% increase in 2008 compared to 1997/8 in the United Kingdom. This increase is again mainly due to a doubling of the computed tomography (CT) examinations performed over this same period. The increase in radiation dose in the UK is thought to be modest and this is partly due to the number of examinations performed. A comparison with other European countries show that the 19 out of 20 most commonly performed examinations, UK shows a lower than European average in frequency (Aroua et al., 2010). Only barium enemas were shown to be more frequent than average compared to the European counterparts. In terms of dose, 17 out of 20 examinations in the UK are less on average dose compared to their European counterparts. This illustrates that even in a country like United Kingdom, the trend for increasing use of medically related ionising radiation continues apace despite the fact that there has been more traditional emphasis on

Country	Annual collective dose per caput (mSv)
Germany	1.52
Belgium	1.39
Switzerland	1.37
Iceland	1.28
France	1.11
Norway	0.94
Sweden	0.59
Netherlands	0.48
Lithuania	0.47
Denmark	0.46
Finland	0.35
United Kingdom	0.30

Table 3. Collective dose per caput in 2008 for the "top 20" examinations in Europe (from Aroua et al, 2010).

dose awareness. Table below shows annual collective dose per capita of European countries. The UK collective dose per caput from all medical and dental X-ray examinations stands at 0.3 mSv. This number can be compared with corresponding values of 2.2 mSv assessed for medical and dental X-rays in the USA in 2006 (NCRP, 2009) and the similar figure of 1.9 mSv (UNSCEAR., 2010) as the average for people living in Healthcare Level 1 (HCL1) countries. The UK collective dose per caput is clearly very low for a HCL1 country.

For CT, the UK annual per caput dose stands at 0.27 mSv. This number can be compared with corresponding levels of 1.5 mSv from CT in the USA in 2006 (NCRP., 2009) and 0.74 mSv from CT in Canada in 2006 (Chen and Moir., 2010). The UK per caput dose from CT is relatively low compared to North American countries.

At the time of writing, there is likely to be an even further increase in the use of CT compared to the level quoted in 2006. This is mainly due to rapid rise in the use of CT in emergency setting (White and Kuo, 2007; Street et al, 2009) and also the increasing use of PET/CT (Elliot A, 2009; Chawla et al, 2010; Devine et al, 2010). However, the dose increase may not be as high as recent concentrated efforts in the medical community have focused on strategies in dose reduction. Recent effective novel innovations have helped to reduce some of the dose burden (see later).

3. Risks

Ionizing radiation has long been known to increase the risk of cancer and is officially classified as a carcinogen by the International Agency for Research on Cancer (IARC). It is now on the official list of carcinogen by the World Health Organization (IARC list of carcinogen, 2011). The exact relationship between dose exposure and cancer induction, however, is complex and several issues merit further discussion.

First, perceived medical-related radiation has traditionally been quantified in comparison with radiation we all receive from background levels. It is worth noting that medical-related radiation exposure is not the same as background radiation. The exposure from background radiation is generally of mixed-energy particles (high and low LET-radiation) while the exposure from diagnostic medical procedures is generally of low-energy x-rays. Low-LET radiation deposits less energy in the cell along the radiation path and is considered less destructive per radiation track than high-LET radiation. Examples of low-LET radiation include X-rays and γ-rays (gamma rays), which are used in medical imaging. Low-LET radiation produces ionizations sparsely throughout a cell; in contrast, high-LET radiation transfers more energy per unit length as it traverses the cell and is more destructive per unit length.

Background radiation are naturally occurring from sources in soil, rocks, bricks and building material, and from radon gas which seeps out to homes. Radon is a colourless, odourless gas that emanates from the earth and as it decays also emits LET radiation. Average annual exposures worldwide to natural radiation sources (both high and low LET) would generally be expected to be in the range of 1–10 mSv, with 2.4 mSv being the present estimate of the central value (UNSCEAR, 2000). Of this amount, about one-half (1.2 mSv per year) comes from radon and its decay products. Some areas will have more radon gas background than others (e.g. Cornwall in the UK generally has twice the amount of background radiation than the rest of the UK).

After radon, the next highest percentage of natural ionizing radiation exposure comes from cosmic rays. This is followed by terrestrial sources, and "internal" emissions. Cosmic rays are particles that travel through the universe and are filtered by the earth atmosphere (therefore this varies from sea level to higher altitudes where less atmospheric filtration occurs). The Sun is a source of some of these particles. Other particles come from exploding stars called supernovas. The amount of terrestrial radiation from rocks and soils varies in different parts of the world. Much of this variation is due to differences in background radon levels. "Internal" emissions come from radioactive isotopes in food and water with uranium and thorium series of radioisotopes present in food and drinking water (UNSCEAR, 2000).

To quantify medical exposures (which are of predominantly low-LET radiation) to background radiation (which are of mixed-LET radiation) is uncertain due to this respect but it is now generally accepted that of the 2.4mSv total average background radiation of mixed-LET, the total average annual population exposure worldwide due to low-LET radiation would generally be expected to be in the range of 0.2–1.0 mSv, with 0.9 mSv being the present estimate of the central value. The pie chart below illustrates proportions of high and low-LET radiation of the background radiation worldwide (adapted from BIER VII, 2006).

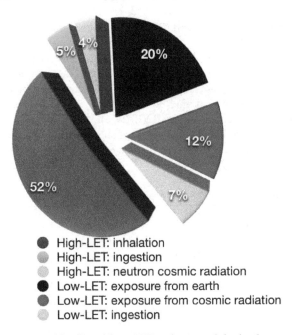

● High-LET: inhalation
◔ High-LET: ingestion
○ High-LET: neutron cosmic radiation
● Low-LET: exposure from earth
● Low-LET: exposure from cosmic radiation
◔ Low-LET: ingestion

Fig. 2. Pie chart proportions of high and low-LET radiation of the background radiation worldwide (adapted from BIER VII, 2006).

Ubiquitous background radiation represents exposure to whole population (all ages, gender, health status) whereas medical radiation distributions are often skewed towards higher age groups and also sicker individuals. There may also be skewed in gender distribution (there is rightly a reluctance to image patients who are pregnant, for example). Therefore, one must acknowledge that the background radiation level is an approximation and will vary from individual to individual living at different locality having varying exposures.

Current radiation protection standards and practices are based on the premise that any radiation dose, no matter how small, can result in detrimental health effects, such as cancer and genetic damage. Further, it is assumed that these effects are produced in direct proportion to the dose received, i.e., doubling the radiation dose results in a doubling of the effect. These two assumptions lead to a dose-response relationship, often referred to as the linear no-threshold model, for estimating health effects at doses of interest. Although, this is of much benefit in practical terms in risk estimation in the context of radiation protection quantification, there is, however, substantial scientific evidence that this model is an oversimplification of the dose-response relationship. In particular, it results in an overestimation of health risks in the low dose range. Biological mechanisms including cellular repair of radiation injury, which are not accounted for by the linear, no-threshold model, reduce the likelihood of cancers and genetic effects. Therefore, it is now generally accepted that quantification of estimated health risk in the dose range similar to that of background radiation should be strictly qualitative and encompass a range of hypothetical health outcomes, including the possibility of no adverse health effects at such low levels (Burk Jr RJ 1996, updated in 2004).

Biological Effects of Ionizing Radiation (BEIR) VII committee defines low doses of ionizing radiation as less than 100mSv and agrees that at doses of 100 mSv or less, statistical limitations make it difficult to evaluate cancer risk. The current best estimation model for risks associated with low dose radiation exposure is that approximately one individual in 100 persons would be expected to develop cancer (solid cancer or leukemia) from a dose of 100 mSv while approximately 42 of the 100 individuals would be expected to develop solid cancer or leukemia from other causes (BEIR VII, 2006).

It is now generally accepted that at effective doses above 50 mSv the risk of cancer induction increases linearly with dose, this dose-response relation has not been demonstrated at doses below 50 mSv. Below 50 mSv no convincing epidemiological evidence exists currently for cancer risk induction.

There is, however, more convincing evidence to support risk of cancer induction using high dose radiation. A recent article from the USA estimated that CT scans performed in 2007 could result in 29,000 future cancers based on current risk estimations (Berrington de Gonzales et al., 2009). The risk to an individual patient is also high. For example a cardiac scan in a 20-year-old female can produce a lifetime cancer risk from that scan of approaching 1% (Einstein et al., 2007).

Specific effects of radiation are thought to be either deterministic and/or stochastic although the exact relationship is difficult to quantify with certainty. There is reasonable epidemiological evidence (though not definitive) from 30,000 A-bomb survivors that organ doses from 5 to 125 mSv result in a very small but statistically significant increase in cancer risk (Preston et al., 2007). Other low dose epidemiological studies from the occupational exposure of radiation workers are also generally in keeping with this trend with increasing cancer risk with increasing radiation dose (Muirhead et al., 2009 and Cardis et al., 2007). Of note, the third analysis of the National Registry for Radiation Workers in the UK (NRRW-3 from Muirhead et al., 2009) provides the most up-to-date and precise information on the risks of occupational radiation exposure based on cancer registrations as well as mortality from an enlarged cohort of 174,541 workers dating back to 1955. Data were available for 26,731 deaths during 3.9 million person-years of total follow-up period. It clearly shows a statistically significant increasing trend with dose in both mortality and incidence for all malignant neoplasms (leukaemia and several solid organs tumours). Table below summarises the findings from NRRW-3. Despite this, one must bear in mind that radiogenic excess cancer risk associated with medical-related radiation is in orders of magnitude smaller than the spontaneous cancer risk (Burk Jr RJ 1996, updated in 2004 and BEIR VII (2006).

	Leukaemia excluding CLL	All neoplasms excluding leukaemia	All malignant neoplasm excluding leukaemia, lung and pleura cancer
3rd NRRW Analysis			
Mortality	1.712 (0.06-4.29)	0.275 (0.02-0.56)	0.323 (0.02-0.67)
Incidence	1.782 (0.17-4.36)	0.266 (0.04-0.51)	0.305 (0.05-0.58)
2nd NRRW Analysis			
Mortality	2.55 (-0.03-7.16)	0.09 (-0.28-0.52)	0.17 (-0.26-0.70)
15-country Nuclear Worker Study			
Mortality	1.93 (0-7.14)	0.97 (0.27-1.80)	0.59 (-0.16-1.51)
Japanese A-bomb survivors			
BEIR VII: Mortality	1.4 (0.1-3.4)	0.26 (0.15-0.41)	-
BEIR VII: Incidence	-	0.43	-

Table 4. Comparison of estimates of ERR per Sv (and 90% CI) for cancer in NRRW, the 15-country nuclear worker and the Japanese A-bomb survivors.

In addition, there appears to be other determinants that also affect the risk of developing cancers. Among these, genetic considerations, age at exposure, sex and fractionation and protraction of exposure appear to play important roles.

Genetic considerations

There are 2 epidemiological studies that suggest that there is a subgroup of population who are more likely to develop cancer when exposed to radiation although in neither case has the genes responsible for the increased radiosensitivity been identified. Ronckers et al (2008) performed a retrospective study looking at 3,010 young women with scoliosis who regularly underwent radiographic follow-up to monitor disease progression between 1912 and 1965. They had found that there was a borderline but significant dose response in a subset of women with a family history of breast cancer in first- or second-degree relatives. Flint-Richter et al (2007) performed a case-control study looking at children who were epilated with x-rays for the treatment of tinea capitis and found that 1% of children developed meningioma with marked clustering in certain families suggestive of genetic susceptibility to the development of tumours after exposure to radiation. A meta-analysis by Jansen-van der Weide (2010) also supports increase risk of radiation in genetically susceptible group. They performed a meta-analysis from seven studies evaluating the effects of low-dose radiation exposure, such as mammography, on cancer risk in women with a familial or genetic predisposition. They had found that there is an increased risk with exposure to radiation that results in 1.3 times increased breast cancer risk. The risk is also higher in women who were exposed before the age of 20 or who were frequently exposed to radiation.

Age at exposure

Following on from the last point, there is convincing evidence to support a relationship between life-time attributable risk of cancer incidence and age at exposure. Graph below show analysis of lifetime attributable risk of radiation-induced cancer incidence derived from BEIR VI committee based on data of A-bomb survivors. This, in general, supports that children are more radiosensitive than adults. However, it is also true that social and environmental factors play a role as for some solid cancers, these risks do not decline with age. When looking at effective risks ratio (ERR), there is little difference in risk between 10 and 40 years of age, while for some cancers such as lung and bladder, there appears to be a significant increase in risk with increasing age of exposure.

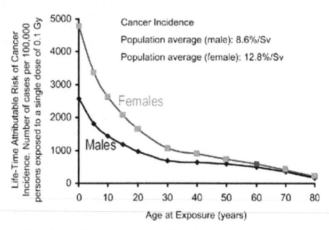

Fig. 3. Lifetime attributable risk of radiation-induced cancer incidence, as a function of age at exposure for males and females (data based on BEIR VII, 2006).
(with permission from Hricak et al, 2011)

Sex

BEIR VI report supports the notion that there is substantially higher lifetime attributable risk of cancer incidence in females compared with males. Figure 5 shows breast cancers risks are higher, but what is more notable is the risk for lung and bladder cancer in women is much higher than in men. This is, in spite of the fact that in 1945 Japan, men were heavy smokers while smoking was deemed as very uncommon in women.

Fractionation and protraction of exposures

It was previously thought that radiation risks per unit dose at low levels and at low dose rates were smaller than that of higher dose and dose rates. This is due to the perceived influence of DNA repair. A suggested dose and dose rate effectiveness factor (this is a multiplication factor use for low dose rates compared to high dose rates) by the ICRP was two, while BEIR VII suggested a value of 1.5. There are a couple of studies that have already been mentioned but are again worth mentioning in this regard. First the International Agency for Research on Cancer 15-country study looked at around 600,000 nuclear workers who were exposed to an average cumulative dose of 19mSv. The estimated ERR for this cohort for developing solid cancers was almost four times larger than that for the A-bomb

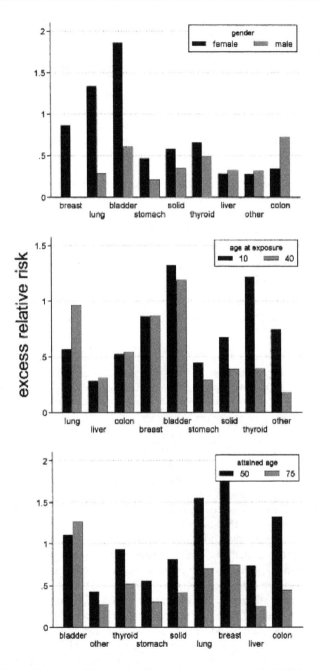

Fig. 4. Comparison of site-specific gender (top), age-at-exposure (middle) and attained-age (bottom) effects on standardised ERR_{1Gy} estimates for selected sites and all solid cancers. (with permission from Preston et al, 2007)

survivors. However, the study noted that there are likely to be important confounders. First, the results are likely to be skewed by the Canadian workers who were relatively few in number with high number of death rates. Second, the predominance of lung cancer suggests possible confounding effect of smoking. More recently, the update from the National Registry for Radiation Workers in the UK (NRRW-3 from Muirhead et al., 2009) followed cohorts with cumulative dose of 25mSv. Cancer risks do increase with dose and the estimated ERR per Sievert was very similar to that for the A-bomb survivor suggestive of a rather small reduction in cancer risk induction with dose protraction which is somewhat surprising. These 2 large studies in themselves suggest that it is not at all clear what the relationship is between dose fractionation and protraction with the risk of cancer induction.

Currently, there are several epidemiological studies following up patients who were exposed to CT at a young age in UK, Australia, Canada, France, Israel and Sweden. Results of such studies will shed more light into the precise influence of radiation and add to the existing body of evidence. This is likely to take time, however, due to the inherent nature of these studies.

4. Strategies in dose optimisations in Computed Tomography (CT)

The fact that CT accounts for most of the ionising radiation used in medical imaging is the reason for its focus on this chapter. Other imaging modalities will also be discussed but in lesser detail.

Dose elimination

Whilst, undoubtedly medical imaging using ionising radiation has several advantages, one must bear in mind that the best way to reduce radiation is to not perform the investigation at all. There is evidence that increasingly CT has been over utilised in various clinical settings meaning that unnecessary scans are being performed, or incorrect examinations are being performed without appropriate justification. The usual practice is to refer to clinical guidelines or appropriateness criteria to direct or justify an examination according to a clinical scenario. The American College of Radiology (ACR), the Royal College of Radiologists in the UK (RCR), and the European Commission all have published decision guidelines for the appropriate use of CT in different clinical scenarios. A retrospective study was performed in a level I trauma centre looking at appropriateness of scans (Hadley et al, 2006). It was found that 44% of the studies ordered would not have been indicated had the guidelines been rigorously followed. One recent innovative approach addressing this has been to incorporate these guidelines into computerised imaging order entry system. Pre-authorisation of CT examinations according to the ACR and RCR guidelines were utilised in an institution which showed significant deferral rate and substantial decrease in the use of CT and MRI. After reauthorisation was implemented, CT annual performance rates decreased from 25.9 examinations per 1,000 in 2000 to 17.3 per 1,000 in 2003 (Blachar et al, 2006). Despite being evidence-based and recommended for routine clinical usage, these guidelines still show poor uptake in general usage (Bautista et al, 2009). When a survey was performed looking at how physicians decide what the best imaging test to use for their patients, the use of ACR Appropriate criteria showed very low uptake in one institution (2.4%) compared with other available resources (e.g. Radiologist consult, specialty journal, UpToDate, Google, Pubmed, etc).

Dose reduction

CT dose reduction can broadly be divided into ways to reduce the total radiation emitted by the X-ray tube and ways to reduce scanning time. Ways to reduce scanning time include ECG gating in cardiac studies, or increasing the pitch of the scanner, for example. Traditionally, the ways to reduce total dose have included X-ray beam filtration, X-ray beam collimation, X-ray tube current modulation and adaptation for patient body habitus (automatic exposure control), peak kilovoltage optimisation, improved detection system efficiency, low dose protocols for specific indications (e.g. CT KUB for renal stones). Since the 1980s, a number of technical innovations have been responsible for dose reduction in CTs including the use of solid state scintillating detectors, electronic circuitry, multi-detector arrays, more powerful x-ray tubes and beam-shaping filters. More recently, a number of newer dose reduction techniques have gained widespread acceptance and these are as follows:

Automatic exposure control

Automatic exposure control (AEC) is one of the most important techniques in clinical practice to reduce radiation dose without compromising image quality. AEC is a broad term that encompasses not only tube current modulation (to adapt to changes in patient attenuation), but also determining and delivering the "right" dose for any patient (infant to obese) in order to achieve the diagnostic quality images. It is technologically possible for CT systems to adjust the x-ray tube current in real-time in response to variations in x-ray intensity at the detector (McCollough et al, 2005), much as fluoroscopic x-ray systems adjust exposure automatically. The modulation may be fully pre-programmed, occur in near-real time by using a feedback mechanism, or incorporate pre-programming and a feedback loop.

Tube current modulation

This is done by maintaining a constant image noise level through longitudinal and/or angular modulation of x-ray tube current according to patient size, shape and the resultant attenuation. This means that the tube current varies across different scan length. This is in contrast to fixed tube current methods where it is constant throughout the scan length meaning in effect that in certain areas, there are wasted radiation as one will not yield increased diagnostic capability.

Longitudinal (z) mA modulation

In the longitudinal (z-axis) modulation (AutomA) technique, the basic strategy is to provide predictable image quality to achieve a reliable diagnosis with the lowest necessary radiation dose depending on patient size and attenuation. This is done along the patient's longitudinal axis (i.e. shoulders to pelvis). For a specific patient anatomy and diagnostic task, a specific parameter (defined as noise index - NI) is prescribed by the user to specify the targeted image quality that represents the average noise in the centre of an image of a uniform water phantom. A 5% decrease in NI demands approximately a 10% radiation increase, whereas a 5% increase in NI decreases radiation dose by approximately 10% (Karla et al, 2004a). Therefore, an appropriate NI selection is imperative to control the balance between radiation dose and image quality. This is often recommended by a combination of manufacturer recommendation and clinical experience. This also varies depending on the type of scan performed. With a given NI, the AutomA automatically adjusts x-ray tube current in the scan to maintain the same noise level in all images regardless of patient size and attenuation. Previous studies have shown that in abdominal CT studies a NI of 10.5 to

15 leads to a reduction in radiation dose by 16.6-53.3% in comparison to that using a constant x-ray tube current (Karla et al, 2004b).

Angular (x,y) mA modulation

Angular (x,y) mA modulation addresses the variation in x-ray attenuation around the patient by varying the mA as the x-ray tube rotates about the patient (e.g. in the A.P. versus lateral direction). The operator chooses the initial mA value, and the mA is modulated upward or downward from the initial value within a period of one gantry rotation. As the x-ray tube rotates between the AP and lateral positions, the mA can be varied according to the attenuation information from the CT radiograph (i.e. Scout image), or in near real-time according to the measured attenuation from the 180° previous projection.

Using both angular and longitudinal mA modulation, significant dose reduction can be achieved. Although an approximately 50% reduction in dose has been found with automatic exposure control, the system is not foolproof. It seems that at the extremes (i.e. smaller and larger patient sizes), there needs to be adjustment of the noise level such that a higher noise level was recommended for a large sized patient to avoid a higher radiation exposure (Karla, 2004a), and a lower noise level for smaller patients was also suggested by a previous investigation (Karla, 2004b). Some of these studies have shown that there is an influence of patient weight on image quality and dose when a constant noise level is chosen for all patient sizes. Kalra et al (2004b) showed that smaller patients (in their study, defined as having weight less than 68 kg and corresponding to smaller transverse and anteroposterior diameters) had subjective image quality scores lower than larger patients (weights greater than 68 kg) despite using a fixed noise index parameter.

Adjusting noise level based on weight alone is also fraught with difficulties. Several studies have shown that weight is not the ideal factor for required dose calculations. This is because two patients with the same weight can have different regional dimensions and tissue attenuation properties on CT scanning, which can affect the image quality significantly (Karla et al, 2003). Schindera et al (2008) found that a phantom with increased anthropomorphic size received significantly increased skin and deep organ dose than a smaller sized phantom for fixed noise level. Some people have proposed a correction factor for patient size to find the optimal noise index using a combination of patient's weight, BMI or information body diameter from CT scout images (Li et al, 2011).

Tube Angle Start Position and Pitch

Various investigators have demonstrated that there are significant dose variations with a sinusoidal pattern on the peripheral of a CTDI 32 cm phantom or on the surface of an anthropomorphic phantom when helical CT scanning is performed, resulting in the creation of "hot" spots or "cold" spots (Svandi et al, 2009). Exploiting this in conjunction with adjustment of pitch can result in dose saving (Zhang et al, 2009). For example, at a pitch of 1.5 scans, the dose is usually lowest when the tube start angle is such that the x-ray tube is posterior to the patient when it passes the longitudinal location of the organ. For pitch 0.75 scans, the dose is lowest when the tube start angle is such that the x-ray tube is anterior to the patient when it passes the longitudinal location of the organ. For organs that have a relatively small longitudinal extent, dose can vary considerably with different start angles. While current MDCT systems do not provide the user with the ability to control the tube start angle, these results indicate that in these specific situations pitch 1.5 or pitch 0.75, small organs and especially small patients, there could be significant dose savings to organs if that functionality adjustment was available.

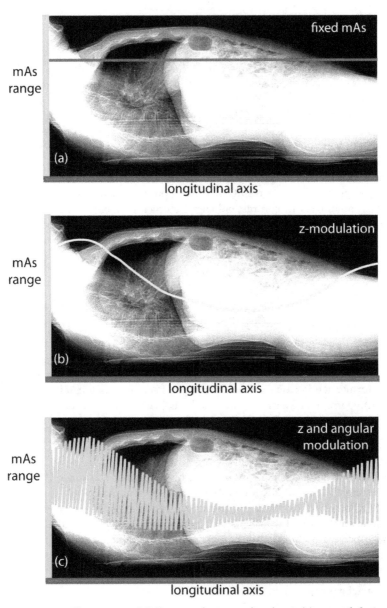

Fig. 5. Diagrammatic illustration of differences between fixed mA (a), z-modulation (b) and combined z and angular modulation (c). Note that the use of fixed mAs means that mA is too high for most parts of the scan. Alternatively, solely relying on z-axis modulation can also mean that tube currents are higher than fixed mA at the edges of the scan where the tube current is increased to account for shoulder girdles and bony pelvis (which inherently have higher attenuation than body scan).
(images are authors' own reproduction, modified from Hricak et al, 2011 with permission)

Tube voltage

There have been various studies looking at the adjustment of tube potential. The most reliable determinant when adjusting the tube voltage at present is the patient's BMI but factors such as fat distribution (e.g. around the thorax) should also be considered. The traditional setting for voltage has been at 120kV. Diagnostic images have been shown to be possible at 100 or even at 80kV (Hausleiter et al, 2010). This can be done according to the patient's size with published reports demonstrating up to 70% reduction in chest scans and 40% reduction in abdominal scans (Yu et al, 2010) by lowering the kV in selected groups to 100kV from 120kV. Lowering the voltage decreases penetration (therefore less useful in patients with high BMI) but also results in increased subject contrast as it approximates towards the k-edge of iodine (33 keV). This means that there is superior enhancement of iodine at lower potentials. As a result, there is improved conspicuity of hypervascular or hypovascular pathologies in contrast-enhanced studies as there is improved contrast-to-noise ratio. There is a trade-off in that there is increase image noise mainly due to higher absorption of low-energy photons by the patient. Therefore, this only works best in patients of smaller sizes and the improved contrast-to-noise is negated in larger patients. Dose is proportional to the current but to the square of tube voltage and so it remains more advantageous to lower the kV than the mAs.

Cardiac CT

Various innovations related to cardiac CT have markedly reduced the dose of this traditionally high dose examination and are worth discussing in more detail (Roobottom et al, 2010). The 2 main methods of data acquisitions in cardiac CT angiographic studies are inherently linked to the cardiac cycle (ECG-gating). These are retrospective and prospective gating. The former tends to be utilised when the heart rate is greater than 65 beats per minute (bpm) or irregular heart rate and the latter with regular heart rate below 65 bpm. Radiation reduction strategies for these are as follows.

ECG-linked tube current modulation

Traditionally, tube current is applied throughout the cardiac cycle in retrospective acquisition. But since the coronary arteries are best image at end diastole where they are most still (usually 75% of the RR interval), the tube current only needs to be maximum at this point. At all other points, the tube currents can be reduced (see Fig. 6 below). Phase data is still available for the rest of the cardiac cycle (although image will be affected by quantum mottle) and functional data can still be obtained. The traditional ECG-gated retrospective spiral acquisition, though results in good image quality, the dose still poses concerns. Thus sequential scanning as an alternative has been developed (Schoenhagen P, 2008).

Prospective axial gating

Axial acquisition (rather than spirally) on a pre-defined block (e.g. 4cm) can be used – so called 'step and shoot' technique (Stolzmann et al, 2008) – is a more radiation-efficient technique in that the tube current is only applied in that block and is zero outside that pre-determined window (see Fig 7, below). This results in a dramatic reduction in dose to a level where 5mSv and below can be regularly achieved. The current reconstruction technique limitation, however, determines that the heart rate needs to be below 65bpm. Therefore aggressive beta blockade should be utilised where possible to achieve this (unless there are contraindications). This technique is also prone to step artefacts. The heart rate also needs to be regular but the use of more 'padding' with centering of the acquisition to include end-systolic (35%) to end-diastolic (85%) phases can combat this to a certain extent. This also allows multi-phase analysis. Expanding phase data, however, increases dose penalty but still remains advantageous as this is still less than the use of retrospective spiral scanning, even with aggressive dose modulation.

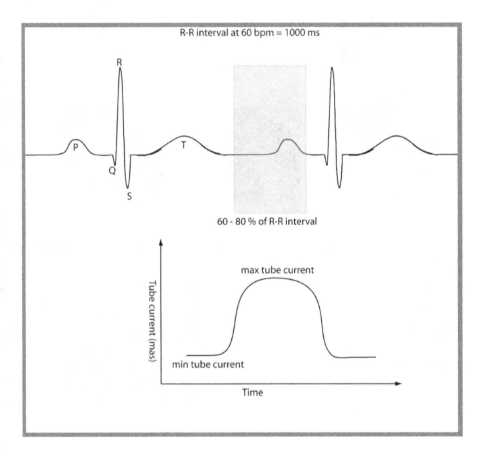

Fig. 6. ECG-linked dose modulation with maximum tube current during diastolic phase and minimum for the rest of the cardiac cycle.
(images are authors' own reproduction)

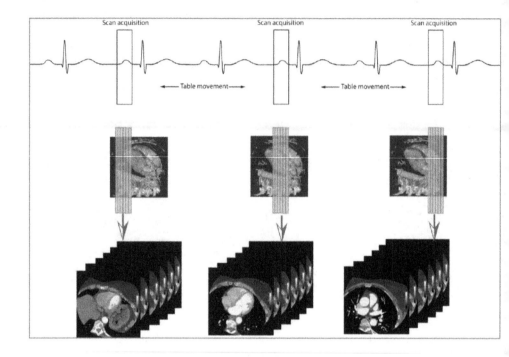

Fig. 7. The "step and shoot" technique with sequential axial scans covering the entire heart. (images are authors' own reproduction)

Scan length

Scan length should be kept at a minimum to lower the dose. In general CT, this is done by the operator from the scout view obtained on initial scanning. In certain studies, increase coverage is required but this is difficult to ascertain on the scanned projection radiograph (i.e. scanogram, scoutview or topogram). For example, increase coverage is needed in patients with vein grafts (e.g. internal mammary artery grafts require coverage from lung apex). At our institution, we performed the initial low dose unenhanced scan to look for coronary calcifications which also allows accurate delineation of the coronary arteries. The added information means that we do not have to perform overcoverage (and therefore eliminate unnecessary radiation) and we now routinely perform this at our centre (Roobottom et al, 2010).

High pitch spiral acquisition

High pitch ECG-gated acquisition using dual source CT scanner is sufficiently fast enough so that in a patient that has a slow heart rate, images of the whole heart can be acquired within a single heart beat. In particular, cardiac imaging have shown great efforts in dose reduction where a dose reduction of up to 90% has been achieved using novel acquisition techniques (Flohr et al, 2009 and McCollough et al, 2009).

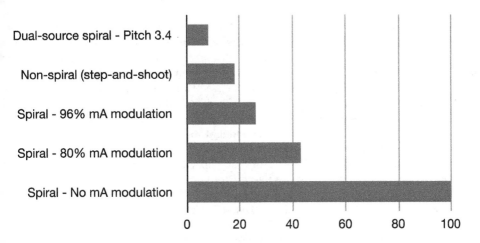

Fig. 8. Graphical illustration showing dose reduction in different techniques of cardiac CT angiography.
(modified from Hricak et al, 2011, with permission)

Displacement and shielding methods

There are several types of selective organ shielding available since 1997 (Hooper et al, 1997) with various studies focusing on the use of shields for the eyes, thyroid breasts and gonads. The evidence shows that they reduce dose (especially surface organ dose), however, whether they affect image quality is still a hugely contentious issue.

Breast shielding has been the most studied shielding method and is worth a more extensive mention for this reason. These are made of latex sheets consisting of bismuth and have the property of attenuating x-ray beam in the energy range used in CT. This can be up to about 50% depending on the tube voltage or filtration. The use of bismuth breast shield does not affect the mean energy of the x-ray beam, but reduces the overall intensity. There is fairly consistent agreement throughout the literature on its effectiveness at reducing radiation dose. Fricke et al (2003) found that bismuth breast shields for paediatric CT reduce dose by 29% in anthropomorphic phantoms. Coursey et al (2008) reported a dose reduction of 52% when the bismuth breast shield was employed after the acquisition of scout view using the automatic tube current modulation in a chest MDCT protocol. Foley et al (2011) showed that the use of breast displacement and lead shielding in a group of women undergoing cardiac CT for investigation of chest pain lead to significant reduction in mean breast surface dose without affecting coronary image quality. This varies with breast size but the mean reduction was around 24% of breast surface dose.

Other organ specific shields have also been studied. Mukundan et al (2007) reported a 42% lens dose reduction when using a 2-ply bismuth shield for axial brain and helical craniofacial paediatric CT protocols. Perisinakis et al (2005) also investigated the orbital bismuth shielding technique in children and found the average eye dose reduction of 38% and 33% for CT scans of the orbit and whole head in their Monte Carlo simulations as well as 34% and 20% for the entire and partial eye globe scans in paediatric patients, respectively.

However, despite some evidence supporting the use of shielding, routine use of these shields has been called into question. Geleijns et al (2006) and Vollmar et al (2008) reported that although dose reduction can be achieved with the use of shields, their use is also associated with significant artefacts and increase in image noise. Their findings were supported by Kalra et al (2009) indicating that as well as increasing noise, there is artifactually increased in attenuation values in the region immediately behind the shields. More recently, Lee et al (2011) showed that the use of thyroid shielding with cotton wool spacer reduces the dose to thyroid by up to 27% without affecting image noise although it was noted that there was noticeably increase in the attenuation values of the superficial neck structures such as the neck muscles. It was thought that this was likely related to a metal artefact caused by bismuth implanted within the shield itself.

Some of the streak artefacts in earlier studies can partly be explained by close apposition of the shields to the skin surface. This was the case for both Geleijns et al (2006) and Vollmar et al (2008). When a spacer or plastic shields are used the streak artefacts are noticeably eliminated (Hohl et al 2006, Karla et al, 2009 and Lee et al, 2011). However, the increase in attenuation appears to be a real issue (Karla et al, 2009, Lee et al, 2011). This could have implications when looking at coronary artery calcifications, renal cyst and adrenal mass characterisation, for example, where increase in attenuation is used specifically to characterise disease entities.

Another contentious issue concerns image quality. Although some of the studies have examined the effect on image quality, the robustness of these assessments has also been called into question. In particular, the image quality was not quantified objectively but rather only in qualitative terms with statements such as "we did not see any differences in quality between the shielded and unshielded lung" (Yilmaz et al, 2007). Some investigators have argued that when noise and image quality were analysed objectively and in a robust manner, studies appear to suggest that there is increase in image noise and deterioration of image quality (Geleijns et al, 2006, Vollmarr et al, 2008, Karla et al, 2009). Although, it must be noted that a more recent study by Lee et al (2011) looking at bismuth thyroid shield did not show significant differences in mean noise values. In general, there needs to be more evidence that robustly assess image quality prior to the routine use of shielding to be universally accepted.

Another issue concerns the effect on image reconstruction and wasted radiation. Some investigators argue that bismuth shielding has very different effects on patient dose for the frontal (AP) projection compared to the dorsal (PA) projection (Geleijns et al, 2010). The attenuation of the incoming and exiting beam ultimately has an effect of image quality and dose efficiency. It is argued that if the x-ray tube was in the dorsal position, the exiting x-ray beam is attenuated prior to reaching the detector and therefore this wastes unnecessary radiation, which may have been useful for image formation process. They argue that if the aim of bismuth shield is to reduce low energy photons that mainly deposit their energies on the surface of the patients, then this can be similarly achieved with lowering the tube current without the added artefacts.

The use of bismuth shielding in combination with automated tube current modulation has also been investigated. This is also fraught with danger as there is the potential for either a dose increase to the patient or unequal noise levels within/between images. The reason for this is that the AEC system may attempt to increase the dose to account for the extra patient attenuation. Some of this can be eliminated by placing the shields after the scanned projection radiograph (i.e. scanogram, scoutview or topogram), however, some manufacturer system performs continual monitoring of patient attenuation and adapts tube current in real time (e.g. when angular modulation is used - see earlier section). If this is the case, then the use of shielding can be detrimental and would actually increase the dose to the patient.

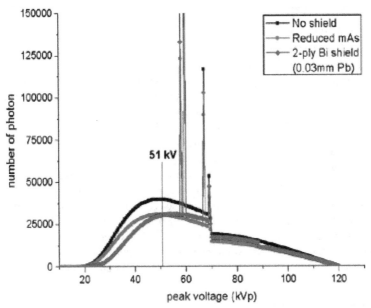

Fig. 9. Comparison of 120 kVp CT beam x-ray spectra with no shield, reduced mAs or 2-ply Bi shield. Note that the reduction of the low-energy photons is substantial in the 2-ply bismuth shield compared to the reduced tube current.
(with permission from Kim et al, 2010)

Patient factors

Increasingly, we are moving towards tailoring our examination according to individual patient to optimise image quality and minimising dose. Previous discussion regarding tube current modulation have already shed light on the benefits of this. In general, factors such as age, chest circumference, body mass index, and specific individual factors such as presence of stents and coronary calcification may influence the type of study being performed as well as dose limitation.

Patient scan length

Patient scan length is also worthy of consideration. Larger length scans produce larger dose. In the PROTECTION I trial, the average dose for a cardiac CT angiography was 12mSv with the average scan volume of 12cm (Hausleiter et al, 2006)). For cardiac scanning (at least for retrospective scanning), an increase in 1cm results in an increase of approximately 1mSv of added radiation. The volume of acquisition must therefore be tailored to suit each individual patient. Both radiologists and radiographers will need to have active role in selecting this.

Special patient groups

Paediatric patients

Usual dose reduction strategies such as tube current modulation, ECG-gating, etc as is used in adults can be similarly applied to children with further dose reduction. Adjustment of kV to 80kv is not an unreasonable approach as well as changing mAs values by employing weight-based specific protocols (Ben Saad et al, 2009, Lee et al, 2006, Young et al 2011).

Women's imaging and pregnancy

Efforts in dose reduction in pregnant patients owing to dose concerns to the foetus have previously been to not perform the study unless it is absolutely paramount. If this was deemed necessary, then various approaches have been adopted but mostly revolve around the use of shielding (see above). Some specific adjustment of protocols have been tried for certain clinical scenario such as for investigation of pulmonary embolism. Litmanovich et al (2009) compared reduced-dose pulmonary CT angiography (200 mA and 100 kV) with matched control group standard protocol (400 mA, 120 kv). The CT dose index, dose-length product, effective dose, image quality, and signal-to-noise ratio were assessed. There was a significant dose reduction of more than 65% using low dose protocols while maintaining diagnostic imaging quality. Though the dose to the chest has been substantially reduced, the dose to the foetus due to scatter radiation still poses concerns. Others have adopted a more simple approach to tackle this. Danova et al (2010) used lead aprons as shielding to the uterus when scanning thoracic CT achieving up to 34% reduction using the wrap around apron to cover for scatter radiation demonstrating that protective aprons are an effective dose reduction technique without additional costs and little effect on patient examination time.

Iterative reconstruction

Out of all the dose reduction strategies discussed, iterative reconstruction shows the most promise. CT workstations have used filtered back projection as the preferred method for producing images from the raw data acquired by the receptors. Filtered Back-Projection (FBP) algorithm has been used as the foundation of commercially available CT reconstruction techniques since the 1970's. Comparatively, it is robust and relatively undemanding on computer processing. It is still widely used today and considered acceptable for clinical diagnosis, but it does not provide optimal results for depiction of the patient as it makes many incorrect assumptions about the data. This is apparent in the inherent level of artefacts and noise in FBP images. To compensate for such noise, larger patient doses are required to overcome this.

Iterative reconstruction algorithms have been put forward as promising recent advances in CT technology but were in fact initially proposed by Shepp and Vardi back in 1982. Only recently, this has been shown to be superior to filtered back projection algorithms for noise reduction (Liu et al, 2007). Even though iterative reconstruction algorithms exhibit great advantages in situations of low signal to noise ratio, their use in real-time CT was previously limited by the time and computing power required to perform the iterations. Recently, however, due to improving computing technology, it is now possible to utilise various facets of iterative reconstruction to reduce the noise and thus achieving significant dose reduction.

- Adaptive Statistical Iterative Reconstruction (ASIR)

Adaptive statistical iterative reconstruction (ASIR) is a post-processing method marketed by GE (General Electric Medical systems, WI, USA) where images are obtained by applying adaptive statistical iterative reconstruction to filtered back projection images. The images are obtained in a low dose mode and the noise in the image is then reduced by applying ASIR. The amount of noise reduction applied can be varied from 1% to 100%. This technique allows modest (up to 40%) dose reduction with similar recorded levels of noise in the image. Several manufacturers have subsequently released similar technology and now all manufacturers are offering iterative reconstruction methods and a means for improved image quality and dose reduction.

There is increasing amount of research to suggest that utilising iterative reconstruction improves image quality, and thereby allows for lowering of kV and mAs and thereby reducing the dose (Prakash et al, 2010). Marin et al (2010) utilized ASIR at low kV and high mAs setting comparing with standard of care FBP technique scanning hepatic organs and found that there is improvement in both image quality and reduction in dose. Similarly, Singh et al (2010) found that abdominal scanning using ASIR compared with FBP yields significant benefit in terms of improved image quality. Due to improved contrast-to-noise ratio, lower radiation parameters can be utilised thereby achieving significant dose reduction. Similar results have been shown using chest CT in comparing ASIR with standard FBP. Leipsic et al (2011) showed that compared with FBP images, ASIR images had significantly higher subjective image quality, less image noise, and less radiation dose with around 30% reduction on average. Studies that have high dose burden such as CT colonography also shows much promise with the use of new ASIR technique. Flicek et al (2010) utilised ASIR in a pilot study using 18 patients undergoing CT colonography using an altered protocol of standard scan 50mAs (supine scan) and 25mAs (prone scan with 40% ASIR). The results show that the radiation dose can be reduced 50% below currently accepted low-dose techniques without significantly affecting image quality when ASIR is used. More recent investigations appear to confirm that using ASIR yields benefits in achieving dose reduction as well as improved image quality. Pontana et al (2011a) studied the utility of iterative reconstruction algorithm in comparison to FBP on 80 patients and found that iterative reconstructions provided similar image quality compared with the conventionally used FBP reconstruction at 35% less dose and also suggested that even higher dose reductions than 35% may be feasible by using higher levels of iterations. Significant noise reduction can also be achieved using the same dose/raw data (Pontana et al 2011b). More and more evidence are appearing in the literature for specific uses of ASIR in specific clinical setting (e.g. CT enterography in Crohn's patients - Kambadakone et al 2011; coronary CT angiography - Leipsic et al 2011) further emphasising its increasing usefulness in quest for improved image quality and dose reduction.

- Model-Based Iterative Reconstruction (MBIR)

Compared with ASIR, newer method so called 'Model-Based Iterative Reconstruction (MBIR)' has now been developed and instead of relying on a single model (as is used in ASIR), performs multiple iterations from multiple models. These models account for a complete three-dimensional assumptions that comprises of focal spot, beam shape, voxel size, and size of detector. In addition, MBIR also accounts for noise from photon flux as well as system noise from the scanner itself.

There is currently no literature research on the practical applications on MBIR as this has only recently been commercialised in December 2010. Preliminary work by the authors on phantoms (unpublished data) have revealed that MBIR shows the most reduction in terms of noise and is superior in terms of objective and subjective image quality and diagnostic confidence compared with ASIR and FBP technique. Dose can be further reduced by up to 80%. Studies are underway to see how image quality compares with traditional methods of FBP and ASIR. From the preliminary work, there is a distinct possibility of achieving body scanning at under 1mSv thus paving a way for significant dose reduction of up to 80% of current levels.

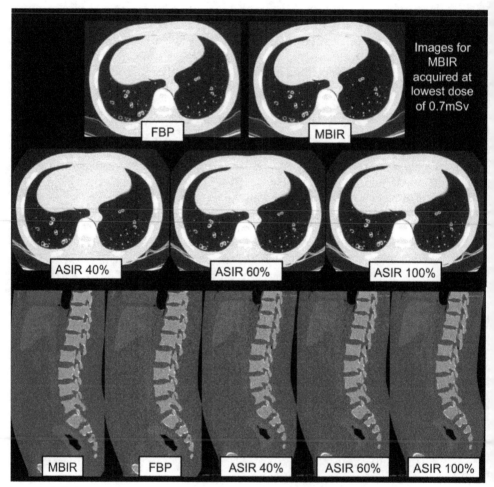

Fig. 10. Side-by-side comparison of scanned torso phantom showing different image quality between traditional filtered back projection, ASIR and MBIR. Note that despite MBIR being acquired at very low dose, this still shows remarkable superior image quality. (images are authors' own work)

Garnet technology

Recently, Garnet-based detectors have been developed which are 100 times faster and have 25% less afterglow compared to the traditional GOS (Gd2O2S)-based detectors. Reduced inherent noise as well as greater contrast resolution should allow for lower voltage/current techniques even for patients with high BMI. Moreover, due to greater contrast resolution, it is hoped that this might aid more accurate detection of in-stent restenosis in cardiac CT angiography – something that is difficult to assess accurately previously due to streak artefact (Cademartiri et al, 2007; Haraldsdottir et al 2010).

320-Row detector

With new 320-row detectors now available, axial volumetric scanning of a 16cm segment range in a single 0.35s rotation with an acquisition configuration of 320x0.5mm is now possible. This has the advantage of preventing overranging, and offers high spatial and temporal resolution. This has the potential to reduce the dose considerably. It has been determined that axial volumetric scanning constitutes a dose saving of up to 55% (Kroft et al, 2010; Al-Mousily et al 2011).

5. Dose reduction in fluoroscopy

At the turn of the century, there is little awareness of the risk of associated with fluoroscopic procedures, and in particular fluoroscopically-guided intervention. This has in some parts been due to the lack of adequate measuring equipment, complex dose relationships, and false sense of perceived lack of risk. The mood has gradually shifted since 2 key publications in 2004 calling for better dose management for fluoroscopically-guided interventions (Hirshfeld Jr et al 2004; Miller et al, 2004). Since then the FDA in the US have included some of the recommended requirements into new safety-related regulations for manufacturers, and these include features such as last image hold, display of cumulative exposure time, cumulative air kerma, and real-time display of the air kerma dose rate.
Dose delivery in fluoroscopic procedure is complex interactions of numerous factors. There are some excellent review articles (Hirshfeld Jr et al 2004; Steckler et al, 2009) but in-depth discussion is beyond the scope of this chapter. The issue of adequate dose monitoring and effective dose estimations have previously been called to question. The latest dose management technology, which is a feature that is made available on all angiographic equipment in the US, is 'cumulative dose at reference point'. This actually refers to the cumulative 'free-in-air' air kerma and quantify total amount of radiation delivered to a specific point located a fixed distance from the x-ray source. Although, actual dose to patient is a complex, procedure-specific, and varies between operators and is dynamic process (patient's movement during procedure will affect dose received, for example), the cumulative dose at reference point has been shown to correlate reasonably well with the absorbed dose at a specific skin site (Miller et al, 2003).

6. Dose reduction efforts in nuclear medicine

In the US, between 1972 and 2005, diagnostic nuclear medicine procedures increased 5- to 6-fold whereas the U.S. population increased by approximately 50%. Between 1996-2005, there was 5% annual growth in the number of nuclear medicine procedures while the growth of the U.S. population has been less than 1% annually. Between 1982 and 2005, the estimated per capita effective dose from in vivo diagnostic nuclear medicine increased by 550% and the collective effective dose increased by 720% (Mettler et al, 2008). In fact, the estimated 2005 per capita effective dose from diagnostic nuclear medicine (0.75 mSv) is greater than the total per capita dose from both diagnostic radiology and nuclear medicine examinations combined in 1982 (0.14 and 0.40 mSv, respectively). As might also be suspected, the estimated 2005 collective effective dose from diagnostic nuclear medicine (220,000 person Sv) is greater than the total per capita dose from both diagnostic radiology and nuclear medicine examinations in 1982 (32,600 and 92,000 person Sv, respectively).

Year	Procedures (Millions)	U.S. Population (Millions)	Exams per 1000 Population	Per-Capita Effective Dose (mSv)	Collective Effective Dose (Person-Sv)
1972	3.3	209.9	15.7		
1973	3.5	211.9	15.6		
1974		213.9			
1975	4.8	216.0	22.2		
1976		218.0			
1977		220.2			
1978	6.4	222.5	28.8		
1979		225.0			
1980	(5.8 to 6.4)	226.5			
1981	7.0	229.5	30.5		
1982	7.55 (7.4 to 7.7)	231.6	32.6	0.14	32,100
1983		233.0			
1984	6.3*	235.8	26.7		
1985	6.2*	237.9	26.2		
1986	6.7*	240.1	27.9		
1987	6.8*	242.3	28.1		
1988	7.1*	244.5	29.0		
1989	7.1*	246.8	28.9		
1990		249.7			
1991		252.1			
1992		255.0			
1993		257.7			
1994		260.3			
1995	10.2†	262.8	38.8		
1996	10.5†	260.3	40.3		
1997	10.9†	262.8	41.5		
1998	11.8†	265.2	44.5		
1999	12.6†	272.7	46.2		
2000	13.5†	282.1	47.9		
2001	14.5†	284.8	50.9		
2002	14.9†	287.9	51.8		
2003	15.7†	290.8	54.0		
2004	16.5†	294.7	56.0		
2005	(19) 17.2†	296.0	58.1	0.75	220,500

*FDA 1985 Radiation Experience Data 1980, Survey of U.S. Hospitals, DHEW Pub FDA 86-8253, National Technical Information Service, Springfield, VA.

†Data from IMV Inc Benchmark Report 2005 are patient visits, not procedures.

Table 5. Number of Nuclear Medicine Examinations Performed in the U.S. between 1972-2005.

(with permission from Mettler et al, 2008)

There has been a marked shift in the type of procedures with the studies of the brain and thyroid decreasing from a combined percentage of more than 56% of all procedures in 1973 to less than 4% in 2005. The most dramatic increase occurred in cardiac procedures increasing from 1% in 1973 to 57% in 2005. Cardiac studies are relatively high dose procedures and account for more than 85% of the effective dose to the patient population. Currently, more than 75% of all studies fall into 2 categories- cardiac and bone and these 2 types of examinations account for almost 95% of the collective effective dose. In 1982, the estimated number of nuclear medicine procedures was about 7.5 million. The per-capita effective dose from nuclear medicine was 0.14 mSv and the collective dose was 32,000 person Sv. By 2005, the estimated number of procedures had increased to about 19.6 million. The per-caput effective dose increased to about 0.75 mSv and the collective dose to about 220,000 person Sv.

	1973*		1982*		2005*	
Procedure	**Number**	**(%)**	**Number**	**(%)**	**Number**	**(%)**
Bone	125	(3.6)	1811	(24.5)	3450	(20)
Cardiac	33	(1.0)	950	(12.8)	9800	(57)
Lung	417	(11.9)	1191	(16.1)	740	(4)
Thyroid	460	(13.1)	677	(9.1)	—	(<2)
Renal	122	(3.5)	236	(3.2)	470	(3)
GI	535	(15.2)	1603	(21.7)	1210	(7)
Brain	1510	(43.0)	812	(11.0)	—	(<2)
Infection					380	(2)
Tumor	14	(0.4)	121	(1.6)	340	(2)
Other	294	(8.4)			—	(<2)
Total	3510	(100)	7400	(100)	17200	(100)

*National Council on Radiation Protection and Measurements.[2]
†IMV is for patient visits. Ratio of visits to procedures is about 1.14.

Table 6. Change (in Thousands) and Percentage of Total Examinations from 1972 in diagnostic nuclear medicine examinations.
(with permission from Mettler et al, 2008)

As with CT, dose reduction efforts have focused on improving technology with innovations such as increased sensitivity of single photon emission computed tomographic (SPECT) instrumentation. This is currently the principal modality for nuclear studies of the heart. Improvement in detectors such as the new cadmium-zinc-telluride detectors will facilitate higher speed/resolution gamma camera and reduce noise (Erlandsson et al, 2009). This equates to higher detector sensitivity, and the hope is that this will generally lead to less administered dose requirement over time (Peterson et al, 2011).

7. General notes on issues relating to dose monitoring

Modality-specific metrics are generally used for dose monitoring. These include administered radioactivity for radiopharmaceuticals, entrance skin kerma for x-rays, CT dose index for CT and dose-area product and cumulative air kerma for fluoroscopy. This can be confusing. In an ideal world, radiation safety should be assessed meaningfully across all imaging modalities using a single standard metric for assessment of radiation risk. This should be done not least because it would be more practical for clinicians, but also for ease of risk stratification for patients who are likely to undergo various imaging investigations of different modalities throughout their lifetime. Today, various methods exists, and more are under development to achieve that very goal but at the same time the validity of such endeavour are being questioned as standardised dose specification with a single quantity/unit may not necessarily encompass for both stochastic and deterministic effects. As it stands, no single unifying unit exists for dose monitoring for all medical-related procedure at present.

8. Conclusion

In summary, recent efforts in dose reduction in the medical community has arisen due to increasing awareness of radiation-related risks. As we perform more imaging, the risk to the population becomes more concerning. Trends to this effect show that this is increasing at an almost exponential rate. Fortunately, the medical community has made great strides both in utilising new technologies but also aided by closer examination of current practices. Some of the improvements can be implemented right away without much efforts and costs but rely on adherence to already available guidelines and decision making systems. Others rely on the use of new technologies that should make it easier for clinicians to both monitor accurately the dose delivered to the patient but also aid in dose reduction/optimization. In the future, medical imaging is likely to become intertwined with individualised medicine to such an extent that each patient will receive an appropriate test, at an appropriate dose level and according to his or her specific characteristics. Tailoring an investigation to the individual patient should become the norm rather than a one-size-fits-all solution. Several advances are on the horizon and further progress in computational power, new tube design, novel detectors and advances in medical physics and engineering will likely overcome current limitations which can only be beneficial.

9. References

Al-Mousily F, Shifrin RY, Fricker FJ, et al. (2011). Use of 320-detector computed tomographic angiography for infants and young children with congenital heart disease. *Pediatr Cardiol*, 32:426-32.

Aroua A, Olerud HM et al (2010). Collective doses from medical exposures: an inter-comparison of the "Top 20" radiological examinations based on the EC guidelines RP 154. Proceedings of the Third European IRPA Congress, June 2010, Helsinki, Finland.

Bautista, A.B., Burgos, A., Nickel, B.J., Yoon, J.J., Tilara, A.A. & Amorosa, J.K., (2009). Do clinicians use the American College of Radiology Appropriateness criteria in the management of their patients? *American Journal of Roentgenology*, 192(6), p. 1581.

BEIR VII report, (2006). Health Risks from Exposure to Low Levels of Ionizing Radiation: BEIR VII Phase 2, U.S. Nuclear Regulatory Commission. Washington, DC: National Academies Press.

Ben Saad M, Rohnean A, Sigal-Cinqualbre A et al (2009) Evaluation of image quality and radiation dose of thoracic and coronary dual-source CT in 110 infants with congenital heart disease. *Pediatr Radiol* 39:668-676

Berrington de Gonzalez A, Mahesh M, Kim K et al. (2009). Projected Cancer Risks From Computed Tomographic Scans Performed in the United States in 2007. *Arch Intern Med*, 169:2071-2077.

Blachar A, Tal S, Mandel A, Novikov I, Polliack G, Sosna J, Freedman Y, Copel L, Shemer J. (2006). Preauthorization of CT and MRI examinations: assessment of a managed care preauthorization program based on the ACR Appropriateness Criteria and the Royal College of Radiology guidelines. *J Am Coll Radiol*. Nov;3(11):851-9.

Burk Jr, R.J. (1996). Radiation risk in perspective: position statement of the Health Physics Society, *Health Physics Society Website. www. hps. org/documents/risk_ps010-1. pdf. Published January*. Updated in 2004.

Cardis, E., Vrijheid, M., Blettner, M., Gilbert, E., Hakama, M., Hill, C., Howe, G., Kaldor, J., Muirhead, C.R. & Schubauer-Berigan, M. (2007). The 15-Country Collaborative Study of Cancer Risk among Radiation Workers in the Nuclear Industry: estimates of radiation-related cancer risks, *Radiation research*, 167(4), pp. 396-416.

Cademartiri F, Schuijf JD, Pugliese , et al. (2007). Usefulness of 64-slice multislice com- puted tomography coronary angiography to assess in-stent restenosis. *J Am Coll Cardiol*, 49:2204-10.

Chawla SC, Federman N, Zhang D, et al. (2010). Estimated cumulative radiation dose from PET/CT in children with malignancies: a 5-year retrospective review. *Pediatr Radiol*. 40:681-686.

Chen J and Moir D (2010). An estimation of the annual effective dose to the Canadian population from medical CT examinations. *J Radiol Prot*, 30 (2), 131.

Coursey C, Frush DP, Yoshizumi T et al (2008) Pediatric chest MDCT using tube current modulation: effect on radiation dose with breast shielding. *AJR* 190:W54-W61

Danova D, Keil B, Kästner B, Wulff J, Fiebich M, Zink K, Klose KJ, Heverhagen JT. (2010). Reduction of uterus dose in clinical thoracic computed tomography. *Rofo*. Dec;182(12):1091-6.

Devine CE, Mawlawi O. (2010). Radiation safety with positron emission tomography and computed tomography. *Semin Ultrasound CT MR*. Feb;31(1):39-45. Review.

Einstein AJ, Henzlova MJ, Rajagoplana S. (2007). Estimating Risk of Cancer Associated With Radiation Exposure From 64-Slice Computed Tomography Coronary Angiography. *JAMA*. 298(3):317-323

Elliott A. (2009). Issues in medical exposures. *J Radiol Prot*. Jun;29(2A):A107-21.

Erlandsson, K., Kacperski, K., van Gramberg, D. & Hutton, B.F. (2009). Performance evaluation of D-SPECT: a novel SPECT system for nuclear cardiology, *Physics in medicine and biology*, 54, p. 2635.

Flicek, K.T., Hara, A.K., Silva, A.C., Wu, Q., Peter, M.B. & Johnson, C.D. (2010). Reducing the radiation dose for CT colonography using adaptive statistical iterative reconstruction: A pilot study, *AJR*. 195(1), pp. 126-31.

Flint-Richter, P. & Sadetzki, S. (2007). Genetic predisposition for the development of radiation-associated meningioma: an epidemiological study, *The Lancet Oncology*, 8(5), pp. 403-10.

Flohr TG , L eng S, Yu L, et al. (2009). Dual-source spiral CT with pitch up to 3.2 and 75 ms temporal resolution: image reconstruction and assessment of image quality. *Med Phys*, 3 6 (12): 5641 -5 653 .

Foley, S.J., McEntee, M.F., Achenbach, S., Brennan, P.C., Rainford, L.S. & Dodd, J.D. (2011). Breast Surface Radiation Dose During Coronary CT Angiography: Reduction by Breast Displacement and Lead Shielding, *American Journal of Roentgenology*, 197(2), pp. 367-73.

Fricke BL, Donnelly LF, Frush DP et al. (2003). In-plane bismuth breast shields for pediatric CT: effects on radiation dose and image quality using experimental and clinical data. *AJR* 180:407-411

Geleijns, J., Wang, J. & McCollough, C. (2010). The use of breast shielding for dose reduction in pediatric CT: arguments against the proposition, *Pediatric radiology*, pp. 1-4.

Geleijns J, Salvado AM, Veldkamp WJ et al. (2006). Quantitative assessment of selective in-plane shielding of tissues in computed tomography through evaluation of absorbed dose and image quality. *Eur Radiol* 16:2334-2340

Hadley, J.L., Agola, J. & Wong, P. (2006). Potential impact of the American College of Radiology appropriateness criteria on CT for trauma, *American Journal of Roentgenology*, 186(4), p. 937.

Hart, Wall, Hillier, Shrimpton. (2010). HPA-CRCE-012 - frequency and collective dose for medical and dental x-ray examinations in the UK, 2008. Health protection agency December 2010: Available from: http://www.hpa.org.uk/Publications/Radiation/CRCEScientificAndTechnicalRe portSeries/HPACRCE012/. Accessed 5 May 2011.

Hausleiter et al. (2010) Image Quality and Radiation Exposure With a Low Tube Voltage Protocol for Coronary CT Angiography Results of the PROTECTION II Trial. *JACC Cardiovasc Imagingvol.* 3 (11) pp. 1113-23

Haraldsdottir et al. (2010) Diagnostic accuracy of 64-slice multidetector CT for detection of in-stent restenosis in an unselected, consecutive patient population. *European Journal of Radiology* vol. 76 (2) pp. 188-94

Hirshfeld JW Jr, Balter S, Brinker JA, Kern MJ, Klein LW, Lindsay BD, Tommaso CL, Tracy CM, Wagner LK, Creager MA, Elnicki M, Hirshfeld JW Jr, Lorell BH, Rodgers GP, Tracy CM, Weitz HH. (2004). American College of Cardiology Foundation; American Heart Association; American College of Physicians. ACCF/AHA/HRS/SCAI clinical competence statement on physician knowledge to optimize patient safety and image quality in fluoroscopically guided invasive cardiovascular procedures. A report of the American College of Cardiology Foundation/American Heart Association/American College of Physicians Task Force on Clinical Competence and Training. *J Am Coll Cardiol.* Dec 7;44(11):2259-82.

Hopper KD, King SH, Lobell ME et al. (1997). The breast: in-plane x-ray protection during diagnostic thoracic CT - shielding with bismuth radioprotective garments. *Radiology* 205:853-858

Hohl C, Wildberger JE, Suss C, et al. (2006). Radiation dose reduction to breast and thyroid during MDCT: effectiveness of an in-plane bismuth shield. *Acta Radiol;* 47:562-567

Hausleiter J, Meyer T, Hadamitzdy et al. (2009). Estimated radiation dose associated with cardiac CT angiography. *JAMA* ;301:500-7.

International Agency for Research on Cancer (IARC)
    Web site: www.iarc.fr
    IARC Carcinogen Monographs: http://monographs.iarc.fr. Accessed 6 Aug 2011.

Israel, G.M., Cicchiello, L., Brink, J. & Huda, W. (2010). Patient size and radiation exposure in thoracic, pelvic, and abdominal CT examinations performed with automatic exposure control, *American Journal of Roentgenology*, 195(6), p. 1342.

Jansen-van der Weide, M.C., Greuter, M.J.W., Jansen, L., Oosterwijk, J.C., Pijnappel, R.M. & de Bock, G.H. (2010). Exposure to low-dose radiation and the risk of breast cancer among women with a familial or genetic predisposition: a meta-analysis, *European radiology*, pp. 1-10.

Kambadakone, A.R., Chaudhary, N.A., Desai, G.S., Nguyen, D.D., Kulkarni, N.M. & Sahani, D.V. (2011). Low-Dose MDCT and CT Enterography of Patients With Crohn

Disease: Feasibility of Adaptive Statistical Iterative Reconstruction, *American Journal of Roentgenology*, 196(6), p. W743.

Kalra, M.K., Maher, M.M., Toth, T.L., Hamberg, L.M., Blake, M.A., Shepard, J.A. & Saini, S. (2004). Strategies for CT Radiation Dose Optimization1, *Radiology*, 230(3), p. 619. - 2004a

Kalra MK, Maher MM, Kamath RS, et al. (2004). Sixteen-detector row CT of abdomen and pelvis: study for optimization of Z-axis modulation tech-nique performed in 153 patients. *Radiology*; 233:241-249. - 2004b

Kalra MK, Maher MM, Prasad SR, et al. (2003). Correlation of patient weight and cross-sectional dimensions with subjective image quality at standard dose abdominal CT. *Korean J Radiol*; 4:234-238.

Kalra MK, Dang P, Singh S et al. (2009). In-plane shielding for CT: effect of off-centering, automatic exposure control and shield-to-surface distance. *Korean J Radiol* 10: 156-163

Kim, S., Frush, D.P. & Yoshizumi, T.T. (2010). Bismuth shielding in CT: support for use in children, *Pediatric radiology*, pp. 1-5.

Kroft LJ, Roelofs JJ, Geleijins J. (2010). Scan time and patient dose for thoracic imaging in neonates and small children using axial volumetric 320-detector row CT compared to helical 64-, 32-, and 16- detector row CT acquisitions. *Pediatr Radiol* ;40:294-300.

Lee, Y.H., Park, E., Cho, P.K., Seo, H.S., Je, B.K., Suh, S. & Yang, K.S. (2011). Comparative Analysis of Radiation Dose and Image Quality Between Thyroid Shielding and Unshielding During CT Examination of the Neck, *American Journal of Roentgenology*, 196(3), p. 611.

Lee T, Tsai IC, Fu YC. (2006). Using multi-detector row CT in neonates with complex congenital heart disease to replace diagnostic cardiac catheterization for anatomical investigation—initial experiences in technical and clinical feasibility. *Pediatr Radiol* 36:1273-1282

Leipsic, J., Nguyen, G., Brown, J., Sin, D. & Mayo, J.R. (2010). A prospective evaluation of dose reduction and image quality in chest CT using adaptive statistical iterative reconstruction, *AJR. American journal of roentgenology*, 195(5), pp. 1095-9.

Leipsic, J., Heilbron, B.G., and Hague, C. (2011). Iterative reconstruction for coronary CT angiography: finding its way. *Int J Cardiovasc Imaging*, Feb 27. (Epub ahead of print].

Leschka et al. (2008). Low kilovoltage cardiac dual-source CT: attenuation, noise and radiation dose. *Eur Radiol* ;18(9):1809-1817.

Li, J., Udayasankar, U.K., Tang, X., Toth, T.L., Small, W.C. & Kalra, M.K. (2011). Patient Size Compensated Automatic Tube Current Modulation in Multi-detector Row CT of the Abdomen and Pelvis, *Academic radiology*, 18(2), pp. 205-11.

Litmanovich D, Boiselle PM, Bankier AA, Kataoka ML, Pianykh O, Raptopoulos V. (2009). Dose reduction in computed tomographic angiography of pregnant patients with suspected acute pulmonary embolism. *J Comput Assist Tomogr*. Nov-Dec;33(6): 961-6.

Liu YJ, Zhu PP, Chen B, et al. (2007). A new iterative algorithm to reconstruct the refractive index. *Phys Med Biol*, 52:L5 – L13

Marin, D., Nelson, R.C., Schindera, S.T., Richard, S., Youngblood, R.S., Yoshizumi, T.T. & Samei, E. (2010). Low-tube-voltage, high-tube-current multidetector abdominal CT:

improved image quality and decreased radiation dose with adaptive statistical iterative reconstruction algorithm--initial clinical experience, *Radiology*, 254(1), pp. 145-53.

McCollough CH, Bruesewitz MR, Kofler JM. (2006). CT Dose Reduction and Dose Management Tools: Overview of Available Options; *Radiographics* March - April, 26:2: 503 - 512.

McCollough CH. (2005). Automatic exposure control in CT: are we done yet? *Radiology* Dec; 237(3): 755-756.

McCollough C H, Leng S, Schmidt B, Allmendinger T, Eusemann C, Flohr TG. (2009). Use of a pitch value of 3.2 in dual-source cardiac CT angiography: dose performance relative to existing scan modes [abstr]. In: Radiological Society of North America scientific assembly and annual meeting program. Oak Brook, Ill: Radiological Society of North America,; 485-486.

Mettler Jr, F.A., Bhargavan, M., Thomadsen, B.R., Gilley, D.B., Lipoti, J.A., Mahesh, M., McCrohan, J. & Yoshizumi, T.T. (2008). Nuclear medicine exposure in the United States, 2005-2007: preliminary results, *Seminars in nuclear medicine*, 38(5), pp. 384-91.

Mettler FA Jr, Bhargavan M, Faulkner K, et al. (2009). Radiologic and nuclear medicine studies in the United States and worldwide: frequency, radiation dose, and comparison with other radiation sources — 1950-2007. *Radiology* 2009 ;2 53 (2): 520-5 31

Miller et al. (2004). Quality improvement guidelines for recording patient radiation dose in the medical record. *J Vasc Interv Radiol*, 15(5):423-429.

Miller DL, Balter S, Cole PE et al. (2003). Radiation doses in interventional radiology procedures: the RAD-IR study. II. Skin dose. *J Vasc Interv Radiol*, 14(8): 977-990.

Monson, R., Cleaver, J., Abrams, H.L., Bingham, E. & Buffler, P.A. (2005). *BEIR VII: health risks from exposure to low levels of ionizing radiation*, Washington DC: National Academies Press.

Muirhead, C.R., O'Hagan, J.A., Haylock, R.G.E., Phillipson, M.A., Willcock, T., Berridge, G.L.C. & Zhang, W. (2009). Mortality and cancer incidence following occupational radiation exposure: third analysis of the National Registry for Radiation Workers, *British journal of cancer*, 100(1), pp. 206-12.

Mukundan S, Wang PI, Frush DP et al. (2007). MOSFET dosimetry for radiation dose assessment of bismuth shielding of the eye in children. *AJR* 188:1648-1650

NCRP (2009). Ionizing radiation exposure of the population of the United States. NCRP Report 160. National Council on Radiation Protection and Measurements, Bethesda MD.

Perisinakis K, Raissaki M, Tzedakis A et al. (2005). Reduction of eye lens radiation dose by orbital bismuth shielding in pediatric patients undergoing CT of the head: a Monte Carlo study. *Med Phys* 32:1024-1030

Peterson, T.E. & Furenlid, L.R. (2011). SPECT detectors: the Anger Camera and beyond, *Physics in medicine and biology*, 56, p. R145.

Preston DL, Ron E, Tokuoka S, et al. (2007). Solid cancer incidence in atomic bomb survivors: 1958-1998. *Radiation research*, 168:1-64.

Pontana, F., Duhamel, A., Pagniez, J., Flohr, T., Faivre, J.B., Hachulla, A.L., Remy, J. & Remy-Jardin, M. (2011). Chest computed tomography using iterative reconstruction vs filtered back projection (Part 2): image quality of low-dose CT examinations in 80 patients, *European radiology*, 21(3), pp. 636-43. A

Pontana, F., Pagniez, J., Flohr, T., Faivre, J.B., Duhamel, A., Remy, J. & Remy-Jardin, M. (2011). Chest computed tomography using iterative reconstruction vs filtered back projection (Part 1): evaluation of image noise reduction in 32 patients, *European radiology*, 21(3), pp. 627-35. B

Prakash, P., Kalra, M.K., Ackman, J.B., Digumarthy, S.R., Hsieh, J., Do, S., Shepard, J.A. & Gilman, M.D. (2010). Diffuse lung disease: CT of the chest with adaptive statistical iterative reconstruction technique, *Radiology*, 256(1), pp. 261-9.

Pflederer T, Rudofsky L, Ropers D et al. (2009). Image quality in a low radiation exposure protocol for retrospectively ECG-gated coronary CT angiography. *AJR AM J Roentgenol*, 192:1045-50.

Preston DL, Ron E ,Tokuoka S, e t al. (2007). Solid cancer incidence in atomic bomb survivors: 1958-1998. *Radiat Res*, 168(1): 1-64 .

Ronckers, C.M., Doody, M.M., Lonstein, J.E., Stovall, M. & Land, C.E. (2008). Multiple diagnostic X-rays for spine deformities and risk of breast cancer, *Cancer Epidemiology Biomarkers & Prevention*, 17(3), p. 605.

Roobottom CA, Mitchell G and Morgan-Hughes G. (2010). Radiation-reduction strategies in cardiac computed tomographic angiography. *Clin Radiol*, vol. 65 (11) pp. 859-67

Schindera ST, Nelson RC, Toth TL, et al. (2008). Effect of patient size on radiation dose for abdominal MDCT with automatic tube current modulation: phantom study. *AJR Am J Roentgenol*, 190:W100-W105.

Schoenhagen P. (2008). Back to the future: coronary CT angiography using prospective ECG triggering. *Eur Heart J* ;29(2):153-4.

Singh, S., Kalra, M.K., Hsieh, J., Licato, P.E., Do, S., Pien, H.H. & Blake, M.A. (2010). Abdominal CT: Comparison of Adaptive Statistical Iterative and Filtered Back Projection Reconstruction Techniques, *Radiology*. Nov;257(2):373-83

Stecker MS, Balter S, Towbin RB, et al. (2009). Guidelines for patient radiation dose management. *J Vasc Interv Radiol*, 20(7 suppl): S263-S273.

Stolzmann P, Leschka S, Scheffel H, et al. (2008). Dual-source CT in step-and-shoot mode: noninvasive coronary angiography with low radiation dose. *Radiology*, 249(1): 71-80.

Savandi, A.S., Demarco, J.J., Cagnon, C.H., Angel, E., Turner, A.C., Cody, D.D., Stevens, D.M., Primak, A.N., McCollough, C.H. & McNitt-Gray, M.F. (2009). Variability of surface and center position radiation dose in MDCT: Monte Carlo simulations using CTDI and anthropomorphic phantoms, *Medical physics*, 36, p. 1025.

Street M, Brady Z, Van Every B, Thomson KR. (2009). Radiation exposure and the justification of computed tomography scanning in an Australian hospital emergency department. *Intern Med J*. Nov;39(11):713-9.

United Nations Scientific Committee on the Effects of Atomic Radiation. (2010). Sources and effects of ionizing radiation. Medical radiation exposures, annex A. 2008 Report to the General Assembly with Annexes. New York, NY: United Nations, 2010.

Vollmar SV, Kalender WA. (2008). Reduction of dose to the female breast in thoracic CT: a comparison of standard-protocol, bismuth-shielded, partial and tube-current-modulated CT examinations. *Eur Radiol* 18:1674-1682.

White CS, Kuo D. (2007). Chest pain in the emergency department: role of multidetector CT. *Radiology*. Dec;245(3):672-81. Review.

Yilmaz MH, Yasar D, Albayram S et al. (2007). Coronary calcium scoring with MDCT: the radiation dose to the breast and the effectiveness of bismuth breast shield. *Eur J Radiol* 61:139- 143.

Young, C., Taylor, A.M. & Owens, C.M. (2011). Paediatric cardiac computed tomography: a review of imaging techniques and radiation dose consideration, *European radiology*, 21(3), pp. 518-29.

Yu L,Li H, Fletcher JG, McCollough CH. (2010). Automatic selection of tube potential for radiation dose reduction in CT: a general strategy. *Med Phys*, 37(1): 234 -2 43.

Zhang, D., Zankl, M., DeMarco, J.J., Cagnon, C.H., Angel, E., Turner, A.C. & McNitt-Gray, M.F. (2009). Reducing radiation dose to selected organs by selecting the tube start angle in MDCT helical scans: a Monte Carlo based study, *Medical physics*, 36(12), pp. 5654-64.

3

Molecular Spectroscopy Study of Human Tooth Tissues Affected by High Dose of External Ionizing Radiation (Caused by Nuclear Catastrophe of Chernobyl Plant)

L. A. Darchuk[1], L. V. Zaverbna[2],
A. Worobiec[1] and R. Van Grieken[1]
[1]*University of Antwerp, Antwerp,*
[2]*National University of Medicine, Lviv,*
[1]*Belgium*
[2]*Ukraine*

1. Introduction

Ionizing radiation remains the most significant environmental factor, causing severe impacts on human health on the vast territory of Ukraine for 25 years.
Medical monitoring of the dental health of more than 1500 people who participated in cleaning work at the Chernobyl power station territory after the catastrophe has demonstrated an increasing of specific dental problems occurred after several years of participating at these works (Leus et al., 1998). The high level of hyperstenzy of hard dental tissues, the increase of enamel abrasion (first and second degrees) and the noncarious dental disease (cuneiform defects, pitting) are typical for around 63%, 36% and 47% of examined patients, respectively (Darchuk et al., 2008). Pathological dental abrasion, cuneiform defects, enamel erosion and enamel scissuras have been shown in Fig.1. Significant damages of parodontium (level II-III) by generalized severe periodontitis were detected for men with an acute radiation syndrome (Revenok, 1998). The main pathology was an intensified degeneration process while carious changes occurred, although these last were not directly related to irradiation.

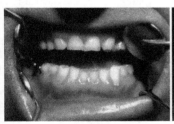
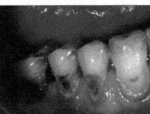
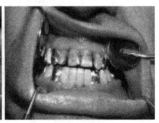

| Pathological dental abrasion | Cuneiform defects, noncaries changes | Enamel erosion |

Fig. 1. Dental defects which are typical of person undergone external ionizing radiation.

The influence of external ionizing radiation on dental tissues was analyzed by several scientists. Zhang et al. (2004) studied with SEM (Scanning Electron Microscopy) the effect of irradiation on the susceptibility of radiation caries, the structures of rats` tooth enamel and dentin. The collagen fibers and the resistance to acid after undergo external radiation were also investigated. Enamel structure changes were found after irradiation of 30 - 70 Gy. It was shown that dentin morphology changed, some collagen fibers vanished and resistance to acid was reduced after irradiation.

A study on the possibility of dose-dependent tooth-germ (for four-week-old dogs) damage produced by ionizing radiation has shown hyperemia and edemata on tooth-germ pulps from 1.3 Gy onward. Both of these dental diseases became more acute as the radiation dose increased from 1.3 Gy to 5.3 Gy. Possible damage to both the dentin and enamel was pointed out (Sobkoviak et al., 1977).

Dental analysis of rats which had undergone an exposure to 13.2 Gy during 100 days showed a temporary halting in tooth formation following exposure. The dentin which was formed immediately after the radiation was architecturally disarranged to the extent that no paralleling of dentinal tubules existed. In some instances, there was actual resorption of tooth structure with a fibrous replacement (English et al., 1954).

The goal of this work was to study structural changes of teeth taken from people which were exposed to high doses of external ionizing radiation. They worked for different periods: from several weeks to several months, on the territory of the Chernobyl reactor IV zone during the first year just after the catastrophe. The teeth were extracted according to medical recommendations because of oral surgery or mechanical damage due to accidents.

Infrared (IR) absorption and micro-Raman spectroscopy (MRS) as the most efficient techniques to analyze molecular structure of biological tissues (Ellis et al., 2006) have been applied to investigate tooth tissues. The main advantage of Infrared spectroscopy is a superior signal-to noise ratio, which is much less for Raman spectroscopy. Anyway Raman spectroscopy offers another benefit. The MRS is very useful for analysis of bone-like objects, which consist of mineral and organic substances, because Raman vibrational bands of crystalline materials are often sharp. So that the vibrational bands typical of organic matrix are more distinguishable and the identification of minerals is less doubtful.

Enamel, dentin, and cement belonging to 17 patients who absorbed high doses (0.5-1.7 Gy, high dose group, HDG) of ionizing radiation were studied. Only patients without the bone marrow affection were investigated. For comparison the same tooth tissues of 10 men who had not been exposed to radiation (control group, CG)) were investigated as well.

Tooth is composed by enamel, dentin, cement and root (Fig. 2) which are bioinorganic materials. The organic part of the dental tissues is represented by collagen (Ten Gate, 1998) which controls the calcification of bone and crystallization of apatite in bones (Castrom et al., 1956). Collagen is rich in glycine, proline, and hydroxyproline. Collagen molecules are composed by 3 polypeptide chains forming tubinal structure. Collagen is rich with glycine, praline and hydroxyproline (Neuberger, 1956). The composition of aminoacids depends on the dental tissues – the ratio of proline:hydroxyproline:glycine is about 11:11:55 for dental enamel and 29:21:62 for dentin (Geller, 1958). The mineral part of enamel, dentin and cement is composed by carbonate hydroxyapatite, with an approximate formula $Ca_{10}(PO_4)_{6-x}(OH)_{2-y}(CO_3)_{x+y}$, where $0 \leq x \leq 6$ and $0 \leq y \leq 2$ (Le Geros, 1991). The carbonate ion CO_3^{2-} can substitute for either PO_4^{3-} (B-substitution, a CO_3^{2-} ion substitutes for a single PO_4^{3-} ion) or OH^- (A-substitution, a CO_3^{2-} ion replaces two OH^- ions) (Le Geros, 1981; Elliott, 1994). Also the carbonate ion can be present in dental tissues as an adsorbed phase in the mineral lattice (Elliott, 1994).

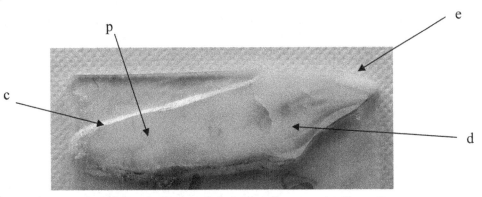

Fig. 2. Cross section of a tooth; e, enamel; d, dentin; c, cement; p, pulp cavity.

Tooth enamel, covering tooth external surface, is the hardest and most mineralized dental tissue. It contains about 96-98 wt.% of mineral matrix and around 2-4 wt. % of protein and water. Protein is rich in proline, glycine and glutamic acid.

Dentin makes up the bulk of the tooth and it is surrounded by cement at the root and by a thin protective layer of enamel. Dentin consists of about 70 wt.% of mineral matter, 20 wt.% of organic materials, and 10 wt.% of water (Ten Gate, 1998). The protein of dentin is composed mainly of Type I collagen and non-collagenous proteins (Le Geros, 1981; 1991). Cement is a bonelike calcified connective tissue covering tooth roots (Ross et al., 2003) and its composition is similar to the composition of dentin. Cement consists of about 60-65 wt. % of mineral matter, around 23-26 wt.% of organic compounds (mainly collage) and about 12 wt.% of water.

Diversities of enamel, dentin and cement are provided by the nature of the tissues' embryogenesis. The tooth dentin and cement are grown up from mesenchyme while the tooth enamel is grown up from ectoderm, so that enamel almost does not consist of organic matrix.

2. Samples and experimental technique

2.1 Infrared absorption spectroscopy

The preparation of samples for infrared spectroscopy has some features because biological tissues consist of some amount of water. So that samples are dried to avoid overlapping of absorption bands arose from analyzed objects with a broad absorption bands typical of water.

For preparing samples teeth were washed with distilled water and then left for several hours in a drying chamber at 50° C. Afterwards enamel, dentin and cement were sawn down with diamond dental drill to get powder of tooth tissues for IR spectral analysis. Then 0.2 g of tooth tissue powder was thoroughly mixed with 50 µl vaseline oil and put in a cuvette of KBr.

All IR spectra were recorded with one-beam infrared spectrometer IKS-31. Ten accumulations of spectra has been applied for each measurement to improve signal-to-noise ratio. Infrared spectral analysis was done in a low vacuum camera (10^{-1} Pa) to minimize overlapping with broad absorption bands typical of atmospheric CO_2 and H_2O. The infrared intensity transmitted by the cuvette containing the sample (I_s) and that transmitted by an empty cuvette (I_o) were measured in the 600-4000 cm^{-1} spectral range with a resolution of 2-4 cm^{-1}. The spectral resolution depends on diffraction grate (number of grating groove per mm) applied for each spectral interval.

The absorption intensity $I_a(v)$, where v is wavenumber in cm^{-1}, of the investigated samples was obtained as $I_a = (1 - I_s/I_c)$, which is approximately proportional to the sample absorption. For a semiquantitative analysis, the $I_a(v)$ spectra were fitted using a standard procedure which describes the $I_a(v)$ spectrum as a sum of Lorentzian components (Burn, 1985).

2.2 Micro-Raman spectroscopy

For micro-Raman spectroscopy washed and dried teeth were put in polymer solution of Technoviz 4004 (Kulzer, Germany), which got solidified after 6-8 hours. Then the tooth samples were cut in slices with a thickness of $1 \div 1.5$ mm with a diamond disk, Fig.2.

Raman spectra have been collected with a Renishaw InVia micro-Raman (Renishaw plc, GB) spectrometer. For vibrational excitation two laser were applied: argon laser (514.5 nm, Spectra Physica) with a maximum laser power of 50 mW as well as a diode laser (785 nm, Renishaw) with a maximum laser power of 300 mW. Raman spectra were obtained in the spectral range 100 – 3200 cm^{-1} at a resolution of 2 cm^{-1}. Measurement times between 10 and 30 s have been used to collect the Raman spectra with a signal-to-noise ratio of better than 100/1.

The slices of tooth were placed on the stage of the microscope, and the transverse cross section was oriented perpendicularly to the incident laser beam. A continuous laser beam was focused on a sample via 50x and 100x microscope objectives. The laser beam spot size depends on objectives. Spectra were obtained using a 100×(NA=0.95; theoretical spot size of 0.36 μm and 0.50 μm for 514.5 nm and for 785 nm laser, respectively) and 50×(NA=0.70; theoretical spot size of 0.45 μm and 0.68 μm for 514.5 nm and for 785 nm laser, respectively) magnification objectives.

Because of the heterogeneous nature of teeth, single point Raman microspectroscopy cannot adequately describe the microstructure of tissues and therefore spatial information is needed. Hence for the analysis of teeth, MRS mapping, with a spatial resolution of ~1 μm, allows to analyze tooth tissue zones such as individual cement lines, individual lamellae. Raman spectral mapping was employed to study the orientation of mineral and collagen components of bone tissues by Morris et al. (2004) and by Kazanci et al. (2006).

A computer controlled micro-Raman mapping was applied to analyze enamel, dentin, and cement. The microscope stage was XY-motorized and computer-controlled for point-by-point scanning with 0.1 μm resolution, 1 μm reproducibility and 90 mm × 60 mm maximum spatial range. Raman mapping data sets are collected by point-by-point scanning with 1 μm as a minimum step. A total of 20 images were obtained by using the motorized translation stage and the pixel reconstruction. Raman images were acquired in different rectangular areas of 60x60, 60x40, 50x50, 50x30 and 40x30 μm². The images were created with step size of 1 or 2 μm. The sample background fluorescence was subtracted with Wire 2.0 softwear (Renishaw, GB).

The Raman spectra, except MRS mapping, were normalized to the intensity of the strongest band at 962 cm^{-1} (v_1-PO_4 mode) of each spectrum.

3. Discussion

3.1 Enamel

Mineral component of dental tissues causes the presence in the IR spectra of bands assigned to the carbonate-ions (CO_3^{2-}) and the orthophosphate-ions (PO_4^{3-}). Collagen is presented by amide groups of proteins - ($CONH_2$). The amide absorption bands are well evident at 1660 cm^{-1} (amide I), 1540-1550 cm^{-1} (amide II), and band at 1240 cm^{-1} (amide III).

Infrared spectra of dental enamel taken from patients of both HDG and CG consisted of strong and wide absorption band at 900-1200 cm^{-1} which arose due to v_3 vibration mode of the phosphate-ion PO_4^{3-} (Fig. 3). This wide absorption band consisted of a set of bands at 950-970, 1000-1020, and 1050-1080 cm^{-1} in the IR spectra of enamel. A comparison of IR spectra of enamel belonging to patients from the control group and people exposed to high doses of ionizing radiation did not show significant changes of absorption band profile typical of phosphate-ion. Nevertheless it was detected that the intensity of the absorption band at 880 cm^{-1} which is typical for CO_3^{2-} ions was stronger in the spectra of irradiated enamel. So that the phosphate-carbon ratio in mineral matrix of enamel changed to increasing of carbonate content in enamel belonging to patients of HDG. Semiquantitative analysis was applied to estimate the area of the band assigned to the carbonate-ions. The area of the band was calculated with Origin softwear according to a procedure described by Burn (1985) and Bebeshko et al. (1998). This analysis showed increasing of the area of the band typical of the anion CO_3^{2-} (at 880 cm^{-1}) up to 0.98 ± 0.04 (a.u.) for people from HDG, in comparison with the area value of 0.84 ± 0.04 (a.u.) for patients of CG. Absorption bands at 1420 and 1482 cm^{-1} arose from the v_3-CO_3^{2-} vibration mode were not distinguished in the obtained IR spectra, because of their overlapping with the strong absorption band at 1440 cm^{-1} typical of vaseline.

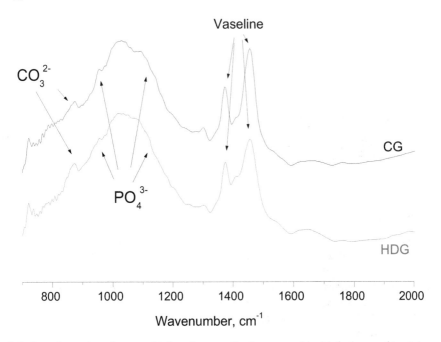

Fig. 3. Infrared spectra of enamel taken from patients exposed to high doses of ionizing radiation (HDG) and people from control group (CG).

Raman spectra of teeth enamel of HDG and CG were more informative than infrared spectra to detect transformations of the enamel mineral matrix. The Raman spectra of enamel obtained with 514 nm were as informative as those obtained with 785 nm laser beam excitation (Fig. 4), but the spectra with excitation at 514 nm often showed an intensive fluorescence background. The fluorescence should be easily subtracted, but if the fluorescence is huge and the intensity of Raman bands is very low compared to the fluorescence signal, the accuracy of analysis will be lower.

Hydroxyapatite single crystals belong to the crystal-class C_6 and they can be described with the Raman-active symmetry tensors (A, E_1, and E_2). The free PO_4^{3-} ion has four internal vibration modes: v_1 at 948 and 962 cm^{-1}, v_2 at 432 and 447 cm^{-1}, v_3 at 1028, 1053 and 1075 cm^{-1}, and v_4 at 580, 593 and 608 cm^{-1} (Fig. 4). Intensities of the Raman bands are invariant to the orientation of the crystal and depend only on the c-axis orientation (Tsuda et al., 1994; Nelson et al., 1982).

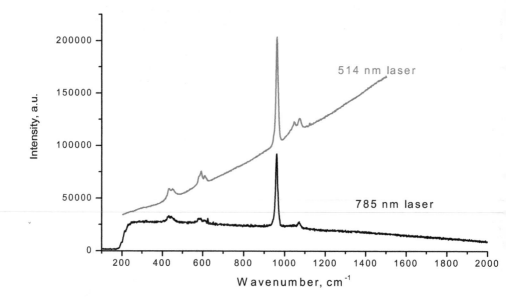

Fig. 4. Raman spectra of enamel recorded with laser beam excitation at 514 nm and at 785 nm.

In the case of enamel taken from patients exposed to ionizing radiation: for the v_4 vibration mode of PO_4^{3-}, the band at 580 cm^{-1} was not detected, the band at 450 and 608 cm^{-1} became very weak, but the band at 593 cm^{-1} did not change (Fig. 5) in comparison with the Raman spectra of enamel belonging to men from control group. These results can be explained as a breaking of the hydroxyapatite crystal symmetry.

The shoulder at 1028 cm^{-1} ($v_3(PO_4)$) was covered by a weak band at 1045 cm^{-1} ($v_3(PO_4)$, out of plane), which is typical of amorphous phosphate and a weak band at 1071 cm^{-1} (Fig. 5) which is assigned to carbonated apatite, type-B substitution (Elliott, 1994; Morris et al., 2004).

The increase of the carbonate content in the enamel is the result of a lack of equilibrium between phosphate and carbonate phases of $Ca_{10}(PO_4)_{6-x}(OH)_{2-y}(CO_3)_{x+y}$, induced by irradiation. The mentioned changes of the mineral matrix lead to a decrease of tooth enamel hardness.

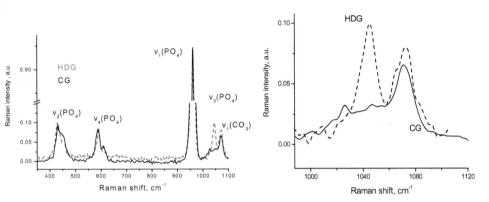

Fig. 5. Raman spectra in the range of internal vibrations of the mineral part of tooth enamel taken from patients of HDG and CG.

3.2 Dentin and cement

Dentin and cement differ from enamel and they look like coarse-fibered bone. IR spectra of dentin and cement were very similar; so only infrared absorption spectra of dentin will be discussed. Infrared spectroscopy of both dentin and cement taken from patients of HDG showed a decreasing content of the mineral matrix (intensities of phosphate bands are lower that those for CG) and changes in organic matrix (Fig. 6). The area of the absorption band at 880 cm-1, typical of the anion CO_3^{2-} increased from 0.97 ± 0.05 (a.u.) for dentin taken from people of control group to 1.05 ± 0.05 (a.u.) for HDG. The absorption band of

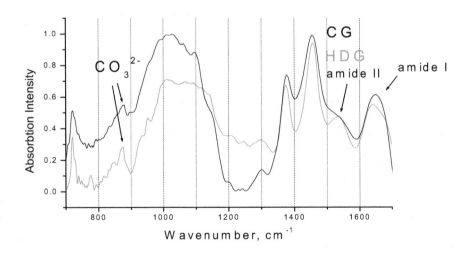

Fig. 6. Infrared spectra of dentin for patients exposed to a high doses of radiation (HDG) and people from control group (CG).

the amide III at 1240 cm-1 and amide II at 1550 cm-1 partly overlapped with strong absorption bands at 1380 and 1440 cm-1 assigned to vaseline oil, so that it was complicated to provide a semiquatitative analysis of areas of the bands assigned to amides. Anyway, reduction of the ratio between amide I and amide II amount was detected – ratio of amide I to amide II bands` area changed from 1.5 ± 0.05 (a.u.) for CG to 1.2 ± 0.05 (a.u.) for HDG. These changes of organic matter were the results of the collagen structure degradation caused by high doses of external radiation.

It can be seen in Fig. 7 that the Raman spectra of dentin and cement were also very similar: therefore mainly Raman spectra of dentin will be discussed.

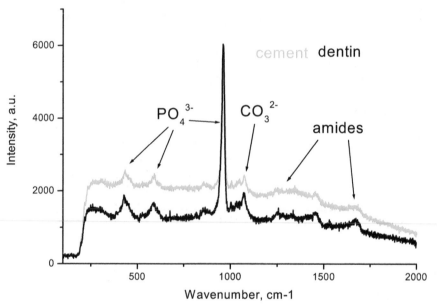

Fig. 7. Raman spectra of dentin and cement (control group) was recorded with 785 nm laser beam excitation.

In order to detect changes of the mineral matrix caused by high doses of external ionizing radiation, the full width at half maximum (FWHM) of the strongest band (at 961 cm-1) assigned to phosphate-ion has been estimated for both HDG and CG. FWHM bandwidth and area S of the PO_4^{3-} v_1 band were determined by curve of it with the procedure describing by Burn, (1985) and Bebeshko et al. (1998). It was found that the band could be well fitted with two Lorentian peaks at 955 cm-1 and 962 cm-1 (Table 1)

Center of peaks	955 cm-1		962 cm-1	
	FWHM (cm-1)	S (a.u.)	FWHM (cm-1)	S (a.u.)
Control group	18.1	29630	11.3	78210
High doses radiation group	18.3	37232	11.5	76138

Table 1. FWHM and area of the strongest band at 961 cm-1 calculated from Raman spectra of dentin belonging to HDG and CG patients

The extra peak at 955 cm⁻¹ had to be involved into the calculation to curve the strong phosphate band and it is an additional v_1 peak assigned as amorphous or crystallographically disordered hydroxyapatite (Marcovich et al., 2004). For HDG this disordering is high and the area of the main peak at 962 cm⁻¹ is less, this could be caused by ionizing radiation. For instance, the strongest band of the PO_4^{3-} in Raman spectra of both CG and HDG could be curved with only one peak at 961 cm⁻¹ with FWHM of 10.1 cm⁻¹, so that crystallography of hydroxyapatite in enamel was less sensitive to high doses ionizing radiation.

The Raman spectra of dentin belonging to patients from HDG showed slight increase of the band at 1071 cm⁻¹ ($v_1(CO_3^{2-})$ mode) which could be explained by substitution of the phosphate ion with carbonate (B-type substitution), Fig. 8.

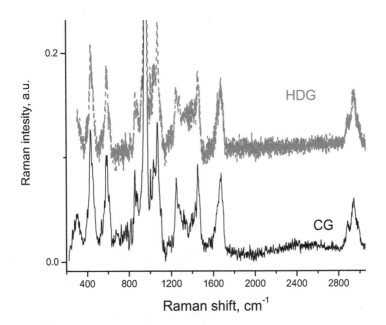

Fig. 8. Raman spectra of dentin taken from patients of HDG and CG.

The feature of dentin organic matrix is the presence of osteonectin which links the bone mineral (hydroxyapatite) and the collagen phases. Osteonectin is a specific noncollagenous protein and consists of two amino acids: glutamine ($R=-CH_2-CH_2-CO-NH_2$) and phenylalanine ($R=-CH_2-C_6H_5$). The band at 1004 cm⁻¹ assigned to phenylalanine was observed in the Raman spectra of the dentine from patients of HDG and CG (Fig. 9). According to the Raman spectra of dentin taken from HDG, phenylalanine is affected by external ionizing radiation as shown by the band at 1004 cm⁻¹ that splits into two narrow bands. The intensities of the bands typical of glutamine (at 815 cm⁻¹ and 1330 cm⁻¹) were significantly lower in case of the HDG, which can be ascribed to the ionizing radiation influence. These results have shown a sensitivity of osteonectin to high doses of external radiation.

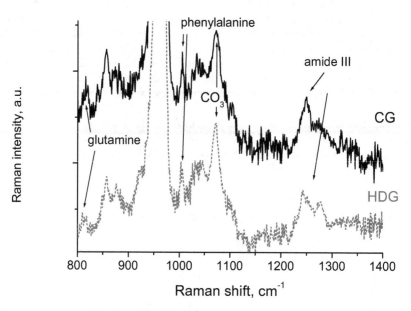

Fig. 9. Raman spectra of dentin taken from patients of HDG and CG in spectral region of osteonectin.

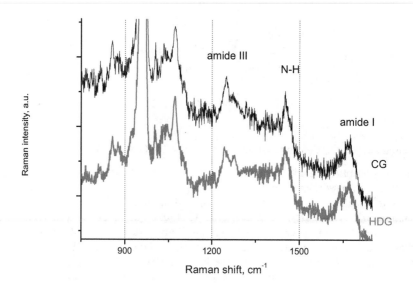

Fig. 10. Raman spectra of dentin belonging to patients of HDG and CG in spectral region of proteins.

The amide absorption bands are well evident in the Raman spectra at 1670 cm⁻¹(amide I) and at 1240 cm⁻¹ (amide III) for both CG and HDG. A strong impact of high doses of ionizing radiation on the amide part of dentin and cement has been demonstrated by changes of amides' bands (Fig.10). The wide band corresponding to amide I vibrations of collagen (peptide carbonyl coupled to the NH-out-of-plane band) is weakly structured with a main wide line at about 1670 cm⁻¹ (poly-L-proline) and a shoulder at 1640 cm⁻¹ for dentin and cement from patients of CG. For dentin and cement from patients of HFG, the wide band of amide I was split into two bands at 1245 and 1270 cm⁻¹. The split of the amide I band could be explained as breakdown of peptide bonds of collagen caused by ionizing radiation. The band at 876 cm⁻¹ could be assigned to hydroxyproline which is an amino acid formed upon hydrolysis of connective-tissue proteins such as collagen (about 14) and elastin but rarely from other proteins.

High doses of radiation also provoked decreasing intensities of the hydroxyproline band at around 855 cm⁻¹ and the proline band at around 875 cm⁻¹ and the proline shoulder at 920 cm⁻¹. Changes of the proline band intensity are more significant compared to those in the hydroxyproline. Loss of nitrogen in collagen has been detected as a decreasing of the band at 1450 cm⁻¹. The changes of Raman bands typical of organic matrix of dentin and cement reflect a degradation of the collagen structure and breaking of collagen bonds affected by high doses of radiation.

So, high doses of external ionizing radiation lead to mineral composition changes and to alteration of the organic matrix of dentin and cement.

3.3 Mapping of dental tissues

Micro Raman analysis offers superior spatial resolution (0.6-1 μm) compared to the infrared microspectroscopy resolution which is not better than 5-10 μm. So that micro-Raman spectroscopy is very useful for the analysis of biological localities such as single cement lines, boundaries around microtracks, and human dentin tubules.

Micro-Raman mapping was performed for enamel, dentin and cement belonging to patients from both HDG and CG. Mapping of Raman spectra did not show any differences in spatial distribution of organic and mineral components, which could be caused by high doses of external ionizing radiation. Due to the Raman mapping it could be seen that osteonectin and amide III formed single lines (Fig. 11) which look like channels going through dental tissues, such as dentin and cement. The diameter of these lines was calculated according to the applied mapping step of 1 μm and was around 3-5 μm. We need to underline that the distribution of ostenectin was similar to the amide III distribution – the same location of maximum and minimum of their Raman bands intensities.

The applied mapping of dentin (from HDG) also found out several spots with size around 1 μm where the intensity of the strongest band typical of phosphate ion was incredibly low. The single Raman spectrum which was extracted from the mapping (Fig.12) showed that at this spot only organic matrix is present. The near absence of the bands typical of the phosphate-ion demonstrated disordering of the hydroxyapatite crystal. It is possible to suggest that this local resorption of mineral matrix was provoked by high doses of ionizing radiation, but only a few similar spots have been found.

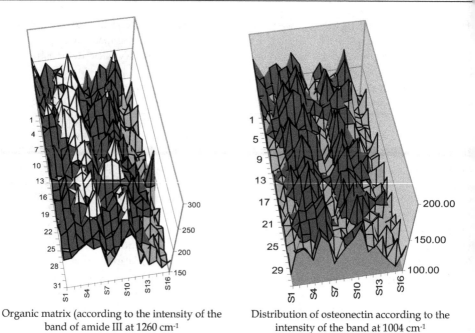

Organic matrix (according to the intensity of the band of amide III at 1260 cm⁻¹

Distribution of osteonectin according to the intensity of the band at 1004 cm⁻¹

Fig. 11. Spatial distribution of amide III and osteonectin in dentin, the mapping has been done with step of 1 μm.

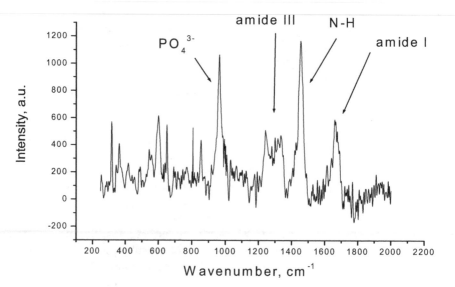

Fig. 12. Raman spectrum of resorbed area of dentin belonging to patient from HDG

Histological dentin consists of a calcified matrix with dental tubules. With Raman mapping it was demonstrated that some of these tubules go through the dentin-enamel junction and terminate in the enamel (Fig.12, taken from Darchuk et al., 2011). It was possible to estimate that the diameter of dentin channels was around 2 µm.

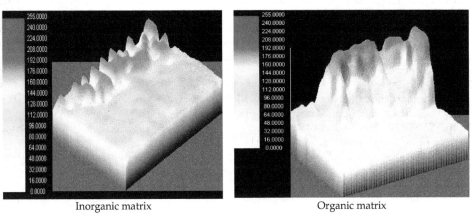

| Inorganic matrix | Organic matrix |

Fig. 12. Spatial distribution of mineral (according to intensities of the band of hydroxyapatite at 962 cm^{-1}) and organic (according to intensities of the band of amide I at 1670 cm^{-1}) components on the boarder between enamel and dentin; area of mapping was 30x50 µm.

4. Conclusions

Thanks to the application of two different spectroscopic analytical methods, complementary information regarding the influence of high dose radiation on the dental tissues was achieved. The results of infrared and micro-Raman spectroscopy analysis of dental tissues taken from patients of CG and HDG were in good agreement but micro-Raman spectroscopy gave a lot of details about the pathological effects of high doses of external radiation on the human dental tissues.

The changes in mineral matrix of enamel and dentin (as well as cement) were caused by substitution of the phosphate ion with the carbonate ion which lead to lack of phosphate-carbonate equilibrium. An increase of amorphous phosphates in dentin and cement showed the breakage of the hydroxyapatite crystal symmetry brought about the decrease of dental tissues hardness.

Both, infrared and Raman spectroscopy have demonstrated changes in organic structure of tooth tissues. High doses of radiation effect resulted in destruction of the collagen chain of dentin and cement: chemical transformations of amino acids, breakdown of peptide bonds of collagen and loss of nitrogen in collagen. Micro-Raman spectroscopy demonstrated sensitivity of osteonectin to high doses of external radiation. Intercrystalline voids, appeared as a result of ionizing radiation effect, have been filling with low differentiation connective tissue which could not fix hydroxyapatite crystals well enough.

5. References

Bebeshko V.G., Darchuk-Korovina L.A., Liashenko L.A., Sizov F.F. & Darchuk S.D. (1998). Spectral diagnostics of the bone structure cnanges in the case of oncohematological disease and acute radiation syndrome, *J. Problems Osteol.* 1, 25-30 (in Russian).

Burn G. (1985). *Solid State Physics*. Burlington, MA: Academic Press, New York, 755 p., ISBN-13: 978-0121460709

Castrom D., Engsstrom A., & Finean J.B. (1956). The influence of collagen on the organization of apatite crystallization in bone. *Chem. Abstr.* 50, 12149 h.

Darchuk L.A., Zaverbna L.V., Bebeshko V.G., Worobiec A., Stefaniak E.A., & R. Van Grieken. (2008). Infrared investigation of hard human teeth tissues exposed to various doses of ionizing radiation from the 1986 Chernobyl accident. *Spectroscopy-Biomedical Application*, 22, 105-111.

Darchuk, L., Zaverbna, L., Worobiec, A., & Van Grieken, R. (2011) Structural features of human tooth tissues affected by high dose of external ionizing radiation after nuclear catastrophe of Chernobyl plant. *Microchemical Journal*, 97, 282-285.

Elliott J.C. (1994). Structure and Chemistry of the Apatites and Other Calcium Orthophosphates. Elsevier, Amsterdam, 387 p. ISBN 0-444-81582-1.

Ellis, D.I. & Goodacre, R. (2006) Metabolic fingerprinting in disease diagnosis: Biomedical applications of infrared and Raman spectroscopy. *Analyst*, 131, 875-885.

English, J.A., Scolack, C.A., & Ellinger F. (1954) Oral manifestations of ionizing radiation II. Effect of 200 kV. X-Ray on rat incisor teeth when administered locally to the head in the 1,500 R dose range. *J. Dent Res.*, 33, 377-388.

Geller J.H. (1958). Metabolism significance of collagen in tooth structure. *J. Dental Res.*, 4, 276-279.

Kazanci, M., Roschger P., Paschalis E.P., Klaushofer K., & Fratzl P. (2006) Bone osteonal tissues by Raman spectral mapping: Orientation–composition. *J. Struct. Biol.* 156, 489-496.

Le Geros R.Z. (1981). Apatites in biological systems. *Prog. Crystal Growth Charact.* 4, 1-5.

Le Geros R.Z. (1991). Calcium Phosphates in Oral Biology and Medicine. Ed. by H. Myers (S. Karger, Basel), New York, N.Y., 200 p., ISBN 978-3-8055-5236-3

Leus P.A., Dmitrieva N.I., & Beliasova L.V. (1998). Dental health of Chernobyl power station accident liquidators, in: *The Second Conf. Further Medical Consequents of Chernobyl Power Station Accident*, Ukraine, Kiev, May 1998, p.98 (in Russian).

Morris, M.D., & Finney, W.F. (2004) Recent developments in Raman and infrared spectroscopy and imaging of bone tissue. *Spectroscopy*, 18, 155-159.

Neuberger A. (1956). Metabolism of collagen under normal conditions. *Chem. Abstr.* 50, 12252f.

Nelson, D.G.A. & Williams B.E. (1982) Low-temperature laser Raman spectroscopy of synthetic carbonated apatites and dental enamel. *Australian Journal of Chemistry*, 35, 715-727.

Marcovic M., Fowler, B.O., & Tung, M.S. (2004). Preparation and comprehensive characterization of a calcium hydroxyapatite reference material. *NIST Journ. Research*, 109 (6), 553-568.

Revenok B.A. (1998) Oral cavity health of people which took part in liquidation of Chernobyl accident consequences, in: *The Second international Conf. Further Medical Consequents of Chernobyl Power Station Accident*, Kiev, Ukraine May 1998, p.352 (in Russian).

Ross, H. Michael, Gordon, I. Kaye, & Wojciech P. (2003) Histology: Text and Atlas. Lippincott Williams and Wilkins, 4th edition, 875 p., ISBN-13: 978-0683302424

Sobkowiak, E.M., Beetke, E., Bienengräber, V., Held, M., & Kittner K.H. (1977) The problem of damage to the tooth germ by ionizing radiation (in Germ). *Zahn Mund Kieferheilkd Zentralbl.*, 65 (1), 19-24.

Ten Cate, A.R. (1998). Oral Histology: Development, Structure, And Function. Saint Louis: Mosby-Year Book, 5th edition. ISBN 0-8151-2952-1

Tsuda, H. & Arends, J. (1994). Orientational Micro-Raman Spectroscopy on Hydroxyapatite Single Crystals and Human Enamel Crystallites. *J.Dent Res*, 73, 1703-1710.

Zhang, X., Li Y.J., Wang S.L., & Xie J.Y. (2004) Effect of irradiation on tooth hard tissue and its resistance to acid (in Chin.) *Zhonghua Kou Oiang Yi Xue Za Zhi*, 39, 463-466.

4

Ultrasound Image Fusion: A New Strategy to Reduce X-Ray Exposure During Image Guided Pain Therapies

Michela Zacchino and Fabrizio Calliada
Fondazione IRCCS Policlinico "San Matteo", Radiology Department,
Pavia - Piazzale Golgi, Pavia,
Italy

1. Introduction

Many pain procedures cannot reliably be performed with a blind technique. Thus, imaging guidance is frequently mandatory, above all when the region of interest is deep and/or difficult to reach. In recent years new imaging techniques have been developed to improve diagnosis and to display greater anatomical details. Both Radiology and Pain Therapy have developed new and more accurate techniques in interventional pain, linked to a better understanding of pathophysiology and mechanisms of pain.

There are many important anesthetic blocks performed under ultrasound guidance, but our experience is manly based on pudendal nerve and sacro-iliac joint infiltration.

2. Pudendal nerve

Chronic perineal pain syndrome, due to pudendal nerve impingement, has a specific etiology and taxonomy among other possible pain sources in the pelvic and perineal area. Typically patients present uni or bilateral pain in the perineum, which may be anterior (urogenital), posterior (anal) or mixed, with a history of local treatments failure (proctologic, urologic or gynecologic). Postural nature of the pain, exacerbated, if not entirely provoked, by the seated position, led to a therapeutic strategy based on peri-truncal anesthetic blocks (Robert et al., 1998). Chronic pelvic pain was estimated to affect approximately 15–20% of women aged 18–50 (Mathias et al., 1996). On the other hand, the prevalence in men is near 8% considering urological examinations, but only 1% during primary care consultations (Schaeffer, 2004). The pudendal nerve, a mixed (sensory and motor) nerve, supplies the anus, the urethral sphincters, the pelvic floor and the perineum, furthermore it provides for genital sensitivity. It arises from anterior rami of the second, third, and fourth sacral nerves on the ventral aspect of the piriformis muscle in the pelvic cavity and crosses the gluteal region, passing through the greater ischiatic foramen, into the infrapiriformis canal, accompanied by its artery. It is also surrounded by veins with plexiform appearance (pudendal neurovascular bundle). This bundle courses around the sacrospinous ligament just before the latter's attachment to the ischial spine, enters the perineum through the lesser ischiatic foramen and courses through the ischiorectal fossa and then through the pudendal

(Alcock's) canal. The Alcock's canal is the fascia tunnel formed by the duplication of the obturator internus muscle under the plane of the levator ani muscle on the lateral wall of the ischiorectal fossa. Subsequently, the pudendal nerve splits into three terminal branches: the dorsal nerve of the penis (or clitoris), the inferior rectal nerve, and the perineal nerve, providing the sensory branches to the skin of the penis (or clitoris), the perianal area, and the posterior surface of the scrotum or labia majora. It also innervates the external anal sphincter (inferior rectal nerve) and deep muscles of the urogenital triangle (perineal nerve) (Labat et al., 2008; Lefaucheur et al.,2007). Pudendal nerve impingements are possible in its proximal segment in the pinch between the sacrospinous and sacrotuberous ligaments at the ischial spine and when it crosses the inner border of the sacrotuberous ligament, which is thickened at the beginning of the falciform process. Another possible entrapment site may occur in the Alcock's canal as a result of a thickening of the obturator internus muscle fascia. Finally the pudendal artery may describe perineural curves or constrict the nerve trunk with its collateral branches. The vessels, often tortuous and dilated, narrow nerve's components within the vascular sheath (Robert et al., 1998; Lefaucheur et al., 2007; Labat et al., 1990).

Then, nerve decompression is made with different therapeutic conservative or surgical strategies (Robert et al., 2004; Amarenco et al., 1991). Peripheral nerve blocks approach was first described in 1908 (Benson and Griffis, 2005) and it is actually used by pain therapists. There are many different ways of placing needle: by a fluoroscopic, electroneuromyography (ENMG), computed tomography (CT) or ultrasound (US) guide. Pudendal nerve block is usually made with conventional fluoroscopic guidance with placement of the needle tip near the apex of the iliac crest (Calvillo et al., 2000) but fluoroscopy is unable to visualize the pudendal nerve in the anatomical plane formed by the sacrospinous and sacrotuberous ligament (interligamentous plane). Moreover, this technique exposed to ionizing radiations both patient and physician. CT guidance, first used 1999 (Thoumas et al., 1999), is also well established and documented (Fanucci et al., 2009; Robert et al., 2005; McDonald and Spigos, 2000). Using CT it is easy to recognize ischial spine, sacrospinous and sacrotuberous ligaments and pudendal bundle. Then, we can not only place needle tip in the interligamentous space (between the sacrospinous and sacrotuberous ligaments, as close as possible to the caudal portion of the ischial spine) but it is also possible to inject pudendal nerve at the entrance of Alcock's canal (a scan, at the level of the pubic symphysis, allows to identify pudendal bundle on the medial aspect of the obturator internus) (Fanucci et al., 2009). Despite all these benefits, this technique is performed without real-time visual control, and it leads to risks of unintended puncture of adjacent vessels. Ultrasonography allows the direct visualization of the ischial spine, sacrospinous and sacrotuberous ligament. Moreover color-Doppler improves pudendal artery's visualization. In a feasibility study (Rofaeel et al., 2008), it was shown that the pudendal nerve could be clearly visualized only in 12% of the patients. Pudendal nerve shows a diameter more or less between 4 mm to 6 mm (Mahakkanukrauh et al, 2005; Gruber et al., 2001; O'Bichere et al., 2000). All the structures of this size are hardly detected by US at a depth of 5.2-1.1 cm. Furthermore, the depth of the ischial spine from the cutaneous plane is usually more than 7 cm and in 30–40% of the cases, the pudendal nerve shows anatomical variants making the nerve visualization more difficult, especially when the pudendal nerve has dense or fatty tissue near itself, with a possible failure of the procedure. For all these reasons US is usually combined with intraoperative fluoroscopy with a concordance of the two methods in 82% of the

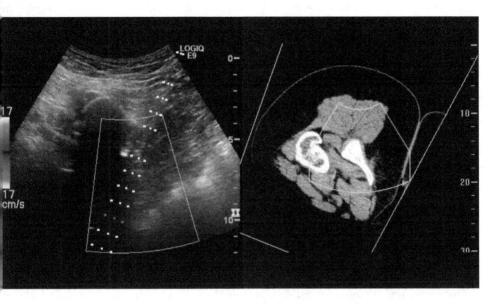

Fig. 1-a. the patient is in prone position. Fusion imaging of left side Alcok canal, pudendal artery (landmark for pudendal nerve) is visible in US side by color Doppler signal medially to ischium bone

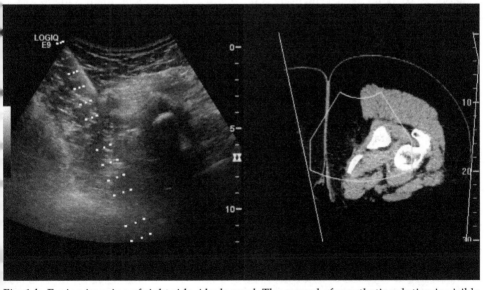

Fig. 1-b. Fusion imaging of right side Alcok canal. The spread of anesthetic solution is visible along the needle tract

procedures (Mahakkanukrauh et al, 2005; Gruber et al., 2001). Some authors describe an easier nerve visualization using a supplementary injection of quiescent solution, as reverse contrast, which should outline nerve's borders (Gray, 2006). Moreover US guidance ensures real-time needle advancement and confirmation of injection spread within the interligamentous plane (Peng and Tumber, 2008). Ultrasound guided block is a simple and reproducible approach avoiding ionizing radiations (Abdi et al; 2004).

3. The sacro-iliac joint

The sacro-iliac joint is a potential source of low back pain and/or buttock pain with or without lower extremity pain. Sacroiliac joint pain may be the result of direct trauma, unidirectional pelvic shear, repetitive and torsional forces, inflammation, or idiopathic onset. The prevalence of sacroiliac joint pain is estimated to range between 10% and 38% with 95% confidence intervals of 0% – 51% (Manchikanti 2009). It's a diarthrodial synovial joint but only the anterior portion and inferior one third of the SI joint is a true synovial joint, because of an absent or rudimentary posterior capsule, the SI ligamentous structure is more extensive dorsally, functioning as a connecting band between the sacrum and ilia. The anterior portion of the sacroiliac joint likely receives its major innervation from posterior rami of the L1-S2 roots but may also receive innervation from the obturator nerve, superior gluteal nerve, and lumbosacral trunk. The posterior aspect of the joint is supplied by the posterior rami of L4-S3, with major contribution from S1 and S2.

For this complex anatomy a clinically guided injection results actually intra-articular only in few cases. Particularly Rosenberg et al. demonstrated a success rate of 22%.

An image guide is required to reach the joint. Since 2000 Dussalt et al proved a fluoroscopy guide as a safe and rapid procedure. They rotated the fluoroscopic tube and reached the posterior aspect of the caudal end of the joint in 97% of the injections. Many authors used a CT guide with subsequent good results.

However radiation exposure under fluoroscopy guided SI joint injection ranges from 12-30 mGy/minute for the skin and 0.1-0.6 mGy/minute for the gonads, and using a CT guidance it may vary from 10 to 30 mGy/minute for the skin. Moreover ionizing radiation exposure leads to stochastic genetic and carcinogenic effects. Some authors have suggested using magnetic resonance (MR) imaging guidance to achieve intra-articular rates of up to 97% (Pereira, 2000). An MRI based procedure was found effective and safe, with the additional advantage to detect bone edema and others inflammatory signs. Even if both CT and MRI imaging are useful technique with an high contrast and spatial resolution, they takes too long with a cumbersome equipment. For all these reasons and to avoid radiation exposure Pekkafali et al. tried to perform SI joint injection using a sonographic guidance. They obtained a successful intra-articular injection rate of 60% for the first half of the procedures and about 93.5%in the last half. Pekkafali, Klauser, Migliore et al. showed that sonographically guided technique may be a valuable alternative in SI joint interventions, safe, rapid and reproducible, with the possibility to detect inflammatory signs using color-Doppler even if the result depends on radiologist's training. However a successful intra-articular injection using a sonographic guide depends not only on the radiologist experience in US performed anesthetic block, but also on the kind of patients and on the disease staging. In fact it's well known that patient's body mass index influences sonographic scans, like the presence of bone spurs or articular space narrowing, due to the pathology itself.

For all these reasons both these anesthetic procedures requires a new strategy in the guidance of the block itself, to reach the region of interest avoiding radiation exposures, like fusion imaging technique.

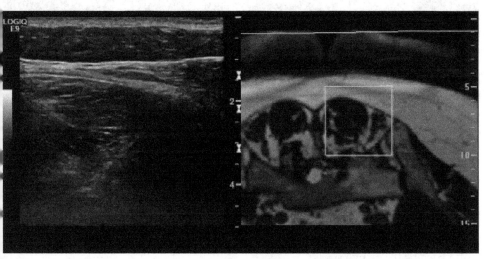

Fig. 2-a. Fusion imaging of right sacroiliac joint at second sacral foramen
Courtesy of Massimo Allegri, Pain Therapy Unit Policlinico S. Matteo Pavia

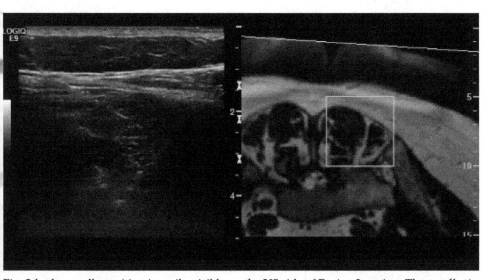

Fig. 2-b. the needle position is easily visible on the US side of Fusion Imaging. The needle tip is approaching the sacroiliac joint
Courtesy of Massimo Allegri, Pain Therapy Unit Policlinico S. Matteo Pavia

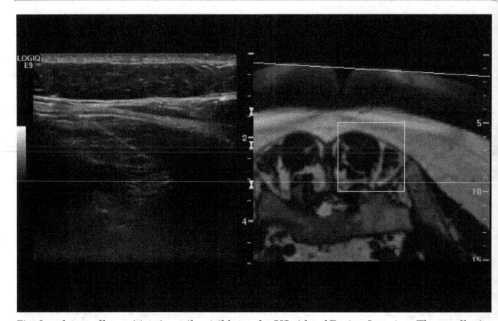

Fig. 2-c. the needle position is easily visible on the US side of Fusion Imaging. The needle tip is approaching the sacroiliac joint
Courtesy of Massimo Allegri, Pain Therapy Unit Policlinico S. Matteo Pavia

4. Fusion imaging

Fusion system can simultaneously show both ultrasound imaging with previous magnetic resonance (MR) and CT study. The basic concept of imaging fusion depends on the idea that an MR/CT study creates a 3D data set that represents the patient's anatomy. The 3D DICOM data set can be stored in any convenient imaging network location. On the other hand, US is the real-time modality, mostly used, that allows to use a feature called "Volume Navigation" to virtually project the patient's previous statics MR/CT data into the space where the clinician is performing a live, real-time, ultrasound. As the operator scans, the ultrasound system shows both the real-time ultrasound and the corresponding slice through the previous MR/CT 3D data. This is a real time reconstruction of an arbitrary slice through the MR/CT 3D data set morphed, with US data into a single unit so that the radiologist or pain therapist can intuitively use both imaging modalities simultaneously to provide diagnosis and treatment guidance. The images are presented in several formats. When presented side by side, the user can see the live ultrasound on one side of the display and the corresponding MR/CT on the other side (split screen image). Similarly when presented side by side the zoom factor of the MR/CT image and US should be varied simultaneously. In this view, the user can see not only the small live field of view (FOV) of ultrasound but also the rest of the MR/CT image that lies outside of the ultrasound scanning area but in the same plane of view. That adds the benefit of the MR/CT's wide FOV to the comparably narrow scope of ultrasound image. The images can also be presented in a mixed mode with MR/CT and US fused data from both modalities. Fusion imaging allows recording CT or

MRI data with US images in real time through an electromagnetic transmitter placed near the patient with two sensors closely bound to the probe. All these components are linked to a GPS system responsible to track the position of the ultrasound probe and to correlate with CT/MRI 3D data set. The critical phase of this procedure is the intermodality registration, involving the precise identification of multiple selected anatomical landmarks of the same patient on the different imaging modalities. Fusion imaging advantages are: 1. Ultrasound suffers from a relatively narrow FOV. MR/CT, by contrast, has an optimal contrast resolution and a wide FOV but lacks real time. Furthermore

- Ultrasound is well suited for many interventional procedures because of its real-time imaging.
- The fusion of CT/MRI and US allows using Color- Doppler or CEUS superimposed with CT/MRI.
- Fusing the images from different modalities could help to avoid multiple exposures to ionizing radiations, improving costs, and quality.
- To date, fusion imaging has been used for many diagnostic and interventional applications.

The first experiences in fusion imaging were carried out on liver imaging, because both biopsy or ablation therapy (radiofrequency) should require an imaging help.

Crocetti's group was the first to test the feasibility of fusion imaging, matching volumetric CT data of the calf livers containing internal targets, which simulated liver lesions, with real time US. The authors tried to reach each target using fusion imaging guidance and finally to perform a radiofrequency ablation. Their protocol consisted of various steps, but the most important is the registration time, during which ultrasound live examination was mixed with the previous CT scan, using radio-opaque markers applied to the calf liver capsule. Every marker was put down with a system of numbering. However during the procedure the authors used others target points (anatomic marker) to confirm the real concordance between both the modalities, such as portal vein. They found an high and consistent level of matching accuracy.

They used very small (1.5 mm) US undetectable target, reproducing a tiny lesion visible only at CT. The navigation system represented therefore the only guidance for the procedures.

However an important limitation of this study was the absence of breath and the absence of a real patient with a subsequent "lesion movement". They suggested solving this problem with the implementation of external electromagnetic position sensors, to patient body and using a breathing motion correction.

Crocetti however proved that fusion imaging could improve liver lesion interventional procedures, with a new possibility for all that lesions that are difficult to see on US B-mode scan.

At the present time, fusion imaging is realized with a new feature called volume navigation technology, which allows to mix US scan with CT/MRI DICOM data, and which uses a magnetic sensor to locate the probe position. Particularly the transmitter is fixed to, or near to, the patient's table as close as possible to the region of interest.

So the registration time becomes very important, like the choice of the target points, and it affects the success of the procedure. There is another important study carried out to analyze the accuracy of fusion imaging between CT and US to guide liver interventional procedures, according to several variables, that can impact the accuracy of this technique in clinical conditions. Hakime et al. measured the different spatial locations of an established target lesion between virtual CT and US examination in real time.

They used some anatomical intra-hepatic and extra-hepatic landmarks with the help of some non-anatomical landmarks (cysts/calcification). On CT scan, they preferred to use portal phase during which tumor is much more visible such as the portal vein, an important anatomical target point. In this study the authors analyzed also breathing influence, that lead to a global distortion along the three axes, and they suggested to display respiration cycle, that could be helpful in non-anesthetized patients.

They found a greater mismatch for anterior-posterior axis versus the lateral one, maybe due to the pressure applied with the probe.

Another important result of this study is the variation in accuracy when patient was under general anesthesia, or when CT scan was performed several days before of the procedure. Both these conditions lead to more errors during registration time, probably because the patients took a different position with a subsequent distortion of the volume body. Moreover for patients under anesthesia was impossible to repeat the same apnea conditions.

In liver pathology diagnosis plays an important rule too. In fact in literature there are some reports using fusion imaging to optimize liver involvement. For example Jung's group first tested liver lesions characterization, vascularisation and perfusion using fusion imaging. Particularly they used contrast-enhanced US mixed with contrast-enhanced MD-CT in seventeen cases and with contrast-enhanced MRI in three cases. They preferred an arterial phase during the examination of Hepatocellular Carcinoma and of neuroendocrine tumors, a portal phase for the metastasis (above all for colo-rectal ones) and finally, for hemangiomas and focal nodular hyperplasia, the phase in which the tumors could be easily visualized. As target points they used above all vascular structure, both artery and veins. An additional registration was done when a lesion was visualized, particularly providing an adjustment of the lesion size.

They found a better characterization of liver lesions by matching different contrast-enhanced modalities, because this new feature could employ the advantages of the different imaging methods . Using an additional registration they marked even small lesions or they reached also a better characterization in cirrhotic livers, where important liver structure changes, due to the pathology itself, influenced a correct diagnosis. Even if one important limitation of this study was the number of cases, few, to prove fusion imaging, actual, diagnostic accuracy. However they found two others important advantages, such as the possibility to detect a lesion in patients with reduced renal function and to reduce radiation exposure during follow-up.

Stang's group too, evaluated colo-rectal liver metastasis using fusion-imaging technique. They found a better characterization of small hepatic lesions, with a higher rate of correctly classified nodules compared with CT imaging alone.

Fusion imaging was tested also in breast biopsy. MR imaging in fact has an high sensitivity in the detection of breast cancer even if it shows a variable specificity. Above all, there is a large overlap of MRI features of the lesions, during the enhancing phase, which influences patient's management. Moreover MRI guidance for biopsy shows many disadvantages above all regarding cost and time. Many groups of research try to reach MR suspicious with a second look US examination, for all these reasons Rizzatto and Fausto tested the feasibility of fusion guidance in breast imaging.

An important problem is the different position of the patient during US examination and during MRI scan, such as the mobility of breast tissue. In fact usually breast is examined using clock position. To solve this problem different approaches were proposed, such as the use of algorithms to develope a model of deformation or the use of a redisegned bed that allows to lie in the prone position both during US and MRI examination. In the first case the registration phase in made using target points that are identified between the two images, obtained with different modalities, and then transformed. After US-MRI imaging coregistration, the software reconstructs a real time multiplanar MR image of the corresponding US examination. Particularly Rizzatto and Fausto performed their first MRI examination in the prone position, using precontrast and postcontrast phases, then they obtained another MRI data set with the patient in the supine position by using external fiducial markers at 9-, 12- and 3- o'clock. So they performed US live examination with the patient in the supine position. Using fusion imaging guidance for breast biopsy, the authors reached an important reduction of the time and costs.

About pain therapy Galiano's group first conducted a research in 2007 that analyzed a kind of fusion imaging. Galiano's group tried to improve US visualization of facet vertebral joint to pain therapists unfamiliar to US scan, that however would like to perform medial branch block with US guidance. They used CT reconstruction images to recognize the anatomical structures during real time US examination of the facet joints. Their research was conducted only on cadavers with subsequent good results. In fact, at the end of the study, they encouraged to start to perform anesthetic blocks in real patients under an US guide.

Klauser and Zacchino's groups published the first papers analyzing fusion imaging in pain therapy in 2010.

Klauser tested fusion imaging to guide sacro-iliac joint-infiltration, using ten cadavers and then ten patients. The basic concept of fusion imaging is always the same, so also Klauser's group chose anatomical target points to match US with CT. Particularly the landmarks were spinous process of the fifth lumbar vertebra, posterior superior and inferior iliac spine, first and second posterior sacral foramen and the sacro-iliac joint. They found a positive success rate, with an increasing learning curve in the second half of patients. Moreover CT scan provides further anatomical details about joint assessment, about eventual bone spurs and so on.

In this work too, registration time takes the longest time, but is the most important time of the entire procedure.

Zacchino's group tested the feasibility of pudendal nerve anesthetic block using fusion imaging. In this work US was matched with CT data, using as anatomical target points ischiatic spine, femoral head and coccyx. They performed anesthetic block in the Alcock's canal, which is difficult to reach using an US or CT guidance alone. Fusion imaging provides further anatomical details for this block, such as the possibility to visualize with the same technique both the pudendal vascular bundle, sacrotuberous and sacrospinous ligaments, independently from pudendal nerve depth. After the procedure the patient was immediately and completely pain free.

All these works carried out on different interventional procedures showed that fusion imaging is helpful to reach the region of interest, and that fusion imaging, combing CT spatial resolution and its wide field of view, with US real time, is essential to perform the procedure, avoiding possible complications (unintended puncture of vessels, hematoma, etc). Fusion imaging improves direct visualization of the region of interest matching CT/MRI advantages plus US advantages avoiding radiation exposure during the procedure, that is always important but above all when repeat the procedure is mandatory and in young patients. It avoids also US disadvantages, first of all, its dependence on patient body mass index. However using fusion imaging too, requires a certain experience that lead to decrease procedure time and it improves the success rate.

Moreover it's important to place the patient in the same position, both during the CT/MRI previous acquisition and during the procedure, as Crocetti proved. Then it's important also to avoid patient's movement both during registration time and during the procedure.

Fusion imaging technique however is now applied to others different kind patology: a new report (Iagnocco 2011) evidences some possibilities in rheumatologic field, where fusion is used to investigate and monitor osteoarthritis and rheumatoid arthritis. This paper showed as fusion imaging should be useful for therapy monitoring matching MRI anatomical landmarks and US details.

5. References

Abdi S, Shenouda P, Patel N, Saini B, Bharat Y, Calvillo O. A novel technique for pudendal nerveblock. Pain Physician 2004;7:319–22.

Amarenco G, Le Cocquen-Amarenco A, Kerdraon J, Lacroix P, Adba MA, Lanoe Y. Perineal neuralgia. Presse Med 1991;20:71–4.

Benson JT, Griffis K. Pudendal neuralgia, a severe pain syndrome. Am J Obstet Gynecol 2005;192:1663–8.

Calvillo O, Skaribas IM, Rockett C. Computed tomography-guided pudendal nerve block. A new diagnostic approach to long-term anoperineal pain: a report of two cases. Reg Anesth Pain Med 2000;25:420–3.

Crocetti L., Lensioni R., De Beni S., Choon See T., Della Pina C., Bartolozzi C. Targeting Liver Lesions for Radiofrequency Ablation An Experimental Feasibility Study Using a CT–US Fusion Imaging System Invest Radiol 2008;43: 33–39

Dussault RG , Kaplan PA , Anderson MW . Fluoroscopy-guided sacroiliac joint injections. Radiology 2000 ; 214 (1): 273 – 277

Fanucci E., Manenti G., Ursone A., Fusco N., Mylonakou L., D'Urso S., Simonetti G. Role of interventional radiology in pudendal neuralgia: a description of techniques and review of the literature. Radiol Med 2009;114:425–36.

Gray AT. Ultrasound-guided regional anesthesia: current state of the art. Anesthesiology 2006;104:368–73.

Gruber H., Kovacs P., Piegger J., Brenner E. New, simple, ultrasound-guided infiltration of the pudendal nerve: topographic basics. Dis Colon Rectum 2001;44:1376–80.

Hakime A. Deschamps F., De Carvalho E. G. M., Teriitehau C., Auperin A., De Baere T. Clinical Evaluation of Spatial Accuracy of a Fusion Imaging Technique Combining Previously Acquired Computed Tomography and Real-Time Ultrasound for Imaging of Liver Metastases Cardiovasc Intervent Radiol 2011; 34:338–344.

Hirooka M, Iuchi H, Kumagi T et al (2006) Virtual sonographic radiofrequency ablation of hepatocellular carcinoma visualized on CT but not on conventional sonography. AJR Am J Roentgenol 186:S255–S260

Iagnocco A. et al. Magnetic resonance and ultrasonography real-time fusion imaging of the hand and wrist in osteoarthritis and rheumatoid arthritis. Rheumatology 2011; 50: 1409-1413

Jung EM, Schreyer AG, Schacherer D, Menzel C, Farkas S, Loss M, Feuerbach S, Zorger N, Fellner C. New real-time image fusion technique for characterization of tumor vascularisation and tumor perfusion of liver tumors with contrastenhanced ultrasound, spiral CT or MRI: first results. Clin Hemorheol Microcirc 2009;43:57–69.

Klauser A.S., De Zordo T., Feuchtner G. M., Djedovic G., Bellmann Weiler R., Faschingbauer R., Shirmer M., Moriggl B. Fusion of real-time US with CT images to guide sacroiliac joint injection in vitro and in vivo. Radiology 2010; 256 (2): 547-553.

Labat J.J., Riant T., Robert R., Amarenco G., Lefaucheur J.P., Rigaud J. Diagnostic criteria for pudendal neuralgia by pudendal nerve entrapment (Nantes criteria). Neurourol Urodyn 2008;27:306–10.

Labat J.J., Robert R., Bensignor M., Buzelin J.M. Les névralgies du nerf pudendal (honteux interne). Considérations anatomo-cliniques et perspectives thérapeutiques. J Urol (Paris) 1990;96:239–44.

Lefaucheur J.P., Labat J.J., Amarenco G., et al. What is the place of electroneuromyographic studies in the diagnosis and management of pudendal neuralgia related to entrapment syndrome? Neurophysiol Clin 2007;37:223–8.

Mahakkanukrauh P., Surin P., Vaidhayakarn P. Anatomical study of the pudendal nerve adjacent to the sacrospinous ligament. Clin Anat 2005;18:200–5.

Manchikanti L., Boswell M. V., Singh V., Benyamin R. M., Fellows B., Abdi S., Buenaventura R. M., Conn A., Datta S., Derby R., Falco F. JE., Erhart S., Diwan S., Hayek S. M., Helm S. II, Parr A. T., Schultz D.M., Smith H.S., Wolfer L.R., Hirsch J.A. Comprehensive Evidence-Based Guidelines for Interventional Techniques in the Management of Chronic Spinal Pain. Pain Physician 2009; 12:699-802

Mathias S.D., Kuppermann M., Liberman R.F., Lipschutz R.C., Steege J.F. Chronic pelvic pain: prevalence, health-related quality of life, and economic correlates. Obstet Gynecol 1996;87:321-7.

McDonald J.S., Spigos D.G. Computed tomography-guide pudendal block for treatment of pelvic pain due to pudendal neuropathy. Obstet Gynecol 2000;95:306–9

Migliore A., Bizi E., Massafra U., Vacca F., Martin-Martin L.S., Granata M., Tormenta S. A new technical contribution for ultrasound-guided injections of sacro-iliac joints. Eur rev Med Pharmacol Sci 2010; 14(5): 465-9

Minami Y, Chung H, Kudo M et al (2008) Radiofrequency ablation of hepatocellular carcinoma: value of virtual ct sonography with magnetic navigation. AJR Am J Roentgenol 190(6): 335–341

O'Bichere A., Green C., Phillips R.K. New, simple approach for maximal pudendal nerve exposure: anomalies and prospects for functional reconstruction. Dis Colon Rectum 2000;43:956–60.

Pekkafahali MZ, Kiralp MZ, Basekim CC, Silit E, Mutlu H. OzturkE, et al. Sacroiliac joint injections performed with sonographic guidance. J Ultrasound Med 2003; 22:553-9.

Peng PW, Tumber PS. Ultrasound-guided interventional procedures for patients with chronic pelvic pain – a description of techniques and review of literature. Pain Physician 2008;11:215–24.

Pereira PL , Günaydin I , Trübenbach J , et al . Interventional MR imaging for injection of sacroiliac joints in patients with sacroilitis . Am J Roentgenol 2000 ; 175: 265 – 266

Rizzatto G, Fausto A. Breast imaging and volume navigation: MR imaging and ultrasound coregistration. Ultrasound Clin 2009;4:261–71.

Robert R, Bensignor M, Labat JJ, Riant T, Guerineau M, Raoul S, Hamel O, Bord E. Le neurochirurgien face aux algies périnéales: guide pratique. Neurochirurgie 2004;50:533–9.

Robert R, Labat JJ, Bensignor M, Glemain P, Deschamps C, Raoul S, Hamel O. Decompression and transposition of the pudendal nerve in pudendal neuralgia: a randomized controlled trial and longterm evaluation. Eur Urol 2005;47:403–8.

Robert R, Prat-Pradal D, Labat JJ, Bensignor M, Raoul S, Rebai R, Leborgne J. Anatomic basis of chronic perineal pain: role of the pudendal nerve. Surg Radiol Anat 1998;20:93–8.

Rofaeel A, Peng P, Louis I, Chan V. Feasibility of real-time ultrasound for pudendal nerve block in patients with chronic perineal pain. Reg Anesth Pain Med 2008;33:139–45.

Rosenberg JM JM, Quint TJ, de Rosayro AM. Computerized tomographic localization of clinically-guided sacroiliac joint injections. Clin J Pain 2000 ; 16 (1): 18 – 21 .

Schaeffer AJ. Etiology and management of chronic pelvic pain syndrome in men. Urology 2004;63:75–84.

Stang et al. Real-time Ulrasonography-Computed tomography fusion imaging for staging of hepatic metastatic involvement in patients with colo-rectal cancer. Initial results from comparison to US seeing separate CT images and to multidetector-row CT alone. Investigative radiology 2010; 45: 491-501

Thoumas D, Leroi AM, Mauillon J, Muller JM, Benozio M, Denis P, Freger P. Pudendal neuralgia: CT-guided pudendal nerve block technique. Abdom Imaging 1999;24:309–12.

Zacchino M., Allegri M., Canepari M., Minella C. E., Bettinelli S., Draghi F., Calliada F. Feasibility of pudendal nerve anesthetic block using fusion imaging technique in chronic pelvic pain. European journal of pain 2010; 4:329-333

5

Ionizing Radiation Profile of the Hydrocarbon Belt of Nigeria

Yehuwdah E. Chad-Umoren
Department of Physics, University of Port Harcourt,
Choba, Port Harcourt,
Nigeria

1. Introduction

Man and his environment are constantly bombarded with naturally occurring ionizing radiation. However, aside from the radiation occurring naturally in the human environment, there are those resulting from man's activities. These anthropogenic activities result in the elevation of the background ionizing radiation levels mainly through the depletion of the ozone layer so that an increase occurs in the amount of cosmic radiation reaching the earth (Foland et al, 1995). One crucial area of human activity is the hydrocarbon industry. Globally, the hydrocarbon industry is a strategic industry, contributing to the wealth of nations and the prosperity of individuals. Global economy is often appreciably impacted by the stability or otherwise of the industry. Furthermore, much of the world's population is dependent on the industry for its energy needs. Significantly also, the processes and techniques of oil and gas exploration, exploitation and usage contribute to the devastation, pollution and degradation of the environment.

In Nigeria, the hydrocarbon industry is reputed to be the highest user of radioactive substances in the country (Chad-Umoren and Obinoma, 2007). And the Niger Delta region of Nigeria plays host to the largest concentration of companies in this sector. The maze of crisis-crossing pipelines in the region (Fig.1) does not only transport oil and gas, but also dangerous radioactive substances so that the environment is not despoiled only during oil spillage, but is also polluted with harmful ionizing radiation. Also abundant in the region are numerous non-oil and gas industries and operations that have been attracted into the region by its oil and gas wealth.

2. History and status of the hydrocarbon industry in Nigeria

Oil was discovered in the Niger Delta of Nigeria in 1956 at Oloibiri in present day Bayelsa state and Shell D'Arcy Petroleum commenced production there in 1958. However, before this period, the German company, the Nigerian Bitumen company, had unsuccessfully drilled fourteen (14) wells in the area between 1908 and 1914 within the eastern Dahomey Basin (NNPC, 2004). By the year 2004, oil production had risen to 2.5 million barrels per day. It is estimated that Nigeria's recoverable crude oil reserves stand at about 34 billion barrels, while the nation's natural gas reserves stand at 163 trillion cubic feet (Tcf) making Nigeria one of the top ten nations with natural gas endowments. The current development plan projects that by year 2020, production will increase to 4million barrels per day.

Though government policy had envisaged that gas flaring will end by 2010, about 50% of the natural gas is still wastefully flared (ECN, 2003), while 12% is re-injected to enhance oil recovery. An estimate by Shell shows that about half of the 2 Bcf/d of associated gas and gaseous by-products of oil exploitation is flared in Nigeria annually, while the World Bank's estimates put Nigeria's gas flaring at 12.5% of the world's total gas flared.

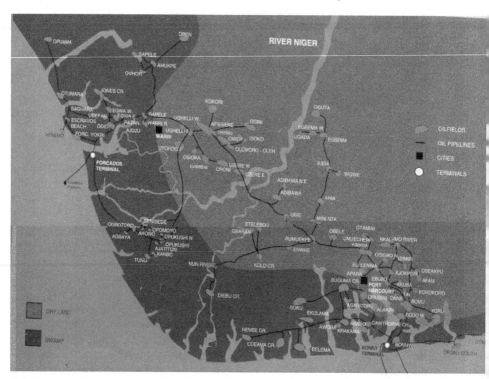

Fig. 1. A map showing network of pipes of a major oil company in the Niger Delta

3. Industrialization and ionizing radiation in the Niger Delta

A number of ionizing radiation surveys have previously been carried out in the Niger delta region to assess the impact of industrialization on the radiation profile of the region and a strong correlation has been established between high industrial activities and elevated background ionizing radiation (BIR) levels for parts of the region (Avwiri and Ebeniro, 1998; Chad-Umoren and Briggs-Kamara, 2010). Ebong and Alagoa (1992 a, b) studied the background ionizing radiation profile at the fertilizer plant at Onne, Rivers State. The study reported a significant increase in the background radiation indicating that the presence of the fertilizer plant had a negative impact on the radiation levels of the area. A survey of the external environmental radiation in the Trans-Amadi industrial area and at other sub-industrial areas of Port Harcourt showed an average value of 0.014 mRh^{-1} (Avwiri and

Ebeniro, 1998). Comparing this result to the standard background radiation level of 0.013mRh^{-1}, the study showed that there was a significant elevation in the radiation status of the surveyed areas which can be attributed to the industrial activities there.

Chad-Umoren and Obinoma (2007) surveyed the background ionizing radiation patterns at the campus of the College of Education at Rumuolumeni in Rivers State. The work gave a low radiation dose equivalence of 0.745±0.085mSv/yr as the average for the campus. It was also suggested that further studies needed to be done to assess the impact of the presence of a nearby cement production plant on the radiation status of the college. Briggs-Kamara et al (2009) studied the ionizing radiation patterns of the Rivers State University of Science and Technology at Port Harcourt. The major focus of the study was to assess the impact of computer and photocopier operations on the environmental ionizing radiation patterns of the campus. Computers and photocopiers are widely used on Nigerian university campuses without any consideration for their health hazards. The study established that in deed these devices contributed to the radiation levels of the campus.

A study was also carried out to determine the radiation levels in solid mineral producing areas of Abia state (Avwiri et al, 2010). Solid minerals have their origin in the earth's crust where the primordial radionuclides such as ^{238}U, ^{232}Th and their progenies are found. Consequently, mining of solid minerals has the potential to impact the environmental ionizing radiation. The radiation exposure rate for the surveyed areas ranged between 14.7 µR/hr to 18.2 µR/hr indicating an elevation over the normal background radiation level of 11.4 µR/hr for the host communities. Considering this elevated radiation level, it was concluded that the mining activity had future radiological health hazards for the miners, the general populace and the environment.

The Niger delta region constitutes the hydrocarbon belt of Nigeria and our aim in this work is to study the impact of oil and gas and its ancillary services on the ionizing radiation profile of the region (Fig. 2). Furthermore, based on international practises, we have also suggested appropriate strategies for the control of the ionizing radiation from the hydrocarbon industry and therefore make the activities of the industry in Nigeria more environmentally friendly. And in conclusion we have suggested possible areas for further work.

4. Geology, physiography and evolution of the Niger Delta

The Niger-Delta forms one of the world's major hydrocarbon provinces. It is situated on the Gulf of Guinea on the west coast of Africa and in the southern part of Nigeria, lying between longitudes 4 – 9°E and latitudes 4 - 9° N (Fig. 1) with an estimated area exceeding over 200,000 square kilometres (Odigie, 2001). It is composed of an overall regressive clastic sequence, which reaches a maximum thickness of about 12km (Evamy et al, 1978) and is divided into three formations: Benin formation, Agbada formation and Akata formation. The Agbada formation consists of sand, sandstone and siltstones and is the principal host of Niger Delta petroleum (Beka and Oti, 1995). The Akata formation forms the base of the transgressive lithologic unit of the delta complex and is of marine origin. It is composed of thick shale sequence (potential source rock), turbidite sand (potential reservoir in deep water) and minor amounts of clay and silt (Avbovbov, 1998). The tectonic framework of the continental margin along the west coast of Equatorial Africa is controlled by Cretaceous fractured zones expressed as trenches and ridges in the deep Atlantic.

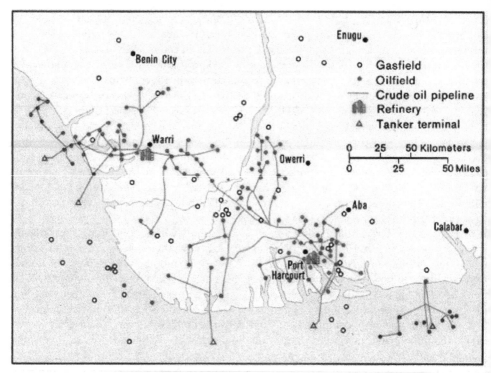

Fig. 2. Map of study area (Niger Delta)

5. Sources of ionizing radiation in the hydrocarbon industry

There are a number of sources for the ionizing radiation associated with the hydrocarbon industry. These include the earth's crust. The earth's crust is a natural source of oil and gas naturally occurring radioactive materials (NORM) as it contains the radionuclides ^{238}U and ^{232}Th. These radionuclides become mixed with the oil, gas and water in the process of oil exploration and exploitation. The radioactive decay chains of these naturally occurring parent radionuclides have very long half-lives and are ubiquitous in the earth's crust with activity concentrations that depend on the type of rock. The radioactive disintegration of ^{238}U and ^{232}Th produce several series of daughter radioisotopes of different elements and of different physical characteristics with respect to their half-lives, modes of decay, and types and energies of emitted radiation. A second source of NORM in the oil and gas industry is scale. Scales consist primarily of insoluble compounds of such elements like barium, calcium and strontium. These compounds precipitate from the produced water as a result of changes in temperature and pressure (IAEA, 2003). The radioactive substances found in scales are the radium isotopes (^{226}Ra and ^{228}Ra). Sulphate scales are the main types of scales in the oil and gas industry (OGP, 2008).

Sludge and scrapings are also sources of NORM in the hydrocarbon industry. Oil field sludge consists of dissolved solids which like scales precipitate from produced water due to variations in temperature and pressure. Generally, sludge is composed of oily, loose materials that may contain both silica compounds and a large amount of barium. The radionuclide found in sludge include radium, while ^{210}Pb and ^{210}Po are found in pipeline scrapings and in sludge that have accumulated in gas/oil separators and liquefied natural gas (LNG) storage tanks. Gas processing facilities are also another source of NORM. The radionuclide in this case is the radon gas.

In the Niger delta region of Nigeria the absence of infrastructural facilities for natural gas utilization has made gas flaring widespread in the region. As previously mentioned, about 50% of the natural gas is flared in the region thereby greatly contributing to the radon in the atmosphere of the region. Seawater injection systems also contain NORM, especially uranium. Seawater is used in the process of oil recovery from reservoirs and incidence of ionizing radiation due to this source is greatly increased in the case where a large volume of seawater is used (OGP, 2003). Incidence of ionizing radiation is further enhanced in the Niger delta of Nigeria as about 12% of the gas is re-injected into wells to improve oil recovery.

6. Effects of ionizing radiation

Radiation monitoring in the hydrocarbon region of Nigeria is important because of the health hazards and the environmental impact of ionizing radiation. Various health hazards of ionizing radiation have been documented. These include the following:

Erythema which is an increased redness of the skin as a result of capillary dilation.

Cancers: The skin is a radiosensitive part of the body and is also its most exposed part. Exposure to ionizing radiation leads to skin cancer. Also, various other kinds of cancers are linked to exposure to ionizing radiation. These include leukaemia, and cancers of the lung, stomach, oesophagus, bone, thyroid, and the brain and nervous system. Not all forms of cancer are traced to exposure to ionizing radiation.

Genetic Effects: Exposure of the reproductive cells to ionizing radiation can lead to miscarriage or genetic mutation. Genetic mutation affects the embryo causing deformity or death.

Sterility: Developing sperm cells known as gonads have very high radiosensitivity, exposure to ionizing radiation can therefore lead to sterility.

Cataracts: Opacities on the surface of the eye lens, called cataracts, result from the denaturing of the lens protein. One effect of ionizing radiation is the denaturing of the lens.

Atrophy of the Kidney: Exposure to ionizing radiation can lead to a condition known as atrophy of the kidney. In this case, the kidney and urinary tract waste or shrink with attendant loss of renal functions.

The level of damage experienced due to exposure to ionizing radiation is determined mainly by the radiation dose received and by such other factors as duration of exposure, nature of radiation and the sensitivity of the part of the body irradiated (ICRP, 1977, UNSCEAR, 1988). Radiation health hazards are dose-dependent. The overall effect of ionizing radiation in man is greatly enhanced when both radionuclide deposition and energy absorption by a specific organ or tissue occur together.

7. Ionizing radiation profile of the Niger Delta, Nigeria

Oil field development and gas exploitation activities have resulted in various forms of unsettling environmental activities in the Niger Delta region that have impacted the ecological, biophysical and socio-economic and political structure of the area (Abali, 2009).

Also, there is ample evidence globally that the hydrocarbon industry contributes to the elevation of the ionizing radiation of the areas where its operations are carried out. The oil and gas industry, like the rest of the global community, places much emphasis on a safe work environment. Various indices may be adopted to assess what constitutes a safe and healthy work environment, among which is the incidence of certain levels of ionizing radiation in the given area. A study was conducted to evaluate the occupational ionizing radiation levels at 30 locations of oil and gas facilities in Ughelli in the Niger delta of Nigeria (Avwiri et al, 2009). The study was carried out during two time periods – during the production periods and during the off-production periods. Not unexpectedly, the study showed that radiation levels were higher during production periods than during the off-production periods (Fig. 3). The mean radiation levels workers in the oil fields were exposed to during both production and off-production periods are shown in Table 1. For example, the mean obtained for the radiation levels during the off-production periods had a range of 13.38±1.69µR/h (0.023±0.003mSv/wk) to 16.29±2.60µR/h (0.027±0.004mSv/wk) while the mean obtained for the radiation levels during the production periods had a range of 15.50± 1.65µR/h (0.026±0.003mSv/wk) to 19.14±3.16µR/h (0.32±0.005mS/wk). For the two periods under investigation, the Eriemu oil field had the highest radiation levels with a mean equivalent dose rate of $7.88\pm1.29\mu Rh^{-1}$, while the Kokori oil field had the highest percentage radiation deviation of 15.19%.

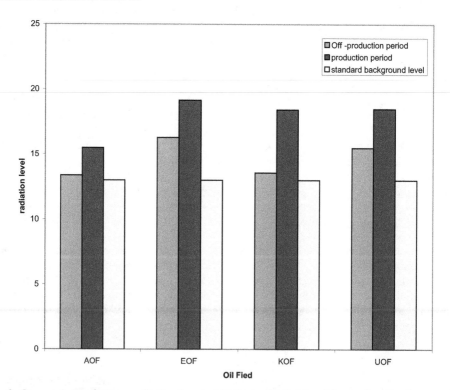

Fig. 3. Comparison of mean radiation levels (µR/h) in the fields with standard background level (Avwiri et al, 2009)

Field code	Surveyed oil and gas field	Mean radiation levels (µR/h)		Mean Deviation (%)	Mean dose equivalent rate (µSv/wk)
		Off - production	Production period		
AOF	Afiesere oil field	13.38 ±1.69	15.50±1.65	7.34	6.43±0.75
EOF	Eriemu oil field	16.29±2.60	19.14±3.16	8.04	7.88±1.29
KOF	Kokori oil field	13.57±1.80	18.43±2.68	15.19	7.12±0.99
UOF	Ughelli oil field	15.50±2.53	18.50±2.68	8.82	7.57±1.16
MEAN		14.69±2.16	17.89±2.54	9.85	7.25±1.65

Table 1. Mean Exposure Rates in the Oil Fields.

We observe that though for both periods, the mean radiation levels are all within the safe radiation limit of 0.02mSv/wk as recommended by the UNSCEAR (1993) there is still a certain measure of occupational radiation health hazard in these locations for the following reasons:

1. The higher radiation levels during production periods compared to those during off-production periods show that the activities at the oil and gas facilities have led to the elevation of the radiation levels at the study locations. It further shows that the activities during the production periods contribute to higher occupational risk for the workers. Workers are therefore exposed to higher radiation risks during production periods and lower risks during off-production periods.

2. The observed disparity in the exposure rates between the production period and the off-production period was seen as implying that there was increased use of radionuclides, increased exposure to flowing crude oil (spillage) and gas flaring activities resulting in elevation of the ionizing radiation levels of the facilities and their immediate neighbourhoods (Avwiri et al, 2009).

3. The exposure rates at all the locations were generally higher than the standard background level of 13.0µR/hr (Fig. 3).

4. Very high exposure levels of 20µR/hr and above were common. The highest level of 26.00±0.5.1µR/h (0.044±0.009mSv/wk) was recorded at the Kokori oil field during production.

5. Even during off-production periods, high exposure levels were also recorded. For example, an exposure level of 22.00µR/h±2.1µR/h was obtained at the Eriemu natural gas compressor station vessel.

6. Generally, the radiation levels at the Natural Gas Compressor (NGC) facilities were found to be higher than the 0.02mSv/week recommended by the UNSCEAR (1993). The NGC facilities were therefore found to be the most unsafe work-environments as the highest radiation levels were obtained there. This may be attributed to the presence of radon at the facility as this radioactive gas is very much associated with the gas exploitation procedures of the hydrocarbon industry (OGP, 2008)

7. A previous study of the radiation around oil and gas facilities in Ughelli showed that for the oil fields the exposure dose rate had a mean range of 12.00± 0.lµRh (5.33±0.35µSv/wk) to 22.00±2.lµRh-1 (9.79±0.16) and 09.00±1.0 to 11.00±0.5µRh-1 in the host communities (Avwiri, et al., 2007)). Compared to the present investigations we observe that the radiation levels have increased appreciably.

A six-year impact assessment of ionizing radiation was published which reported the effect of oil spillage on the ionizing radiation profile of the oil spillage environment and the host communities in parts of Delta state (Agbalagba and Meindinyo, 2010). The study used a geographical positioning system (GPS) and a digilert nuclear radiation monitor. For the actual survey, 20 sites were chosen along with 6 host communities spread across the affected area and a control sample. The mean values obtained for the location ranged from 0.010 mRh^{-1} (0.532 mSv y^{-1}) to 0.019 mRh^{-1}(1.010mSv y^{-1}). In the area affected by oil spillage, the study gave a yearly exposure rate that ranged from 0.013±0.006 mRh^{-1}(0.692±0.080 $mSvy^{-1}$) to 0.016±0.005 mRh^{-1} (0.851±0.100 $mSvy^{-1}$); the values for the host communities ranged from 0.011 mRh^{-1}(0.585 $mSvy^{-1}$) to 0.015 mRh^{-1}(0.798 $mSvy^{-1}$). The exposure rate for the control was 0.010 mRh^{-1}(0.532 $mSvy^{-1}$).

The study established the impact of oil spillage on the radiation levels within the area and the host communities showing that the radiation levels of both the oil-spilled area and the host communities were elevated by the oil spillage. The radiation exposures were 55% and 33.3% respectively above the normal background level of 0.013 mRh^{-1}. Also, the mean equivalent dose rate for the study area although within the safe limit of 0.05 $mSvy^{-1}$ recommended by ICRP (1990) and NCRP (1993) was higher than the 0.0478 mSv/y normal background level.

Chad-Umoren and Briggs-Kamara (2010) assessed the ionizing radiation levels in parts of Rivers State, one of the states in the Niger delta region. The major thesis of that work was that since there is a near homogeneous distribution of oil and gas operations in the state, the radiation profile of the state will correlate with this, such that the state will exhibit a homogeneous radiation profile. The state was then delineated into three zones, namely, an upland college campus; a collection of rural communities and a group of industrial establishments forming the industrial zone. A Comparative analyses of the results for the three sub-environments showed that the highest dose equivalent of 1.332±0.076mSv/yr occurred in the industrial zone while the lowest value of 0.57±0.16mSv/yr was obtained in the rural riverine sub environment. The computed mean dose equivalent for the three zones also showed that the industrial sector had the highest mean value of 1.270±0.087mSv/yr, while the mean for the upland college campus environment was 0.745±0.085mSv/yr with the rural riverine communities having the lowest mean dose equivalent of 0.690± 0.170mSv/yr.

It is to be observed that the very high exposure rate of 0.0168mR/hr was recorded in this study and that it was obtained in an oil-activity-related environment.

This study revealed the following: Firstly, anthropogenic activities have great impact on the radiation levels of the environment. Secondly, the industrial environment of the state contributes the most to the radiation levels of the state (Tables 2, 3 and 4). Thirdly, the dose equivalent for the different components of the industrial environment are all higher than the European Council for Nuclear Research (CERN) recommended value of 1.0mSv/yr for the general population who are not engaged in nuclear radiation related occupations (CERN, 1995). Fourthly, the dose equivalents obtained for the other two environments are within the CERN regulations.

A national energy policy articulated by the national government of the federation of Nigeria encourages the efficient utilization of the nation's abundant oil and gas reserves (NEP, 2003). Recently the impact of this energy policy on the radiation profile of Ogba/Egbema/Ndoni, a Local Government Area in the central Niger delta state of Rivers state was studied (Ononugbo et al, 2011). The study area was divided into six zones (Table 5). To ensure that the original environmental characteristics of the samples were not tampered with, an *in situ* approach of background radiation measurement was used. The

Station	Name
1	Port Harcourt Refinery Company (PHRC) PHRC Junction, Alesa
2	Nigeria Ports Authority, Onne NAFCON(Fertilizer company), Onne
3	Bori/Onne Junction
4	Rumuokoro Junction
5	Choba (Wilbros)
6	Nkporlu Village
7	Arker Base
8	Mobil Area.

Table 2. Stations for experiment (Industrial environment) (Avwiri and Ebeniro, 2002)

Station	Counter	*mR/hr	S	μ	T	Tc	Remarks
	3	0.0149	0.0008	0.0144	1.98	2.26	t<tc
1	2	0.0154	0.0006	0.0144	5.27	2.26	t>tc
	1	0.0128	0.0008	0.0144	-6.32	-2.26	t<tc
	3	0.0148	0.0008	0.0143	1.93	2.26	t<tc
2	2	0.0159	0.0007	0.0143	7.23	2.26	t>tc
	1	0.0122	0.0010	0.0143	-6.04	2.26	t<tc
	3	0.0145	0.0006	0.0146	-0.53	2.26	t<tc
3	2	0.0156	0.0011	0.0146	6.04	-2.26	t>tc
	1	0.0138	0.0022	0.0146	-1.15	-2.26	t>tc
	3	0.0151	0.0007	0.0145	2.71	2.26	t>tc
4	2	0.0159	0.0011	0.0145	4.02	2.26	t>tc
	1	0.0124	0.0007	0.0145	-4.50	-2.26	t<tc
	3	0.0150	0.0006	0.0143	3.69	2.26	t>tc
5	2	0.0150	0.0006	0.0143	3.69	2.26	t>c
	1	0.0121	0.0007	0.0143	-9.94	-2.26	t<tc
	3	0.0148	0.0009	0.0147	0.032	2.26	t<tc
6	2	0.0160	0.0008	0.0147	4.30	2.26	t>tc
	1	0.0136	0.0019	0.0147	-2.00	-2.26	t>tc
	3	0.0147	0.0003	0.0147	-0.109	-2.26	t>tc
7	2	0.0155	0.0008	0.0147	3.35	2.26	t>tc
	1	0.0138	0.0006	0.0147	-4.688	-2.26	t<tc
	3	0.0141	0.0009	0.0144	0.94	-2.26	t>tc
8	2	0.0155	0.0012	0.0144	3.00	2.26	t>tc
	1	0.0135	0.0015	0.0144	-1.83	-2.26	t>tc
	3	0.0130	0.0006	0.0140	-5.47	-2.26	t<tc
9	2	0.0154	0.0006	0.0140	7.79	2.26	t>tc
	1	0.0135	0.0004	0.0140	-3.25	-2.26	t<tc
	3	0.0155	0.0007	0.0152	1.26	2.26	t<tc
10	2	0.0168	0.0013	0.0152	3.86	2.26	t>tc
	1	0.0132	0.0006	0.0152	-10.89	-2.26	t<tc

Table 3. T-test for stations at 5% confidence level and (n-1) degrees of freedom showing mean counter rate (*), mean background radiation (μ), standard deviation(s), computed t and critical t (tc) with sample size (n) of 10 (Industrial environment) (Avwiri and Ebeniro, 2002)

Station	Mean exposure rate, R (mR/hr)	Dose equivalent, D (mSv/yr)
1	0.0144+0.0007	1.261+0.064
2	0.0143+0.0008	1.253+0.073
3	0.0146+0.0013	1.279+0.114
4	0.0145+0.0008	1.270+0.073
5	0.0143+0.0019	1.253+0.166
6	0.0147+0.0012	1.288+0.105
7	0.0147+0.0006	1.288+0.053
8	0.0144+0.0012	1.261+0.105
9	0.0140+0.0005	1.226+0.044
10	0.0152+0.0009	1.332+0.076

Table 4. Dose equivalent, D computed for the data of Table 3 (Industrial environment) (Avwiri and Ebeniro, 2002; Chad-Umoren and Briggs-Kamara, 2010)

study used a digilert 50 and a digilert 100 nuclear radiation monitoring meters containing a Geiger-Muller tube capable of detecting α, β, γ and x-rays within the temperature range of -10^{0}C to 50^{0}C, alongside a geographical positioning system (GPS) which was used to measure the precise location of sampling. The survey was carried out between the hours of 1300 and 1600 hours, because the exposure rate meter has a maximum response to environmental radiation within these hours (Louis et al, 2005). At each of the selected sites and within the host communities four readings were taken at intervals of 5 minutes and the average value determined. The tube of the radiation meter was held at a standard height of 1.0m above the ground with its window facing first the oil installations and then vertically downwards (Chad-Umoren et al, 2006). A ^{137}Cs source of specific energy was used to calibrate the instrument to read accurately in Roentgens. Estimate of the whole body equivalent dose rate was done using the National Council on Radiation Protection and measurements (NCRP, 1993) recommendation:

$$1mRh^{-1}=(0.96\times24\times365/100)mSvy^{-1} \tag{1}$$

Table 5 compares the mean radiation exposure rates for the hydrocarbon industry facilities and the mean radiation exposure rates for the host communities. The lowest mean radiation exposure rate of 0.014mRh^{-1} was obtained at Obite/Ogbogu. This low level may be attributed to the concrete shielding of the gas plant in the area and the distance of the host community from the Idu flow station (about 2km). For the surveyed oil and gas facility locations, the highest mean exposure rate of 0.018$\pm$0.002mRh^{-1} was obtained at the gas plant; while for the host communities, the highest mean exposure rate of 0.017$\pm$0.001mRh^{-1} was recorded in Ebocha and Obrikom communities of zones A and B. This value is well above the mean field/site radiation exposure level. The percentage exposure rate difference is minimum at the treatment plant and maximum at the gas turbine in Obrikom/Omoku.

At location 5 of zone C, a high exposure rate of 0.034mRhr^{-1} was recorded. This could be because of both the presence of sharp sand (asphalt) along the river bank and the industrial waste discharged into the river.

A comparison of the measured radiation levels for the six zones with the normal background level (Fig. 4) shows that the computed mean effective equivalent dose rate for the locations in all the six zones are above the dose limit of 1mSvyr^{-1} for the general public and far below the dose limit of 20mSvyr^{-1} for radiological workers as recommended by the

Area Code	Industrial Site Surveyed	Host community	Mean Site Radiation levels (mRh⁻¹)	Mean Host Community Radiation levels (mRh⁻¹)
A	Ebocha Oil Gathering Center	Ebocha/Mgbede	0.016 ± 0.001	0.017 ± 0.001
B	OB/OB Gas Plant	Obrikom/ Ndoni	0.016 ± 0.001	0.017 ± 0.001
C	Gas Turbine	Obrikom /Omoku	0.018 ± 0.002	0.016 ± 0.003
D	Treatment plant (Arrival Manifold)	Obrikom /Ebegoro	0.014 ± 0.001	0.014 ± 0.001
E	Obite Gas Plant	Obite /Ogbogu	0.015 ± 0.001	0.014 ± 0.001
F	Idu Flow Station	Idu /Obagi	0.016 ± 0.001	0.015 ± 0.001

Table 5. Comparison of Exposure rate at Industrial sites and Host Communities (Ononugbo et al, 2011)

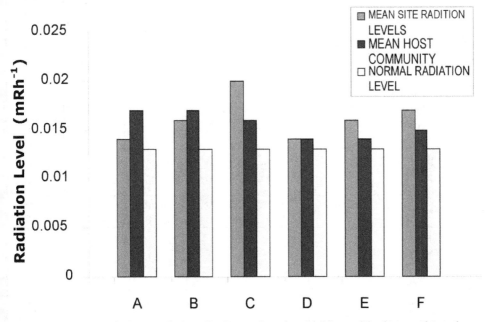

Fig. 4. Comparison of Measured Mean Radiation Levels with Normal Background Level (Ononugbo et al, 2011)

International Commission on Radiological Protection (ICRP, 1990). This indicates that while those exposed occupationally are safe, members of the host communities are not.

The results also show that although the values are within the range of values previously reported for the Niger Delta region (Arogunjo et al, 2004; Agbalagba and Avwiri, 2008 and Chad-Umoren and Briggs-Kamara, 2010), the radiation exposure rate for 71.7% of the sampled area (43 locations) exceeded the accepted ICRP background level of 0.013mRhr[-1].

S/N	Oil and Gas Facilities	Flow station installation radiation levels (μRh^{-1})					
		Afiesere Delta	Ughelli West Delta	Etelebou Bayelsa	Kolo-Creek Bayelsa	Adibawa Rivers	Egbema Rivers
1	Flow station entrance	08.00± 0.70	14.00± 2.30	12.00± 1.00	09.00± 1.10	19.00± 1.10	13.00± 1.10
2	Manifold	15.00± 2.00	19.00± 3.00	19.00± 2.80	12.00± 0.90	17.00± 2.20	16.00± 2.10
3	Flare monitoring meter units	12.00± 0.60	15.00± 2.10	16.00± 1.50	16.00± 1.00	16.00± 1.40	15.00± 2.20
4	Flare knock-out drum/vessel	14.00± 1.70	23.00± 4.20	18.00± 2.10	14.00± 1.60	25.00± 4.20	24.00± 3.20
5	Stack gas vent	15.00± 1.00	15.00± 1.00	21.00± 3.70	13.00± 0.80	19.00± 2.00	22.00± 3.20
6	Flare stack pathway	11.00± 0.80	17.00± 3.20	17.00± 1.80	24.00± 4.3.10	16.00± 2.20	17.00± 2.50
7	Flare stack point	19.00± 3.20	14.00± 1.40	14.00± 1.00	17.00± 2.10	14.00± 1.40	21.00± 3.60
8	Inflow delivery crude oil pipes	18.00± 1.60	19.00± 2.60	17.00± 3.00	24.00± 4.20	20.00± 2.00	21.00± 3.10
9	Inflow gas delivery pipes	15.00± 3.00	20.00± 4.00	13.00± 2.00	22.00± 3.00	20.00± 3.40	18.00± 2.40
10	Associate gas control meter	15.00± 2.00	20.00± 1.80	16.00± 3.10	20.00± 3.70	14.00± 0.60	19.00± 3.00
11	Natural gas compressor station (NGC)	21.00± 2.40	22.00± 2.10	Nil	Nil	Nil	Nil
	MEAN	14.82± 1.74	18.00± 2.54	16.30± 2.20	17.10± 2.15	18.00± 2.02	18.60± 2.64
Mean Rate	Dose equivalent ($\mu Sv/wk$)	6.70± 0.78	8.01± 1.13	7.25± 0.98	7.61± 0.96	80.01± 0.90	8.28± 1.17

Table 6. Facilities exposure rate (Agbalagba et al, 2011)

A recent survey studied the gamma radiation profile of oil and gas facilities in six selected low stations in the Niger delta region (Agbalagba et al, 2011). Of the different types of onizing radiations, gamma rays are the most penetrating and therefore very hazardous. They emanate from radionuclides containing radon and may be ingested or inhaled by personnel in the course of routine repairs and maintenance of oil facilities. When inhalation occurs, the dust particles and aerosols containing radon become attached to the lungs so that the presence of the gamma rays emitted in the decay increases the risk of lung cancer and other hazards such as eye cataracts and mental problems to personnel and host communities (Laogun et. al., 2006).

The results are shown in Tables 6 and 7. The radiation levels for the facilities (Table 6) range from 08.00±0.70μRh⁻¹ in Afiesere flow station entrance to 25.00±4.20μR⁻¹ in Adibawa flare knockout vessel. The high value at the Adibawa knockout vessel can be attributed to the spill of associated crude and the exposure of the environment to effluent. The mean exposure rate for the flow stations range from 14.82±1.74μRh⁻¹ (6.70±0.78μSv/wk) at Afiesere flow station to 18.60±2.64μRh⁻¹ (8.28±1.17μSv/wk) at Egbema flow station, with field mean radiation value of 17.14±2.22μRh⁻¹. The high radiation levels recorded at the natural gas compressor stations (NGC) at Afiesere and Ughelli West flow stations may be attributed to the high concentration of radon in the natural gas and gas production facilities.

Table 7 compares the radiation exposure rate for both the flow stations and the host communities. The lowest radiation exposure rate of 10.00±0.70μRh⁻¹ was obtained at Joinkrama 4 in Rivers State. This low radiation level may be attributed to the geology (underlying rock) of the area and to the distance of the host community from the flow station (~2.5km) which is the farthest among the six host communities. The highest average exposure rate (21.00±2.10μRh⁻¹) was obtained in Emeregha community in Afiesere field. This value is much higher than the mean field radiation exposure level.

S/N	Oil field/flow station	Host community	Mean fields Radiation levels μRh⁻¹	Mean Host community Radiation level μRh⁻¹	Exposure rate diff (%)
1	Afiesere	Emeragha	14.82.±1.74	21.00±2.10	17.3
2	Ughelli west	Ekakpamre	18.00±2.54	17.00±2.00	2.9
3	Etelebou	Nedugo	16.30±2.20	15.00±1.40	4.2
4	Kolo-creek	Imirigin	17.10±2.15	18.00±1.60	2.6
5	Adibawa	Joinkrama 4	18.00±2.02	10.00±0.70	28.6
6	Egbema	Egbema	18.60±2.64	14.00±1.10	14.1

Table 7. Comparison of Flow station and Host Communities (Agbalagba et al, 2011)

The presence of the radioactive gas, radon which is produced by the radioactive disintegration of radium–226, may be responsible for the variation of the radiation levels within the facilities. This is so because radon is known to be present in crude oil and gas products (IAEA, 2003). The dispersal of the gas is also favoured by wind direction at the point of liberation.

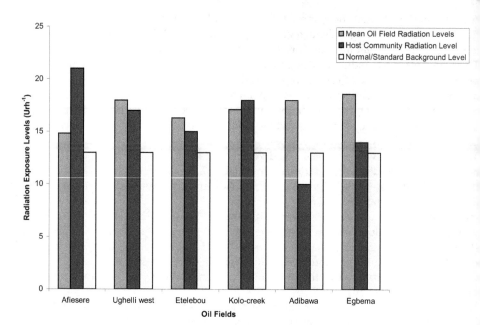

Fig. 5. Comparison of Mean Oil Fields and Host Communities Radiation Levels with Normal background Radiation Level (Agbalagba et al, 2011)

Fig.5 is a comparison of the mean radiation levels for the oil fields and host communities with the normal background ionizing radiation level. Overall, the results show that in all the six flow stations and their host communities except Joinkrama 4, the exposure rates exceed both the standard background levels and values reported previously by Ebong and Alagoa (1992a); Avwiri and Ebeniro (1998), but are in agreement with results reported in similar environments (Arogunjo et al., 2004; Laogun et al., 2006; Agbalagba et al., 2007). The differences in the exposure rates between host communities and flow stations can be attributed to the input materials and output substances (effluents) associated with the activities of the operating companies and dilution of radon as it transits to the host communities. The only exception, the Emeragha community, may be due to the existence of oil wells (wells 7, 10 and 14) within the community which may have enhanced radon concentration and hence the radiation level in the community.

Furthermore, the radiation levels recorded for gas flare facilities and natural gas compressor stations are fairly higher than those for other facilities. This is in agreement with the study of occupational radiation patterns during production and off-production periods (Avwiri et al, 2010). Also, the observation implies the presence of a high concentration of radon gas and heavy metals normally associated with natural and associated gas (Arogunjo et al. 2004; Laogun et al., 2006). However, this study indicates that the radon concentrations in the natural gas at these facilities in the region are low when compared to those obtained in countries like the USA, Great Britain and Canada (Laogun et al., 2006) where radon concentrations (radiation levels) pose enormous environmental challenges requiring government legislation for the control of naturally occurring radiation material (NORM) contamination in their petroleum industries.

The results of the survey did not indicate that there is in the short term the possibility of any health hazard either for the flow station attendants (staff) or the residents of the host communities since the highest average radiation exposure levels of $18.60\pm2.64\mu Rh^{-1}$ $(8.28\pm1.17\mu Sv/wk)$ recorded at Egbema flow station and $21.00\pm2.10\mu Rh^{-1}(9.35\pm0.93\mu Sv/wk)$ obtained at Emeragha community are within the UNSCEAR recommended permissible limit of $200\mu Sv/wk$ for public health and safety (UNSCEAR, 1993). However, the possibility of future health complications for the staff and host communities does exist as a result of prolonged cumulative dose intake from both direct and indirect radionuclides. Also, it is observed from Fig. 5 that the mean radiation exposure in all the oilfields is higher than the normal background which implies that the oil exploration and gas exploitation activities have impacted significantly on the background ionizing radiation levels of the area. According to Agbalagba et al (2011), this impact can be attributed to the underlying crude oil and gas being contaminated by radionuclide bearing rocks/elements (uranium, thorium and radium) which when drilled to the surface and exposed to the terrestrial environment at the oil installations, releases radon gas and other heavy metals which enhances the background levels of the area and other input materials."

S/N	SAMPLED AREA	GEOGRAPHICAL LOCATION	RADIATION LEVEL mRh⁻¹		AVE. RAD. VALUE mRh⁻¹	EQ. DOSE mSvy⁻¹
			RAD 50	RAD 100		
1	Crude Flow Pipe	NO5 32.297' E005 53.780'	0.019	0.018	0.0185±0.004	0.9843±0.32
2	Natural Gas Compressor	NO5 26.021' E005 52.940	0.025'	0.019	0.0220±0.008	1.170±0.43
3	Flow station entrance	NO5 26..057' E005 52.926'	0.017	0.018	0.01759±0.007	0.931±0.37
4	Well 7	NO5 25.918' E005 53.014'	0.021	0.024	0.2230±0.010	1.186±0.53
5	Pegging Manifold	N05 26.062' E005 52.901''	0.019	0.021	0.020±0.008	1.064±0.43
6	Well 10	N05 25.671' E005 52.930'	0.016	0.018	0.0170±0.006	0.9041±0.32
7	Flare Stack Site	N05 26.141 E005 52.653	0.024	0.025	0.0245±0.011	1.303±0.58
8	Well 5	NO5 25.701' E005 52.608'	0.018	0.020	0.0190±0.009	1.011±0.48
9	Otorogu Gas Plant	NO5 25.701' E005 52.608'	0.028	0.034	0.0310±0.010	1.649±0.53
10	*Otujeremi Town*	*NO5 25.865' E005 52.567'*	*0.022*	*0.020*	*0.0210±0.007*	*1.117±0.37*
	MEAN FIELD LEVELS				0.0213±0.008	1.134±0.44

Table 8. Otorogu Oil and Gas Field

Tables 8 - 11 give the results of ionizing radiation monitoring for oil and gas fields located at Otorogu, Evwreni, Oweh, Uzere East and West, while Table 12 compares the ionizing radiation profile of the study fields with those of the host communities (Agbalagba et al, 2011). The radiation status of various areas and parts of the oil and gas facilities are given such as the crude flow pipe, flow station, gas plants, crude oil control valve, manifold, oil wells and host communities. In each specific area two Radalerts were used – Radalert 50 and Radalert 100 and the average radiation level computed.

S/N	SAMPLED AREA	GEOGRAPHICAL LOCATION	RADIATION LEVEL mRh^{-1}		AVE. RAD. VALUE mRh^{-1}	EQ. DOSE mSvy^{-1}
			RAD 50	RAD 100		
1	Camp site	NO5' 22.720' E006 02.962'	0.011	0.011	0.0110± 0.003	0.585± 0.16
2.	Well 13	NO5' 22.615' E006' 02 640'	0.015	0.014	0.0145± 0.005	0.771± 0.27
3	Manifold	NO5' 22.405' E006' 02.405'	0.019	0.013	0.0160± 0.006	0.851± 0.32
4	Well 1	N05' 22.327' E006' 02.410'	0.017	0.014	0.055± 0.005	0.825± 0.27
5	Flow station Gate	N05' 22.445' E006. 02.470'	0.015	0.016	0.0155± 0.006	0.825± 0.32
6	L & S Tanga crude flow pipe	N05' 22.428' E006" 02 500	0.015	0.014	0.0145± 0.005	0.771± 0.27
7	Gas vent (knockout drum)	N05' 22.432' E006 22.482'	0.020	0.022	0.0210± 0.009	1.117± 0.48
8	Flare stock site	N05" 22.361' E006" 02.451'	0.021	0.018	0.0195± 0.008	1.0371± 0.83
9	Well 11	N05" 22 .394' E006' 02.439'	0.014	0.014	0.0140± 0.005	0.771± 0.27
10	*Evwreni Community*	*N05' 24.243' E006" 03.451'*	*0.017*	*0.014*	*0.0155± 0.007*	*0.8258± 0.32*
	MEAN FIELD LEVEL				0.0160± 0.006	0.839± 0.34

Table 9. Evwreni Oil and Gas Field

S/N	SAMPLED AREA	GEOGRAPHICAL LOCATION	RADIATION LEVEL mRh⁻¹		AVE. RAD. VALUE mRh⁻¹	EQ. DOSE mSvy⁻¹
			RAD 50	RAD 100		
1	Flow Station Gate	NO5′ 29.271′ E006 08.101′	0.016	0.012	0.0140±0.005	0.745±0.27
2.	Crude oil control valve	NO5′ 08.101′ E006′ 08′	0.019	0.019	0.0190± 0.007	1.011±0.37
3	Gas vent (knockout drum)	NO5′ 29.289′ E006′ 08.201′	0.017	0.016	0.0165±0.006	0.878±0.32
4	Flare stack site	N05′ 29.304′ E006′ 08.244′	0.016	0.018	0.017±0.005	0.904±0.27
5	NGC Station	N05′ 29.216′ E006. 08.132′	0.022	0.020	0.0210±0.008	1.117±0.43
6	L & S tango Crude flow pipe	N05′ 29.285′ E006″ 28 185′	0.016	0.014	0.0150± 0.006	0.798±0.32
7	Manifold	N05′ 28.185′ E006 07.720′	0.019	0.018	0.01850±0.008	0.984±0.43
8	Well 12	N05″ 29.666′ E006″ 06.567′	0.020	0.018	0.0190±0.007	1.011±0.37
9	Well 2	N05″ 29 .219′ E006′ 08.128′	0.018	0.023	0.0205±0.010	1.091±0.53
10	*Otor-Oweh community*	*N05′ 29.614′ E006″ 06.248′*	*0.012*	*0.014*	*0.0130±0.005*	*0.692±0.27*
	MEAN FIELD LEVEL				**0.0178±0.007**	**0.949±0.37**

Table 10. Oweh Oil and Gas Field

S/N	SAMPLED AREA	GEOGRAPHICAL LOCATION	RADIATION LEVEL mRh⁻¹		AVE. RAD VALUE mRh⁻¹	EQ. DOSE mSvy⁻¹
			RAD 50	RAD 100		
1	Manifold	NO5' 20.080' E006 14.865 '	0.016	0.015	0.0155±0.006	0.525±0.32
2.	Buster station	NO5' 20.162' E006' 14.781''	0.017	0.014	0.0155± 0.005	0.825±0.27
3	NGC Station	NO5' 19.751' E006' 14.762'	0.016	0.019	0.0175±0.006	0.931±0.32
4	Flow station Gate	N05' 19.627' E006' 14.655'	0.027	0.028	0.0275±0.013	1.463±0.69
5	L & S Tango crude flow pipe	N05' 19.167' E006. 14.642'	0.022	0.024	0.230±0.010	1.224±0.53
6	Flare knock out down	N05' 19.601' E006'' 14. 633'	0.020	0.018	0.01900± 0.008	1.011±0.43
7	Flare stack site	N05' 19.584' E006' 14.566'	0.017	0.021	0.0190±0.007	1.011±0.37
8	Well 6	N05'' 19.251' E006'' 15.960'	0.019	0.023	0.0205±0.009	1.277±0.64
9	Well 2	N05'' 19 .421' E006' 15.862'	0.022	0.026	0.0240±0.012	1.277±0.64
10	*Uzere community*	*N05' 20.268' E006'' 14.338'*	*0.016*	*0.019*	*0.0175±0.007*	*0.931±0.27*
	MEAN FIELD LEVEL				0.0202±0.008	1.075±0.45

Table 11. Uzere East and West Oil and Gas Field

Area Code	Oil and Gas Field	Host Community	Mean field dose rate (mSvy⁻¹)	Host Community dose rate (mSvy⁻¹)	Difference (%)
OUT	Otorugu	Otujeremi	1.134±0.31	1.117±0.37	1.51
EVN	Evwreni	Evwreni	0.839±0.34	0.612±0.16	22.70
OWT	Oweh	Otoweh	0.949±0.37	0.692±0.27	37.14
OLO	Olomoro-Oleh	Olomoro	0.943±0.37	0.931±0.27	1.29
UZE	Uzere West & East	Uzere	1.075±0.45	0.931±0.37	14.4

Table 12. Comparison of Studies fields and Host Communities Radiation Data

able 12 shows that the highest mean dose rate for the oil fields, 1.134±0.31mSvy⁻¹ was ecorded at the Otorugu field and its host community, Otujeremi, recorded the highest dose ate, 1.117±0.37mSvy⁻¹, for the host communities. It would appear that the preponderance of ;as operations in Otorugu as compared to the other fields has resulted in an increased elease of radon gas into the atmosphere over the oilfield and its host community leading to ignificant impact on the radiation level of the oilfield and its host community.

. Water and soil analyses

The Niger delta has a plethora of interconnected underground and surface waterways and his can have some impact on the dispersal of ionizing radiation in the environment as they have the potential to transport particulate radioactive matter. Sediments from different vater bodies and industrial effluents discharged into the water bodies are transported from ocation to location. Furthermore, evaluating the water in the region for radioactivity becomes especially important as the water from creeks, streams and rivers in much of the egion is also used for such activities as washing, bathing, cooking and drinking! Through any of this means radionuclide ingestion and consequent radiation contamination of the populace can attend epidemic proportions. Another route for radionuclide ingestion by the human population is through the consumption by man of fish and other sea foods that have been contaminated by radiation.

A study conducted in Delta state, another Nigeria Niger delta state, using Exploranium - The-Identifier GR-135 model, assessed the natural radioactivity concentration in river Forcados (Avwiri et al, 2008). 20 water samples were collected from 20 different locations separated a distance of 100m from each other. The study identified three radionuclides, namely ^{40}K, ^{226}Ra and ^{232}Th with respective average specific activity of 13.94±1.97Bq/l, 12.80±2.84Bq/l and 34.62±3.71Bq/l. The mean absorbed dose rates and dose equivalent were computed to be 9.90±1.61nGy/h and 0.084±0.003mSv/y respectively. The survey did not indicate any radionuclide concentration gradient as the values in the survey were found to be randomly distributed. This can be attributed to the water transportation mechanism and effluent discharge.

Also, the results show low concentration of the identified radionuclides in the water and may therefore not pose any radiological effects on the populace that use the water from this river. Furthermore, the values are comparable to previous surveys within the Niger delta region (Arogunjo et al, 2004), however, they are higher than the value of 1.2mBq/l (^{226}Ra) reported for Lake Ontario (Baweja et al, 1987) and 28 x 10⁻⁶ Bq/l (^{232}Th) for Lake Michigan. This disparity can be attributed to the activities of the hydrocarbon industry in the Niger delta.

In a survey of the gross alpha and beta radionuclide activity in the Okpare Creek in Delta State, Nigeria, Avwiri and Agbalagba (2007) classified the creek into three zones and reported average alpha activities in the classified zones as 1.003±0.097, 4.261 ± 0.109 and 10.296±0.489Bql⁻¹ respectively, and the beta activities as 0.129±0.100, 0.523+0.003 and 0.793+ 0.010Bql⁻¹ respectively. These values are far above the 0.1Bq/l for alpha and 1.0Bq/l for beta WHO maximum recommended level for screening for drinking water (WHO, 2003). A previous study by Avwiri et al (2005) assessed the radionuclide contents of soil, sediments and water in Aba River in the Niger delta state of Abia state. In addition to these, the radionuclide content of the fish in the river was also evaluated. The work employed a rigorously calibrated sodium iodide (NaI) detector. The radionuclides identified were [226]Ra, [228]Ra and [40]K for all three study samples. For the soil samples the respective average specific activity levels obtained were 4.73±0.71Bq/kg, 4.26±0.9Bq/kg, and 118.75±10.54Bq/kg. The activity levels of the radionuclides for the water samples were 3.34±0.43Bq/l, 3.69±0.17Bq/l and 111.39±10.04Bq/l respectively. In the case of the sediments the study gave 11.56±2.35Bq/kg, 17.50±1.77Bq/kg and 253.42±21.10 respectively. And for the catfish from the river, the activity levels were 19.14±4.33Bq/kg, 22.58± 5.58Bq/kg and 63.42±14.09Bq/kg respectively. These results show significantly higher concentrations compared to a previous study (Jibiri et al, 1999). The operations of the Industrial Zone in Aba account for these significantly high values as they discharge their effluents directly into the river. An analogous study to assess the radiological impact of the petrochemical industry on the Aleto Eleme River showed a significant elevation of the ionizing radiation levels at the point of effluent discharge into the river (Avwiri and Tchokossa, 2006).

A study carried out to assess the level of natural radionuclides in borehole water in some selected wells in Port Harcourt, Rivers State showed that the mean specific activity of the resulting annual effective doses for [226]Ra, [228]Ra and [40]K were 3.51±2.22, 2.04 ± 0.29 at 23.03 ±4.37 and 0.36 ± 0.12, 0.51± 0.02 and 0.05 ±0.01mSy/y respectively (Avwiri et al, 2006). The results of this survey are within the range obtained elsewhere. Generally, public places showed the highest activity concentration due to poor sanitation.

Waste management in urban centres is important for environmental health. In many parts of the Niger delta region, wastes are sometimes collected and heaped up and no further action taken until they are scattered again by wind; disposed of indiscriminately or they may be disposed of in landfills. Landfills are openings in the soil used for waste disposal. They could be purpose-made or an abandoned pit or quarry. Many of the landfills in the Niger delta region are usually not differentiated and therefore will hold a mixture of different types of wastes, including those from the hydrocarbon industry. Landfills are an important source of groundwater pollution and may also result in the elevation of the ionizing radiation profile of the environment due to its radionuclide content. One survey conducted in the city of Port Harcourt, Rivers State, assessed the radionuclide content, the ionizing radiation level and associated dose rates of a landfill around the Eliozu area of the city (Avwiri et al, 2011). The Eliozu landfill is composed of different types of wastes including industrial wastes, chemical wastes, medical wastes, scraps, metals and other debris.

10 samples each of soil and water were collected from different parts of the landfill and analyzed for their radioactivity and radionuclide content using the gamma-ray spectrometer NaI (Tl) detector system. The results are shown in Table 13 for the soil samples and in Table 14 for the water samples.

The radionuclides found in both the soil and water samples were [232]Th, [238]U and [40]K. The study reported that for the soil samples, the mean activity concentration were 27.41 ± 9.97 Bq/kg for [238]U, 19.27 ± 8.14 Bq/kg for [232]Th and 326.08 ± 66.74 Bq/kg for [40]K (Table 7). And for the water samples, the mean activity concentration were 7.92 ± 2.69 Bq/l, 6.96 ± 2.37 Bq/l and 24.77 ± 8.33 Bq/l for [238]U, [232]Th, [40]K respectively (Table 8). The calculated absorbed dose rates for the soil had a range of 23.53 nGy.h^{-1} to 50.39 nGy.h^{-1} and a mean of 38.17 ±12.45 nGy.h^{-1} while for the water the range was 6.62 nGy.h^{-1} to 10.71 nGy.h^{-1} with a mean of 9.03±3.07 nGy.h^{-1}. The mean absorbed dose rate for the area is lower than the world's average of 55 nGy.h^{-1}for soil. It should however be observed that the upper limit of 50.39 nGy.h^{-1} is quite close to the world average.

S/No	Sample	Soil (Bq/kg)				
		K-40	U-238 (Ra-226)	Th-232 (Ra-228)	Absorbed dose rates (nGy/hr)	Equivalent dose rate (mSv/yr)
1	A1	570.08 ± 87.6	20.52 ± 5.2	18.95 ± 9.9	45.25 ± 12.5	0.3964 ± 0.1
2	A2	105.57 ± 24.9	18.94 ± 8.4	16.62 ± 6.7	23.53 ± 9.1	0.2061 ± 0.1
3	A3	404.95 ± 99.8	34.81 ± 13.6	19.33 ± 6.5	44.68 ± 14.3	0.3914 ± 0.1
4	A4	140.49 ± 35.8	23.87 ± 9.7	17.84 ± 6.5	27.92 ± 10.0	0.2446 ± 0.1
5	A5	254.55 ± 79.5	38.28 ± 11.4	29.54 ± 11.5	46.60 ± 15.8	0.4082 ± 0.1
6	A6	256.22 ± 68.6	26.54 ± 9.7	22.46 ± 8.7	36.97 ± 12.8	0.3239 ± 0.1
7	A7	527.91 ± 89.5	35.91 ± 11.4	19.41 ± 9.9	50.39 ± 15.2	0.4414 ± 0.1
8	A8	545.13 ± 87.5	26.41 ± 11.1	18.63 ± 7.8	46.52 ± 13.6	0.4075 ± 0.1
9	A9	323.64 ± 75.9	29.36 ± 9.3	13.43 ± 7.3	35.05 ± 12.0	0.3070 ± 0.1
10	A10	132.23 ± 18.7	19.44 ± 9.9	16.52 ± 6.5	24.80 ± 9.3	0.2173 ± 0.1
Mean Values		326 ± 66.7	27.41 ± 10.0	19.27 ± 8.1	38.17 ± 12.5	0.3344 ±0.1

Table 13. Radionuclide Concentration of Soil Samples (BqKg^{-1}) (Avwiri et al, 2011)

S/N	Sample	Water (Bq/l)				
		K-40	U-238 (Ra-226)	Th-232 (Ra-228)	Absorbed dose rates (nGy/hr)	Equivalent dose rates (mSv/yr)
1	W1	16.40 ± 7.3	8.24 ± 2.8	5.87 ± 2.0	7.80 ± 2.8	0.0683 ± 0.03
2	W2	26.74 ± 7.7	7.48 ± 3.3	7.44 ± 2.2	9.24 ± 3.2	0.0809 ± 0.03
3	W3	24.98 ± 9.8	9.32 ± 2.3	8.32 ± 2.5	10.54 ± 3.0	0.0923 ± 0.03
4	W4	19.84 ± 7.0	7.89 ± 3.0	7.89 ± 2.1	9.43 ± 3.0	0.0826 ± 0.03
5	W5	23.51 ± 9.7	8.04 ± 2.1	6.78 ± 2.3	8.91 ± 2.9	0.0781 ± 0.03
6	W6	32.08 ± 8.2	9.06 ± 3.2	8.29 ± 3.1	10.71 ± 3.8	0.0938 ± 0.03
7	W7	16.12 ± 7.2	9.41 ± 2.7	7.89 ± 2.5	9.92 ± 3.1	0.0869 ± 0.03
8	W8	27.16 ± 9.0	6.94 ± 3.5	5.20 ± 2.0	7.55 ± 3.2	0.0661 ± 0.03
9	W9	21.67 ± 7.4	5.99 ± 2.1	4.76 ± 2.1	6.62 ± 2.6	0.0580 ± 0.02
10	W10	39.15 ± 10.0	6.85 ± 1.8	7.34 ± 2.9	9.68 ± 3.1	0.0848 ±0.03
Mean Values		24.77± 8.3	7.92±2.7	6.96±2.4	9.03 ± 3.1	0.0791 ± 0.03

Table 14. Radionuclide Concentrations of Water Samples (Bql^{-1}) (Avwiri et al, 2011)

The determination of the equivalent radiation exposure (effective dose rate) for the immediate neighbourhood of the dumpsite indicate that the population in the vicinity of the site receives a dose that lies in the range of 0.2061 to 0.4414 mSv.y^{-1} with a mean of 0.3344±0.1091 mSv.y^{-1} from the soil and a dose that ranges from 0.0580 to 0.0938 mSv.y^{-1} with a mean of 0.0791±0.0269 mSv.y^{-1} from the water. The dumpsite therefore appears to pose minimal ionizing radiation risk to the environment and the population.

9. Ionizing radiation control strategies for Nigeria's hydrocarbon industry

The following strategies are proposed for the control and management of ionizing radiation in the region:

9.1 Identification of areas of radiation risks
Activities and locations with high ionizing radiation risks or potential for radiation hazards should be delineated. The reason for this is that radiation pollution and risks cannot be controlled if the location of the risk is unknown. Therefore, regular radiation monitoring in the region is to be sustained. Such radiation monitoring should include baseline surveys of oil and gas facilities and host communities, water and soil analyses, monitoring of well heads, production manifolds, storage tanks, flow stations, etc.

9.2 Establishment and enforcement of ionizing radiation limits for the hydrocarbon industry
Here such government agencies as the Nigeria Nuclear Regulatory Authority (NNRA) must enforce compliance with relevant laws by industry operators. Where existing laws are inadequate or obsolete they should be reviewed and appropriate legislations enacted.

9.3 Education and public enlightenment
Education of industry personnel and proper enlightenment of members of the public should form critical components of any ionizing radiation control strategy. Industry personnel, especially those most likely to be exposed to radiation, need to be trained to work safe and ensure that ionizing radiation is not spread.
Also, the general public, especially residents of communities that host oil and gas installations should be educated on the hazards of ionizing radiation.

9.4 Handling of ionizing radiation contaminated waste
Ionizing radiation contaminated waste generated by the hydrocarbon industry include sludge, scale, drilling pipes, storage tanks, etc. In the absence of appropriate control mechanisms, unintended spreading of ionizing radiation contamination with consequent contamination of areas of land and the environment and eventual exposure of the public can occur in the course of the handling, storage and transportation of ionizing radiation contaminated equipment or waste.
Used oilfield equipments should therefore be evaluated first to determine their radioactivity status before transportation; otherwise unintended spreading of ionizing radiation can occur when such materials as contaminated pipes and tanks are transported. Possible strategies for handling radioactive waste generated in the oil and gas industry include among others disposal in salt caverns, underground injection, and smelting (OGP, 2003).

Industry operators will have to study the Niger delta environment to determine the most appropriate strategies for the region.

9.5 Decommissioning of oil and gas production facilities

At some point in its productive cycle, an oil and gas facility such as a reservoir will become economically unviable for further exploitation and must therefore be properly disposed of rather than just abandoning it. This is the process of decommissioning. For example, available statistics (NNPC, 2004) show that there are 606 oil fields in the Niger Delta region, 251 of which are offshore, while the balance of 355 wells are on-shore. Also, of these, 193 wells are currently being exploited while 23 have been shut in or abandoned as no more viable and would therefore need to be properly decommissioned.

Among other reasons for decommissioning is the desire to limit the hazards of radioactive contamination. The steps in oilfield facility and equipment decommissioning include the following (IAEA, 2003):

i.   Radiological assessment of equipments and facilities to ascertain radiation levels and therefore level of risk to personnel, the general populace and the environment.

ii. Submission of decommissioning plans, surveys and radiological reports on the equipments and facilities to relevant regulatory authorities such as the Nigeria Nuclear Regulatory Authority (NNRA) and such other agencies responsible for environmental health and safety especially as it relates to ionizing radiation.

iii. Decontamination of equipments and facilities to levels set by relevant regulatory bodies, especially if such equipments and facilities will become available for unrestricted public use.

iv. Radioactive waste management. This is the last step in the decommissioning procedure, where all hazardous radioactive wastes and the remaining contaminated items are disposed of in designated and approved radioactive waste disposal facilities.

10. Conclusion and suggestions for future research

It is obvious from this study that the Niger delta region of Nigeria is inundated with ionizing radiation as a result of the activities of the hydrocarbon industry and its subsidiary services. Though the profile in general does not indicate any immediate health complications both for the industry personnel and the general populace, the ionizing radiation patterns of the region is still crucial for the following reasons:

10.1. Low level radiations have long-term cumulative effects (Chang et al, 1997). Moreover for the hydrocarbon industry evidence exists on long term effect of low radiation dose. It has been observed that there is prevalence among hydrocarbon industry retirees and host community members of such diseases as eye cataracts, various kinds of cancer such as lung and bone cancer, leukaemia and mental disorder (UNDP, 2006; Otarigho, 2007). In the long term therefore, the current low radiation exposure could become a serious health hazard. It is therefore apparent that the present low radiation level in Nigeria's hydrocarbon belt should not be ignored.

10.2. Various research findings presented in this study show that in many specific locations, radiation levels exceed the expected normal background levels. The implication of this is that the ionizing radiation levels in the region are rising progressively so that the environment is gradually becoming unsafe for the general populace in the region.

10.3. Though the overall ionizing radiation profile for the region is that of low exposure rates, however evidence from the study indicates that in some of the locations the radiation levels exceed internationally recommended standards, especially for the general populace.
10.4. The activities of the industry are not abating. On the contrary, there appears to be more aggressive activities aimed at exploiting the abundant resources of the region with attendant increased introduction into the region of ionizing radiation. Increased oil and gas exploration, exploitation and production activities will result in increased radiation pollution due to increased radioactive waste generation and increased generation of such hydrocarbon industry sources of ionizing radiation as scales, sludge, formation water, etc.
10.5. The current practise of gas flaring poses the danger of radionuclide build-up in the atmosphere. As the work has shown, areas of high gas-related activities in the region have tended to experience marked ionizing radiation levels. Furthermore, when the flared gas comes in contact with rain water, it results in the radioactive pollution of the rain water. And like the water from the creeks and rivers, rain water in many parts of the Niger delta is an important source of water for various activities like bathing, washing, cooking and drinking.
10.6. It will be helpful to investigate the health challenges of the workers at the oil and gas facilities in the Niger delta to see if there is any correlation between ionizing radiation exposure and the health complaints of the workers.
10.7. Further research in the area should include:
i.   Mathematical modelling of oil spill phenomenon in the area
ii.   Mathematical modelling of impact of gas flaring in the area
iii. More extensive soil and water analyses of areas affected by oil spillage in the region so as to properly establish the impact of oil spillage especially considering that much of the population of the region are fishermen and rural subsistence farmers
iv.   A study of the health challenges of oil and gas industry personnel and host community residents to determine any possible correlation between ionizing radiation and the prevailing health issues of these respective groups

11. References

Abali,B.K. (2009). Oil and Gas Exploration: What ONELGA suffers. Port Harcourt, B'Alive publications co.

Agbalagba, E. O , Avwiri, G. O and Chad-Umoren, Y. E. (2011). Radiological Impact of Oil and Gas Activities in Selected Oilfields in Production Land Area of Delta State, Nigeria (In press).

Agbalagba, O. E. and and Meindinyo, R. K. (2010). Radiological impact of oil spilled environment: A case study of the Eriemu well 13 and 19 oil spillage in Ughelli region of delta state, Nigeria. Indian Journal of Science and Technology, Vol. 3(9), 1001 – 1005

Agbalagba E.O., Meindinyo. R.K., Akpata, A.N. and Olali, S.A (2011). Assessment of Gamma-radiation Profile of Oil and Gas Facilities in Selected Flow Stations in the Niger- Delta Region of Nigeria. Indian Journal of Emerging Science and Technology (in press)

Arogunjo, M.A., I.P. Farai and I.A. Fuwape (2004). Impact of oil and gas industry to the natural radioactivity distribution in the delta region of Nigeria. Nigerian Journal of Physics, vol.16, 131-136.

Avbovbo, A. A. (1998). Tertiary Lithostratigraphy of Niger Delta: American Association of Petroleum Geologists Bulletin, vol. 62(2), 295-300

Avwiri, G. O. and Agbalagba, E.O. (2007). Survey of gross alpha and gross beta radionuclide activity in Okpare – Creek, Delta State, Nigeria. Asian Journal of Applied Science, vol. 7 (22), 3542 – 3542.

Avwiri, G.O., Chad – Umoren, Y.E., Enyinna, P.I. and Agbalagba, E.O. (2009). Occupational radiation Profile of oil and gas facilities during and off – production periods in Ughelli, Nigeria. Factia Universititatis, Series Working and Living Environmental Protection, Vol. 6(1).

Avwiri, G. O; Enyinna, P. I and Agbalagba, F. O.(2007). Terrestrial Radiation Around Oil and Gas Facilities in Ughelli, Nigeria. Journal of Applied Science, vol. 7, 1543 – 1546.

Avwiri, G. O., Owate, I. O. and Enyinna (2005). Radionuclide Concentration Survey of Soil, Sediments and Water in Aba River, Abia State, Nigeria. Scientia Africana, vol. 4(1&2), 67-72.

Avwiri, G.O. and Ebeniro, J.O. (1998). External Environmental Radiation in an industrial area of Rivers State, Nigeria. Nigerian Journal of Physics, vol. 10, 103 – 107.

Avwiri, G.O., Enyinna, P. I. and Agbalagba, E.O. (2008). Assessment of natural radioactivity concentration and distribution in river Forcados, Delta State, Nigeria. Scientia Africana Vol. 4(1 & 2), 128-135

Avwiri, G.O., Enyinna, P.I and Agbalagba, E.O. (2010). Occupational radiation levels in solid minerals producing areas of Abia State, Nigeria. Scientia Africana. Vol. 9 (1).

Avwiri, G.O., and Ebeniro, J.O. (2002). A survey of the background radiation levels of the sub-Industrial areas of Port Harcourt, Global Journal of Pure and Applied Sciences. Vol. 8(2), 111 – 113.

Avwiri, G.O. and Tchokossa, P. (2006). Radiological Impacts of Natural radioactivity along Aleto River due to a Petrochemical Industry in Port Harcourt, Rivers state, Nigeria. Journal of Nigeria Environmental Society (JNES), Vol. 3 (3), 315 – 323.

Avwiri, G.O., Mokobia, C.E. and Tchokossa, P. (2006). Natural radionuclide in borehole water in Port Harcourt, Rivers state, Nigeria. Radiation Protection Dosimetry, Vol. 123(4), 509 – 514, Oxford Journals UK

Avwiri, G.O., Nte, F.U. and Olanrewaju, A.I. (2011). Determination of radionuclide concentration Of land fill at Eliozu, Port Harcourt, Rivers State. Scientia Africana (in press).

Baweja A. S, Joshi S.R. and Demayo A . (1987). Report on the National Radionuclide monitoring program, (1981-1984) Environment Canada, Ottawa, Scientific services No.156.

Beka, F. T., and Oti, M. N., (1995). The distal offshore Niger Delta: frontier prospects of a mature petroleum province, in, Oti, M.N., and Postma, G., eds., Geology of Deltas: Rotterdam, A.A. Balkema, p. 237-241.

Briggs-Kamara, M.A; Sigalo, F.B; Chad-Umoren, Y.E. and Kamgba, F.A. (2009). Terrestrial Radiation Profile of a Nigerian University Campus: Impact of Computer and Photocopier Operations. Journal Facta Universitatis: Working and Living Environmental Protection, vol 6 (1), 1 – 9

CERN (1995). Safety Guide for Experiments at European Council for Nuclear Research, CRN, Part III – Advice, 40, Ionizing Radiation. (http://cern.web.cern.../40).

Chad-Umoren, Y. E and Briggs-Kamara, M. A (2010). Environmental Ionizing Radiation Distribution in Rivers State, Nigeria. Journal of Environmental Engineering and Landscape Management, vol. 18(2), 154–161

Chad-Umoren, Y. E and Obinoma, O (2007). Determination of Ionizing Radiation Level of the Main Campus of the College of Education, Rumuolumeni, Rivers State, Nigeria. Int'l Journal of Environmental Issues, vol. 5 (1 and 2), pp 5 -10

Chad-Umoren, Y. E; Adekanmbi, M and Harry, S. O. (2006). Evaluation of Indoor Background Ionizing Radiation Profile of a Physics Laboratory. Journal Facta Universitatis: Working and Living Environmental Protection., Vol 3 (1), pp 1 – 8

Chang, W.P; Hwang, BF; Wang, D. and Wang, JD (1997). Cytogenetic Effect of Chronic Low-Dose, Low-Dose-Rate Gamma Radiation in Residents of Irradiated Buildings. Lancet. 350:330 – 333.

Ebong I.D.U. and Alagoa, K. D. (1992a). Estimates of gamma – ray background air exposure at a fertilizer plant. Discovery innovate, 4, 25-28.

Ebong, I.D.U. and Alagoa, K. D. (1992b) Fertilizer Impact in ionization radiation background at a production plant. Nig. J. Phys., vol. 4, 143 – 149.

Evamy, B.D., Haremboure, J., Kamerling, P., Knaap, W.A., Molloy, F.A., and Rowlands, P.H., (1978). Hydrocarbon habitat of Tertiary Niger Delta: American Association of Petroleum Geologists Bulletin, vol. 62, p. 277-298.

Foland, C.K., T.K. Kirland and K. VinniKoov (1995). Observed climate variations and changes (IPCC scientific Assessment) Cambridge University Press New York, Pp. 101 – 105.

IAEA (2003). Radiation Protection and the Management of Radioactive Waste in the Oil and Gas Industry. International Atomic Energy Agency Vienna, 2003 Safety Reports Series No.34

ICRP (1990). Recommendations of International Commission on Radiological Protection., Annals of ICRP-60, Pergamon press, Oxford.

ICRP (1977). Recommendation of the International Commission on Radiological Protection. ICRP Publication 261 (3).

Jibiri, N.A.O., Mbawonku, A.O., Oridale A.A. and Ujiagbadion C. (1999). Natural Radionuclide concentration levels in soil and water around cement factory, Ewekoro, Ogun State. Nigeria Journal of Physics, Vol. II, 12 – 16.

Laogun, A.A., N.O Ajayi and S.A. Agaja (2006). Variation in well head gamma radiation levels at the Nigeria Petroleum development company oil field, Ologbo Ede State, Nigeria. Nig. J. Phys, 18(1), 135-140.

Louis, E. A; Etuk, E. S. and Essien, U (2005). Environmental Radioactive Levels in Ikot Ekpene, Nigeria. Nig. J. Space Res. Vol 1, 80 – 87.

ECN(2003).National Energy Policy. The Presidency: Energy commission of Nigeria. Federal Capital Territory, Abuja, Nigeria, pp: 82

NNPC Yearly Bulletin (2004): History of the Nigerian Petroleum Industry, Nigeria pp 54 – 64

OGP (2008) Guidelines for the management of Naturally Occurring Radioactive Material (NORM) in the oil & gas industry, International Association of Oil & Gas Producers

Ononugbo, C. P; Avwiri, G. O. and Chad-Umoren, Y. E. (2011): Impact of Gas Exploitation on the Environmental Radioactivity of Ogba/Egbema/Ndoni Area, Nigeria. Energy and Environment (In press).

Otarigho, M.D., (2007). Impact of oil spillage on the people of Ughelli South Local Government Area, Delta State. J. Environ. Res. & Policies, vol. 2, 44 – 50.

United Nations Development Programme (UNDP) (2006). Niger Delta Human Development report: Environmental and Social Challenges in the Niger Delta. UN House, Abuja, Nigeria.

United Nations Scientific Committee on the Effects of Atomic Radiation (UNSCEAR, 1993). United Nations sources and effects of atomic radiation 1993, Report to the general Assembly with scientific Annexes, United Nations, New York 1993.992.

Part 2

Medical Uses

6

Radiosensitization with Hyperthermia and Chemotherapeutic Agents: Effects on Linear-Quadratic Parameters of Radiation Cell Survival Curves

Nicolaas A. P. Franken et al.[*]
Academic Medical Centre, University of Amsterdam,
Laboratory for Experimental Oncology and Radiobiology (LEXOR),
Centre for Experimental Molecular Medicine, Department of Radiation Oncology,
The Netherlands

1. Introduction

Radiosensitization effects of hyperthermia and chemotherapeutic agents as currently exploited in the clinic are discussed with respect to the linear quadratic parameters of dose-survival curve presentations. Studies of different human tumour cell lines show that a synergistic interaction can be obtained between hyperthermia, chemotherapy and radiation and that this interaction is more likely to occur in cell lines which are relatively sensitive to chemotherapy. The influence of modifying agents on radiation dose survival curves can adequately be analysed with the use of the linear-quadratic model: $S(D)/S(O)=$ exp-$(\alpha D + \beta D^2)$. The linear parameter, α, represents lethal damage from single particle events and describes the low dose area while the quadratic parameter, β, indicating sub lethal damage (SLD) dominates the effectiveness in the high dose region (Barendsen, 1990, 1994, 1997). The linear-quadratic model is based on well accepted biophysical concepts, involving the assumption that lethal damage can be induced by single-particle tracks and by interaction of damage from multiple particles. It has been found to describe the low-dose region of the survival curves up to 6 Gy rather accurately. Furthermore the LQ-model has been shown to describe adequately dose fractionation effects for normal tissue tolerance and for experimental tumours. The LQ-model has also the advantage that it requires only two parameters to describe radiation dose-survival curves. It allows the separate analysis of changes in effectiveness in the low dose range, mainly determined by the linear term and in the high dose range determined mainly by the quadratic term (Barendsen, 1982; Joiner &

---

[*] Suzanne Hovingh[1], Arlene Oei[1], Paul Cobussen[1], Judith W. J. Bergs[3], Chris van Bree[1],
Hans Rodermond[1], Lukas Stalpers[1], Petra Kok[1], Gerrit W. Barendsen[1,2] and Johannes Crezee[1]
Academic Medical Centre, University of Amsterdam,
[1]*Laboratory for Experimental Oncology and Radiobiology (LEXOR),*
Centre for Experimental Molecular Medicine, Department of Radiation Oncology, The Netherlands
[2]*Department of Cell Biology and Histology, Amsterdam, The Netherlands*
[3]*Institute of Molecular Biology and Tumor Research,Philipps-University Marburg, Emil-Mannkopff-Str. 2,*
Marburg, Germany

Kogel, 2009). An additional advantage of the LQ model is that its parameters can be discussed in terms of specific mechanisms of cell inactivation by radiation (Barendsen, 1990, 1994). Linear-quadratic analyses of hyperthermia- or chemotherapy-induced radiation sensitisation have been reported for exponentially growing and plateau phase human tumour cells in culture and for different experimental rodent cell lines (Franken et al., 1997a, 1997b; van Bree et al., 1997, 2000; Castro Kreder et al., 2004; Bergs et al., 2006, 2007abc). When the additional treatment results in increases of the value of the α-parameter, this indicates that this treatment radiosensitizes at clinically relevant doses. When the additional treatment influences the value of the β-parameter it indicates that the additional treatment has an effect on the repair of sublethal damage.

In order to determine the linear and quadratic parameters and the effects of the different agents on these parameters clonogenic assays (Franken et al., 2006) were conducted of cells after ionizing radiation only and after combined hyperthermia or chemotherapy with radiation treatment. Radiation dose survival curves have been obtained and analysed according to the LQ formula: $S(D)/S(0)= \exp(-(\alpha D+\beta D^2))$. (Barendsen, 1982; Barendsen et al., 2001; Franken et al., 2004, 2006). The effects of the different agents on the linear parameter, α and quadratic parameter, β will be described. If significant changes are derived, the values of α/β ratio's are also of interest, because these ratio's show whether the influence of the dose per fraction, dose fractionation and dose rate in radiation treatments is larger or smaller due to the combined treatments. These ratio's may increase with increasing α or decreasing β. A change in the α/β ratio might be clinically of interest for the selection of fractionation schedules.

2. Hyperthermia

Hyperthermia refers to heat treatments if cells or malignancies in which the temperature is elevated in the range of 39°C to 45°C. It is used in combination with chemo- and/or radiotherapy since it is has been shown to enhance the anti-cancer effects of both therapies (Gonzalez Gonzalez et al., 1995; Van Der Zee et al., 2002; Crezee et al. 2009). Many *in vitro* studies on the combination of hyperthermia and radiation have shown a synergistic interaction between the two modalities, especially at higher temperatures (above 42°C) (Dewey et al., 1978; Roti Roti, 2004; Raaphorst et al., 1991). This interaction is believed to result from inhibition of repair of radiation-induced DNA damage by hyperthermia (Kampinga et al., 2001; Hildebrandt et al., 2002). The sequence of combined radiation and hyperthermia treatment is important. Optimal sensitization is obtained when radiation and hyperthermia are applied simultaneously or with a short interval (Hall & Giaccio, 2006). In the clinic this is not always possible. In our experiments hyperthermia was applied immediately after radiation treatment.

Despite the clinical goal to reach (cytotoxic) temperatures as high as 43 °C, tumour temperature distributions are in practice heterogeneous. In large areas of the tumour temperatures are often lower than 43°C. Nonetheless, good results have been obtained in locally advanced cervical cancers with tumour temperatures below 43 °C (van der Zee et al., 2000). Mild temperatures have more subtle effects than high temperatures, such as tumour-reoxygenation (Dewhirst et al., 2005; Bergs et al., 2007abc). Recently it has been shown that hyperthermia (42 °C for 1h) transiently breaks down the BRCA2 protein (Krawczyk et al., 2011). In this paragraph the effects of hyperthermia treatment for 1h at 41 or 43 °C on the linear quadratic parameters are summarized. Several different cell types have been studied.

.1 Effect of hyperthermia treatment on radiosensitivity of RKO cells

ʰe RKO cells, derived from human colon cancer, are relatively sensitive to hyperthermia ʳeatment. Hyperthermia treatment for 1h at 43°C decreases the relative survival to less than ₀.01 and combination with radiation doses in excess of 5 Gy always resulted in a situation in ᵥhich no colony formation was observed. Treatment of cells with 41°C hyperthermia (1h) ᵈlone had little effect and resulted in a surviving fraction of 0.8 ± 0.4 in immediately plated ᵢp) cells and of 0.9 ± 0.1 in delayed plated (dp) cells. When cells were treated at 41°C for 1h ₘmediately prior to irradiation, a significant (p < 0.001) enhancement of cellular ᵃdiosensitivity was observed both in ip (figure 1A) and dp (figure 1B) cells.

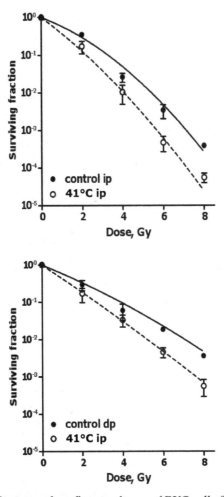

Fig. 1. Radiation survival curves of confluent cultures of RKO cells (human colon cancer ɔells) plated immediately after irradiation, ip (top) or 24h after irradiation, dp (below) with ɔr without hyperthermia pre-treatment at 41°C for 1h. Means with standard errors of at least ₜhree experiments are shown

The effects of hyperthermia on the LQ parameters are summarized in table 1. The value of the linear parameter α increased by a factor 1.7-1.8 while the value of the β parameter ever increased with a factor as high as 2.5-7.0. One must bear in mind that the quadratic component in this cell line is very small and small changes can have a large effect on the numerical values of β.

2.2 Effect of hyperthermia treatment on radiosensitivity of SW-1573 cells

SW-1573 cells are derived from a human lung tumour and are much less sensitive to hyperthermia treatment than RKO cells. Studies were performed to evaluate whether pretreatment with hyperthermia at 41°C or at 43°C in SW-1573 cells was able to enhance the radiosensitivity of these cells. Hyperthermia treatment at 41°C for 1h without radiation did not result in a decrease of the surviving fraction for ip and dp cells as compared to radiation alone. One hour hyperthermia treatment at 43 °C decreased survival to 0.5 ± 0.1 for ip and to 0.4 ± 0.2 for dp cells. Pre-treatment of cells at 41°C for 1h did not alter cellular radiosensitivity of both ip and dp cells (figure 2A). However, 1h treatment at 43°C resulted in a significant (p < 0.001) radiation enhancement both in ip and dp cells (figure 2B). In table I the values of the linear-quadratic parameters for radiation alone and for combined treatments are given. Hyperthermia treatment for 1 h at 41°C did result in an increase of the value of β by a factor 1.3-1.8 while the value of α even decreased. Hyperthermia treatment for 1 h at 43°C result in an increase of the value of α by a factor 2.3-4.4 while the value β increased with a factor 1.8-2.0.

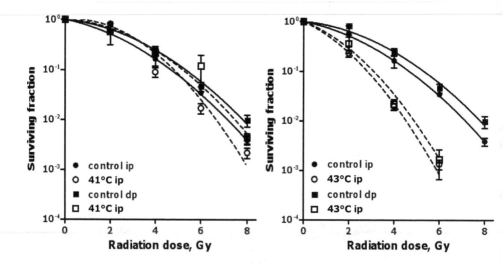

Fig. 2. Radiation survival curves of confluent cultures of SW-1573 cells (human lung tumour cells) plated immediately after irradiation (ip) or 24h after irradiation (dp) with or without hyperthermia pre-treatment at 41°C (left) or at 43°C (right) for 1h. Means with standard errors of at least three experiments are shown.

Radiosensitization with Hyperthermia and Chemotherapeutic Agents: Effects on Linear-Quadratic Parameters
of Radiation Cell Survival Curves

143

Cells	Treatment	α (Gy⁻¹)	β (Gy⁻²)	α/β	α-enhanc factor	β-enhanc factor
RKO ip	sham	0.55 ± 0.09	0.02 ± 0.01	27.5 ± 14.1		
	HT 41 1h	$0.93 \pm 0.09^*$	0.05 ± 0.02	18.6 ± 7.7	1.7 ± 0.3	2.5 ± 1.6
RKO dp	sham	0.47 ± 0.09	0.01 ± 0.01	47.0 ± 47.6^1		
	HT 41 1h	$0.83 \pm 0.08^*$	0.07 ± 0.02	11.9 ± 3.6	1.8 ± 0.4	7.0 ± 7.3
SW1573 ip	sham	0.21 ± 0.02	0.06 ± 0.02	3.5 ± 1.2		
	HT 41 1h	0.06 ± 0.02	0.11 ± 0.03	0.6 ± 0.2	0.3 ± 0.1	1.8 ± 0.8
	HT 43 1h	$0.49 \pm 0.04^*$	0.12 ± 0.03	4.1 ± 1.1	2.3 ± 0.3	2.0 ± 0.8
SW1573 dp	sham	0.09 ± 0.02	0.06 ± 0.02	1.5 ± 1.6		
	HT 41 1h	0.05 ± 0.02	0.08 ± 0.02	0.6 ± 0.6	0.6 ± 0.3	1.3 ± 0.6
	HT 43 1h	$0.40 \pm 0.04^*$	0.11 ± 0.03	3.6 ± 1.1	4.4 ± 1.1	1.8 ± 0.8

Sham= is radiation only; ip=immediately plated; dp=delayed plated. * Significant from sham p<0.05.
[1]The α/β has a large variation because of the high uncertainty of the value of β.

Table 1. Values of the linear-quadratic parameters α and β, α/β and enhancement factors from cells treated with ionizing radiation only and after combined radiation and hyperthermia treatment

3. Cisplatin

Cisplatin is a widely used anti-cancer drug, often combined with radiotherapy (Gorodetsky et al., 1998). Chemo-radiation application based on cisplatin has now become the standard treatment for, among others, locally advanced cervical carcinoma (Duenas-Gonzalez et al., 2003) and locally advanced non-small cell lung cancer (NSCLC) (Loprevite et al., 2001). There have been many studies on the radiation sensitizing effect of cisplatin, but results vary from a clear cisplatin-induced radiosensitization (Begg et al., 1986; Nakamoto et al., 1996; Bergs et al., 2006, 2007ab) to only an additive effect on cell survival (Fehlauer et al., 2000). Cisplatin and radiation have in common that their cellular target is DNA (Rabik & Dolan, 2007).

Cisplatin causes DNA damage by the formation of inter- and intrastrand adducts (Crul et al., 2002). The cisplatin-DNA adducts can cause cell cycle arrest, inhibition of DNA replication and transcription, and eventually apoptosis (Myint et al., 2002). Repair inhibition of DNA has also been implicated (Lawrence et al., 2003) The most important repair pathways reported to be involved in cisplatin-induced DNA damage repair are nucleotide excision repair (NER) and/or homologous recombination (HR) (Haveman et al., 2004; De Silva et al., 2002). An additional route for the repair of cisplatin-DNA interstrand adducts is the post-replication/translation repair pathway which helps the cell to tolerate or bypass the lesion (Dronkert & Kanaar, 2001). Irradiation causes repairable (potentially lethal) and non-repairable (lethal) lesions to the DNA which are induced independently. The ultimate effect of the repairable lesions depends on competing processes of repair and misrepair. The repair of the potentially lethal damage (PLDR) is reflected by the difference in survival between immediately and delayed plated cells. Inhibition of PLDR is implicated to play a role in cisplatin-induced radiation sensitization (Bergs et al., 2006). More specifically, cisplatin-induced radiation sensitization has been shown to occur through inhibition of the non-homologous end joining (NHEJ) pathway and recombinational repair (Myint et al., 2002; Haveman et al., 2004; Dolling et al., 1999).

In this paragraph the radiation sensitization of cisplatin on the lung tumour cell line SW1573 and the cervical tumour cell line Siha is described as changes in linear and quadratic parameters. In figure 3 survival curves are shown for SW1573 lung tumour cells after radiation alone and after radiation combined with cisplatin treatment (1 µM for 1 h). Cisplatin was added to the cultures just before radiation. The survival curves are obtained directly (ip) and 24 h after (dp) treatment to determine potentially lethal damage repair. A slight, but statistically significant effect of cisplatin on the radiosensitivity was only

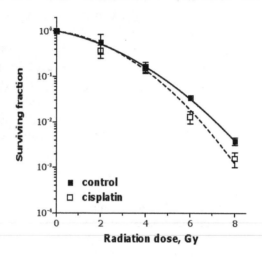

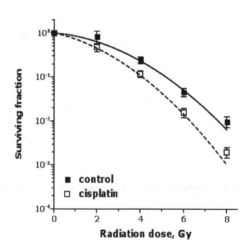

Fig. 3. Radiation survival curves of confluent cultures of SW-1573 cells plated immediately after irradiation, ip (top) or 24 h after irradiation, dp (below) with or without 1µM cisplatin treatment for one hour. Means with standard errors of at least three experiments are shown.

Radiosensitization with Hyperthermia and Chemotherapeutic Agents: Effects on Linear-Quadratic Parameters
of Radiation Cell Survival Curves

145

observed in delayed plated cells (p = 0.02). This was also described by an increase in the α- and β-value (table 2). Only for the delayed plated cells an increase with a factor of 2.5 for the value of α was obtained by cisplatin treatment. For both plating conditions an increase with a factor of 1.2 was obtained for the value of β. In the table 2 also the effects on the linear and quadratic parameters of different plating conditions are presented as well as a 1 h incubation with 1 or 5 μM cisplatin and a continuous incubation with cisplatin during the complete duration of the clonogenic assay. It is obvious that the cervical tumour cells Siha are more radiosensitzed with 1 μM continous cisplatin incubation than the SW1573 lung tumour cells.

Cells	Treatment	α (Gy^{-1})	β (Gy^{-2})	α/β	α-enhanc factor	β-enhanc factor
SW1573 ip	sham	0.21 ± 0.09	0.061 ± 0.016	3.4 ± 1.7		
	1 μM cisplatin (1h)	0.21 ± 0.08	0.072 ± 0.018	2.9 ± 1.3	1.0 ± 0.6	1.2 ± 0.4
SW1573 dp	sham	0.10 ± 0.09	0.063 ± 0.016	1.6 ± 1.5		
	1 μM cisplatin (1h)	0.25 ± 0.09*	0.077 ± 0.017	3.3 ± 1.4	2.5 ± 2.4	1.2 ± 0.4
SW1573 ppi	sham	0.37 ± 0.12	0.014± 0.034	26.4 ± 64.8[1]		
	1 μM cisplatin (cont)	0.41 ± 0.08	0.019 ± 0.025	21.6 ± 28.7[1]	1.1 ± 0.4	1.4 ± 3.8
	5 μM cisplatin (cont)	0.58 ± 0.20*	0.030 ± 0.008*	19.3 ± 8.4	1.6 ± 0.7	2.1 ± 5.2
Siha ppi	sham	0.41 ± 0.04	0.01 ± 0.01	41.0 ± 41.2[1]		
	1 μM cisplatin (cont)	0.81 ± 0.12*	0.02 ± 0.02	40.5 ± 41.0[1]	2.0 ± 0.4	2.0 ± 2.8

Sham is radiation only; ip=immediately plated; dp=delayed plated; ppi=plated prior to irradiation. * Significant from sham p<0.05. α/β values show that with SW1573 cells the quadratic term is affected more than the linear term, while with Siha cells only the linear term is significantly increased. [1] The α/β has a large variation because of the high uncertainty of the value of β.

Table 2. Values of the linear-quadratic parameters α and β and enhancement factors from SW1573 and Siha cells treated with ionizing radiation only and after combined radiation cisplatin (1 μM for 1h; 1 μM continuously; 5 μM continuously) treatment

4. Gemcitabine

Gemcitabine (dFdC, Difluorodeoxycytidine) is a deoxycytidine analogue with clinical activity in non-small cell lung cancer (NSCLC) and pancreatic cancer (Fosella et al., 1997; Manegold et al., 2000; Castro Kreder et al., 2004). It requires phosphorylation to its active metabolites, gemcitabine-diphosphate (dF-dCTP) and gemcitabine-triphosphate (dF-dCTP), with the initial phophorylation by deoxycytidine kinase (dCK) being the rate limiting step (Heinemann et al., 1992; Shewach et al., 1994). The dF-dCTP inhibits ribonucleotide reductase which regulates the production of deoxynucleotides necessary for DNA synthesis and repair (Plukett et al., 1995). The depletion of the deoxynucleotides leads to an increased incorporation of the dF-dCTP into DNA, blocking DNA synthesis (masked chain termination). After incorporation of the dF-dCTP into the DNA an increase in the number of DNA single-strand breaks, chromosome breaks and micronuclei has been observed (Auer et al., 1997).

Both in vitro and in vivo studies have shown that gemcitabine is a potent radiosensitizer (Shewach et al., 1994; Rockwell & Grindey, 1992; Lawrence et al., 1996; Latz et al., 1998; Gregoire et al. 1998; Milas et al., 1999; van Putten et al., 2001, Wachters et al., 2003, Castro Kreder et al., 2003). However, in an early study in non-small cell lung cancer patients, concurrent gemcitabine and radiotherapy resulted in unacceptable pulmonary toxicity related to the large volume of radiation delivered to the lung (Scalliet et al., 1998). More recent ongoing phase I trials show that concurrent gemcitabine at lower doses and radiotherapy is feasible without severe pulmonary toxicity (Manegold et al., 2000; Blackstock et al., 2001). Its unique mechanism of action, its lack of overlapping toxicity and its favourable toxicity profile define gemcitabine as an ideal candidate for combination therapy (Manegold et al., 2000). Currently many randomized studies are ongoing in which gemcitabine is combined with radiotherapy.

Gemcitabine radiosensitization is studied in a gemcitabine sensitive and resistant human lung tumour cells, SWp and SWg, resp., and in gemcitabine sensitive and resistant human ovarian tumour cells, A2780 and AG6000, resp. (Van Bree et al., 2002; Bergman et al., 2000). Gemcitabine was given 24 h before radiation treatment. The SWp is in fact similar to the SW1573 cell line which has been described above. It is called here SWp to distinguish it from SWg, the gemcitabine resistant counterpart which has been developed by van Bree et al. (2002). The lung tumour cells have different sensitivities to radiation alone as compared to the ovarian cancer cells (Van Bree et al., 2002; Bergman et al., 2000).

In table 3 the linear and quadratic parameters of the different cell lines obtained after analyses of the radiation dose survival curves for radiation alone and after combined radiation and gemcitabine treatment are summarized. SWp and SWg were almost equally sensitive to ionizing radiation alone with respect to the low dose region described by the α-value of the linear quadratic formula (Table 3). A slight increase in survival was observed in SWg cells in the high dose region which was reflected by a slightly lower β-value of the linear-quadratic formula (0.040 ± 0.006 vs 0.055 ± 0.008). The human ovarian carcinoma cell line A2780 and its gemcitabine-resistant variant AG6000 were equally sensitive to ionizing radiation. The surviving fractions of the different cell lines after incubation with gemcitabine alone are: SWp 10 nM: 0.52 ± 0.06; SWg 10 µM: 0.95 ± 0.03, 100 µM: 0.24 ± 0.11; A2780 2 nM: 0.82 ± 0.08, 10 nM: 0.21 ± 0.08; AG6000 20 µM: 0.62 ± 0.07, 50 µM: 0.22 ± 0.04.

As can be observed in figure 4 and table 3 radiosensitization is observed with gemcitabine-sensitive as well as in gemcitabine resistant cells. For the resistant cells much higher gemcitabine doses are needed for the radiation sensitization to result in similar cytotoxicity. Both gemcitabine-sensitive cell lines SWp and A2780 are sensitized by incubation with 10 nM of gemcitabine for 24 h before irradiation while the SWg and AG6000 are not sensitized with this dose of gemcitabine. The sensitization is described by an increase in the α-values with factors of 3 and 1.4 respectively, whereas the β-values are not significantly altered. Higher concentrations of Gemcitabine (50 and 100 nM resp.) are required to sensitize gemcitabine-resistant AG6000 and SWg cells to irradiation. For the SWg cells, the radiosensitization was reflected by an increase by a factor of 2.25 in the value of β, whereas in the AG6000 only the α-value was increased by factor of 1.3.

Radiosensitization with Hyperthermia and Chemotherapeutic Agents: Effects on Linear-Quadratic Parameters
of Radiation Cell Survival Curves

147

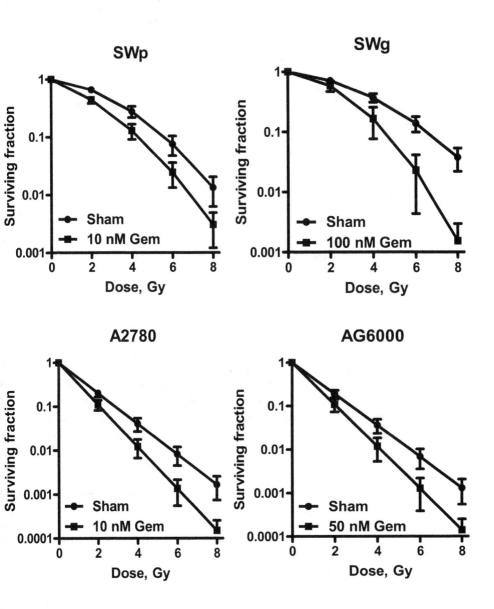

Fig. 4. Radiation sensitization after 24 h incubation with different concentrations of gemcitabine in gemcitabine sensitive SWp and resistant SWg lung tumour cells and in gemcitabine sensitive A2780 and resistant AG6000 ovary cancer cells. Surviving fractions are corrected for gemcitabine toxicity alone (for values see text). Cells are plated immediately after irradiation. Means with SEM of at least three separate experiments are shown.

Cells	Treatment	α (Gy^{-1})	β (Gy^{-2})	α/β	α-enhanc factor	β-enhanc factor
SWp	sham	0.10 ± 0.03	0.055 ± 0.008	1.8 ± 0.6		
	10nM gemcitabine	0.30 ± 0.06*	0.053 ± 0.007	5.7 ± 1.4	3.0 ± 2.8	0.96 ± 0.2
SWg	sham	0.09 ± 0.02	0.040 ± 0.006	2.3 ± 0.6		
	100 μM gemcitabine	0.09 ± 0.03	0.090 ± 0.041†	1.0 ± 0.6	1.0 ± 0.5	2.3 ± 1.1
A2780	sham	0.80 ± 0.10	na			
	10nM gemcitabine	1.10 ± 0.15*	na		1.4 ± 0.3	
AG6000	sham	0.83 ± 0.13	na			
	50μM gemcitabine	1.11 ± 0.20†	na		1.3 ± 0.3	

significant difference with *P<0.01, †P<0.05, na is not applicable

Table 3. Values of the linear-quadratic parameters α and β and enhancement factors from cells treated with ionizing radiation only and gemcitabine-sensitized radiation dose survival curves of gemcitabine-sensitive (SWp and A2780) and gemcitabine-resistant (SWg and AG6000) cells.

5. Halogenated pyrimidines

Incorporation of halogenated pyrimidines (HPs), chloro-, bromo- and iodo-deoxyuridine (CldUrd, BrdUrd, IdUrd) into DNA is known to sensitise cells to ionizing radiation (Franken et al. 1997ab, 1999ab; van Bree et al. 1997; Iliakis et al. 1999; Miller et al. 1992ab). The induced radiosensitisation increases with the degree of thymidine-replacement. The mechanism of radiation sensitisation by the HPs has been suggested to be either an increase in the amount of DNA damage induced by radiation, an influence on repair of sublethal damage (SLD), and/or an enhanced expression of potentially lethal damage (PLD) (Jones et al. 1995; Franken et al. 1997). Since different processes are involved in these phenomena several mechanisms might contribute to the radiosensitisation.

HPs have been suggested to provide an advantage in radiotherapy as radiosensitisers of cells in rapidly growing tumours, in particular in clinical conditions in which critical normal tissues show limited proliferation and as a consequence take up less HP. Labelling depends on the growth fraction, cell loss, cell cycle time and potential doubling time. Of special importance for sensitisation is the rate at which non-cycling cells are recruited into the proliferative compartment during exposure to HPs and a course of radiotherapy. However, even in rapidly growing tumours, cells may, after proliferative cycles, move into a non-proliferative stage. This might compromise the degree of radiation sensitisation if resting cells are less affected by HPs, or are better able to cope with additional damage by repair of PLD.

Here the results of radiosensitization after incubation with 4 μM IdUrd for 72 h are presented. IdUrd-induced radio sensitisation was obtained in all studied cell lines, SW1573, RUCII (Rat urether carcinoma), R1 (Rat rhabdomyosarcoma) and V79, in exponentially growing and in plateau-phase cells. Values of α and β derived by linear-quadratic analyses

Radiosensitization with Hyperthermia and Chemotherapeutic Agents: Effects on Linear-Quadratic Parameters of Radiation Cell Survival Curves

149

of survival curves of exponentially growing cells and plateau-phase cells are presented in table 4. Survival curves of SW cells and V79 cells are given in figure 5. The plating conditions of the V79 cells, i.e. exponentially growing cells plating before or after irradiation (ppi or dp resp.), and plateau phase cells plated immediately or 6-24 h delayed after irradiation (ip or dp resp.) had no influence on the factor of increase of the α-value. It is shown that the value of the linear parameter, α can be enhanced by a factor of 1.9 to 7.5 and that in general low values of α are enhanced more than higher values of α. The value of β is less enhanced and the enhancement factor ranges from 0.7 to 2.4.

The direct comparison between immediate and delayed plating of plateau-phase cells and between plateau phase and exponentially growing cells shows significant quantitative differences. The data on the linear and quadratic parameters described here provide various new insights in the interpretation of radiosensitisation of delay plated plateau-phase cells. It is demonstrated that in delay plated HP-sensitized plateau phase cells PLD is not abolished.

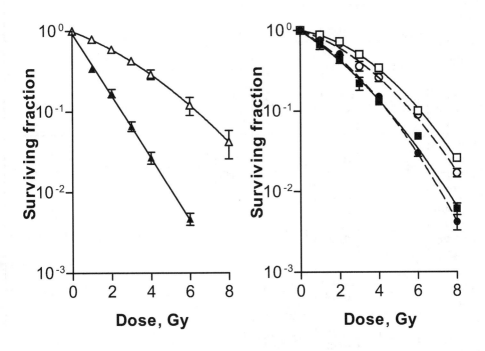

Fig. 5. Radiation dose-survival curves of exponentially growing SW-1573 cells (left) without IdUrd (open triangles) and after incubation with 4 μM IdUrd (closed triangles) and plateau-phase cells (right) plated immediately after irradiation (dashed lines) and plated 24 h after irradiation (solid lines) without IdUrd (open symbols) and after incubation with 4 μM IdUrd (closed symbols). Each point represents the mean value of 3 different experiments ± sem.

Cell line	α (Gy^{-1}) control	β (Gy^{-2}) control	α (Gy^{-1}) IdUrd-sens	β (Gy^{-2}) IdUrd-sens	α/β control	α/β IdUrd-sens	α-enhanc factor	β-enhanc factor
SW 1573 cells Exp growing ip	0.22 ± 0.01	0.022 ± 0.001	0.83 ± 0.06	na	10.0 ± 0.6	na	3.8 ± 0.3	na
SW 1573 cells Plateau phase ip	0.17 ± 0.03	0.042 ± 0.004	0.31 ± 0.03	0.047 ± 0.005	4.1 ± 0.8	6.6 ± 1.0	1.8 ± 0.4	1.1 ± 0.2
SW 1573 cells Plateau phase dp	0.09 ± 0.02	0.046 ± 0.002	0.37 ± 0.04	0.033 ± 0.006	2.0 ± 0.4	11.2 ± 2.4	4.1 ± 1.0	0.7 ± 0.1
RUCII cells Exp growing ppi	0.008 ± 0.007	0.025 ± 0.001	0.06 ± 0.02	0.026 ± 0.001	0.3 ± 0.3	2.3 ± 0.8	7.5 ± 7.0	1.0 ± 0.1
R1 cells Exp growing ppi	0.23 ± 0.01	0.068 ± 0.003	0.44 ± 0.05	0.075 ± 0.016	3.4 ± 0.2	5.9 ± 1.4	1.9 ± 0.3	1.1 ± 0.2
V79 cells Exp growing ip	0.18 ± 0.02	0.017 ± 0.003	0.38 ± 0.04	0.023 ± 0.007	10.6 ± 2.2	16.5 ± 5.3	2.1 ± 0.3	1.4 ± 0.5
V79 cells Exp growing ppi	0.15 ± 0.02	0.013 ± 0.003	0.29 ± 0.03	0.016 ± 0.004	11.5 ± 3.1	18.1 ± 4.9	1.9 ± 0.3	1.2 ± 0.4
V 79 cells Plateau phase ip	0.09 ± 0.03	0.026 ± 0.004	0.17 ± 0.02	0.062 ± 0.005	3.5 ± 1.3	2.7 ± 0.4	1.9 ± 0.7	2.4 ± 0.4
V 79 cells Plateau phase dp	0.07 ± 0.02	0.020 ± 0.002	0.30 ± 0.03	0.024 ± 0.004	3.5 ± 1.1	12.5 ± 2.4	4.3 ± 1.3	1.2 ± 0.2

Means with SEM of at least three separate experiments are shown. ip=immediately plated after irradiation; dp=delayed plated after irradiation; ppi=plated prior to irradiation; na=not applicable.

Table 4. Values of the linear-quadratic parameters α and β and enhancement factors of several cell lines treated with ionizing radiation only and after sensitization with iododeoxyuridine (incubation with 4 µM IdUrd for 72 h)

6. PARP-1 inhibitors

The effect of the Parp-1 inhibitor NU-1025 on the linear and quadratic parameters was tested in Mouse embryonic fibroblasts. Parp1 also known as Poly (ADP-ribose) polymerase is an enzyme which is involved in the single strand break (SSB) repair of the DNA. The DNA SSB induced by ionizing radiation are mostly repaired by the base excision repair system, BER, wheras the DNA DSB are repaired by non homologous endjoining NHEJ or homologous recombination, HR. Inhibiting Parp-1 activity reduces the single strand break repair (Bouchard, 2003). Besides its role in BER Parp-1 is involved in many nuclear processes like DNA replication, transcription, DSB repair, apoptosis and genome stability (Rouleau et al. 2010; Bouchard et al. 2003, Löser et al. 2010). Recently it was hypothesized that cells deficient in BRCA2 or BRCA1 are particularly sensitive to inhibition of Parp-1 (Rouleau et al. 2010, Krawczyck et al. 2011). During DNA replication SSBs are induced. In the absence of Parp-1 these SSB are transformed in DSB. These DSB are repaired with homologous recombination (HR). Therefore cells deficient in HR (e.g. BRCA1 or BRCA2 tumours) might be sensitive to Parp-1 inhibitors. As Parp-1 is involved in many DNA repair processes, Parp-1 inhibitors might work well as radiosensitizers (Löser et al. 2010). As can be observed in figure 6 a modest sensitization effect by the Parp-1 inhibitor NU-1025 was obtained and increase of the value of α in the repair deficient cell line was larger than in the repair proficient cell line, 1.4 vs 1.2 respectively (table 5). The radiation dose survival curves of these MEF cells did not show a shoulder and therefore the quadratic parameter, β, could not be determined.

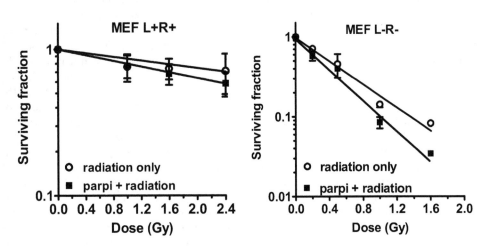

Fig. 6. Radiation dose-survival curves of Mouse embryonic fibroblasts (A: LigIV+/+, Rad54+/+) A and LigaseIV deficient and RAD54 deficient (B: LigIV-/-, Rad54-/-) mouse embryonic fibroblasts. Open circles radiation only curves, closed squares: radiation + Parp inhibitor. Cells were treated with 100 μM Nu-1025 for 24 h before irradiation.

MEF Cells	Treatment with Parp-i	α (Gy^{-1})	β (Gy^{-2})	α -enhanc. factor
LigIV+/+,Rad54+/+	No	0.28 ± 0.01	na	
LigIV+/+,Rad54+/+	yes	0.33 ± 0.03	na	1.2
LigIV-/-,Rad54-/-	no	1.59 ± 0.18	na	
LigIV-/-,Rad54-/-	yes	2.28 ± 0.42	na	1.4

Na= not applicable.

Table 5. Values of the linear parameter, α, and the enhancement factors. The quadratic parameter β could not be determined in these MEF cells.

7. Discussion and conclusion

In most cases an increase of the α-component was observed which corresponds to an enhanced (potentially) direct lethal damage (PLD) at low doses. The β-component, which is assumed to depend on the interaction of sublethal lesions (SLD), was rarely affected by the studied radiosensitization agents. Moreover, it appeared that more radioresistant cell lines were more sensitised than the radiosensitive lines. Furthermore it can be concluded that radiosensitization is also dependent on cell cycle stage like plateau or exponentially growing phase or post treatment plating conditions.

Hyperthermia is an excellent radiosensitizer which can already be effective at mild temperatures. One hour hyperthermia treatment at 41°C without radiation had only a small cytotoxic effect in both the heat sensitive and the heat resistant cell line. This is in agreement with the general idea of cell kill induction at temperatures ≥42°C for 1h or more (Dewhirst, 2005). Hyperthermia treatment at 43°C for 1h did not have a large cytotoxic effect in heat resistant SW-1573 cells. Radiosensitization by 41°C temperature hyperthermia was observed in RKO, but not in SW-1573 cells. The ability of mild temperatures (in the range of 40-42°C) hyperthermia to increase radiosensitivity of human tumor cells has been shown to be cell line dependent (Ruy et al., 1996; Franken et al., 2001; Bergs et al., 2007ab; Larsson & Ng, 2003; Murthy et al., 1977). In a study by Xu et al. (1999) 41.1°C pre-treatment of cells for 1h did not induce radiosensitization whereas treatment for 2h or more resulted in radiosensitization, in the hyperthermia resistant, but not in the hyperthermia sensitive cell line (Xu et al., 1999]. However, simultaneous treatment of the sensitive cell line with 1h 41.1°C hyperthermia and radiation did increase cellular radiosensitivity (Xu et al., 2002). An important mechanism of mild hyperthermia induced radiosensitization *in vivo* is the reoxygenation of tumors by an increase in blood flow (Vujaskovic et al., 2004; Oleson & Robertson, 1995; Song et al., 1995). Recently it was demonstrated that the BRCA-2 protein is transiently inhibited by mild hyperthermia (Krawzcyk et al., 2011). Also translocation of the Mre11 DSB repair protein from the nucleus to the cytoplasm has been implicated (Xu et al., 2002, 2007). However, disappearance of Mre11 protein foci at the sites of irradiation induced DNA double strand breaks by 41°C pre-incubation of cells was not observed (Krawzcyk et al., 2011; Bergs, 2007a). A role for mitotic catastrophe occurring as a result of G2/M checkpoint abrogation has also been suggested (Mackey & Ianzini, 2000). It has been shown that radiosensitization by 41-43°C hyperthermia correlates with an increased number of chromosomal fragments, but not of color junctions, at 24h after treatment compared to radiation alone (Bergs et al., 2008).

It is shown that cisplatin causes radiosensitization as measured by clonogenic survival, but only after allowing a potentially lethal damage repair (PLDR) time of 24 hours. These results are in agreement with those of Wilkins et al. (1993) who investigated the effect of cisplatin and radiation on PLDR in confluent cultures of two different brain tumor cell lines. Wilkins et al. (1993) also observed no radiosensitization by cisplatin in immediately plated cells whereas a cisplatin-induced radiosensitization was seen in cells plated eight hours after irradiation. Their results indicate that the radiosensitizing effect of cisplatin occurs through the inhibition of post-irradiation recovery. The strongest inhibition of PLDR was achieved when cisplatin was administered shortly before or after irradiation (Wilkins et al., 1993). In our experiments, cells were irradiated while cisplatin was present in the medium. Results from studies using exponentially growing cell cultures vary from a cisplatin-induced radiosensitization (Loprevite et al., 2001; Begg et al. 1986; Nakamoto et al. 1996; Huang et al. 2004) to only an additive effect (Gorodetsky et al., 1998; Loprevite et al., 2001; Britten et al., 1996; Monk et al., 2002). The effect of cisplatin treatment on radiosensitivity may depend on the cell type used. Loprevite et al (2001) observed synergism in a squamous lung carcinoma cell line when exposed to cisplatin, whereas an adenocarcinoma of the lung was not sensitized by cisplatin. Even cell lines derived from a single biopsy can differ in the response to cisplatin and radiation combination therapy (Britten et al., 1996). Although dependence on cell cycle phase (Meyn et al., 1980; Krishnaswamy & Dewey, 1993), cisplatin incubation time and the sequence of treatment modalities have been implicated (Gorodetsky et al., 1998; Meyn et al., 1980; Krishnaswamy & Dewey, 1993), there is currently no consensus to account for the varying response of cells to cisplatin and radiation.

The mechanism of cisplatin induced radiosensitization might be due to the inhibition of the DNA repair, NHEJ and HR, pathways (Myint et al., 2002; Dolling et al., 1998). The Ku protein complex, which plays an important role in NHEJ, was demonstrated to show a reduced ability to translocate on DNA containing cisplatin-DNA adducts compared with undamaged DNA. This resulted in a decreased interaction between Ku and DNA-PK$_{cs}$ (Turchi et al., 2000) However, the biochemical processes that cisplatin undergoes in the cell are complex and the intracellular fate of cisplatin may be linked to copper transport (Muffia & Fojo, 2004). Therefore, other processes such as the formation of peroxy complexes inside the cell might be involved in cisplatin-induced radiosensitization (Dewit, 1987). Bergs et al. (2006) demonstrated an increase in the induction of apoptosis after combined treatment as compared to radiation or cisplatin alone at 24 h after treatment. This was confirmed in several other studies (Kumala et al., 2003; Guchelaar et al. (1998). These apoptotic effects observed by Bergs et al. (2006) correlated with clonogenic survival. Fujita et al. (2000) also observed an inhibitory effect of the combination of cisplatin and radiation on the survival of lung tumor cells and ascribed this effect on the induction of tumor cell apoptosis.

In conclusion, a radiosensitizing effect of cisplatin on cell survival is observed in confluent cultures when cells were replated after a 24 hour incubation period during which PLD repair could take place. In contrast, cisplatin did not induce a significant radiosensitization after immediate plating.

Several studies have shown that gemcitabine is a potent sensitizer of ionizing radiation (Shewach et al., 1994; Gregoire et al., 1999; Ostruszka & Shewach, 2000). Among other proposed mechanisms of action, the effect of gemcitabine on cell cycle distributions may be the most important (Milas et al., 1999; Van Putten et al., 2001). In our studies, both gemcitabine-sensitive cell lines SWp and A2780 could be sensitized to irradiation when cytotoxic gemcitabine-treatments were given. The radiosensitization was accompanied by a clear arrest of cells in early S phase which has been argued to be vital for gemcitabine-induced radiosensitization (Latz et al., 1998). Both cell lines showed an increase in α-value indicating the efficacy of gemcitabine-induced radiosensitization in the clinically relevant dose range. Although the gemcitabine resistant cells still could be sensitized only much higher gemcitabine doses were necessary to reach an effect. In the resistant ovarian carcinoma cell line AG6000 this was demonstrated by an increase in the value of α. In contrast with this change, in the gemcitabine resistant lung tumour cell line an increase in the β-value was obtained, the α-value was not affected. In both gemcitabine-resistant cell lines the sensitivity to ionizing radiation alone was not altered. It is reported that gemcitabine resistant tumours are cross-resistant to related drugs like Ara-c (Ruiz van haperen et al., 1994; Peters et al., 1996). In both gemcitabine-resistant cell line, AG6000 and SWg, this was indeed the case (Van Bree et al., 2002). Moreover, the AG6000 cells were also more resistant to cisplatin and taxoids (Bergman et al., 2000). However, no altered sensitivity was found in SWg cells for cDDP, paclitaxel, MTX and 5 FU, while AG6000 cells were 2.5-fold more sensitive to MTX (Bergman et al., 2000). These findings indicate that patients previously treated with gemcitabine may receive additional radiotherapy with or without cDDP or paclitaxel.

The HP-induced-radiosensitisation is mainly due to an increase in the linear parameter α. The quadratic parameter, β, is only rarely influenced. Different mechanisms involved in the radiosensitisation induced by halogenated pyrimidines have been described. Wang et al. (1994) suggested that in exponentially growing cells increased DNA damage production was the major component of radiosensitisation while in plateau-phase cells radiosensitisation occurred through inhibited repair and/or enhanced fixation of potentially lethal damage. The increase of the α values for exponentially growing cells as found in our study, indicates an increase in the number of directly lethal events due to the HPs. This is in agreement with observations of Webb et al. (1993) and Jones et al. (1995) which suggest that an important mechanism of radiosensitisation involves an increase of effective DNA double strand breaks. Miller et al. (1992ab) have suggested that radiation-induced damage in cells which have HPs incorporated into the DNA after low-LET irradiation resembles the damage produced by high-LET radiation. In plateau-phase cells plated immediately after irradiation the increase of α might be due to the same mechanism as involved in exponentially growing cells. In these cells also an increase of β was observed indicating that accumulation of sublethal lesions contributed significantly (Barendsen 1990). Due to the immediate plating after irradiation this sublethal damage might be fixated.

Greatest increases in α were found in delayed plated plateau-phase cells. This radiosensitisation can be interpreted as an enhanced fixation of potentially lethal damage due to immediate DNA damage and/or to damaged DNA repair function in these cells expressed during the interval before delayed plating. The value of β in these cells returned to values as found in cells not containing HPs. This demonstrates that sublethal damage has been repaired in HP-containing plateau-phase cells.

Because Parp-1 is implicated in several DNA repair processes, Parp-1 inhibitors might be good radiosensitizers. Several studies have already demonstrated the radiosensitizing effect of Parp-1 inhibitors (Albert et al. 2007; Löser et al. 2010; Krawczyk et al. 2011). Loser et al concluded that the effects of Parp-1 inhibitors are more pronounced on rapidly dividing

Radiosensitization with Hyperthermia and Chemotherapeutic Agents: Effects on Linear-Quadratic Parameters of Radiation Cell Survival Curves

155

and/or DNA repair deficient cells. In our study at time of treatment most of the cells in culture had accumulated in G1 phase. Therefore radiosensitization effects are modest. However, the increase of value of the linear parameter, α, of the repair deficient cells was more increased after the parp-1 inhibition than of the repair proficient cells.

8. Acknowledgements

We thank Dr. G. Iliakis for sending us (with permission from Dr. F. Alt) the mouse embryonic fibroblast cells. The authors are thankful for financial support from several foundations. The Maurits and Anna de Kock and the Nijbakker Morra foundations are acknowledged for sponsoring laboratory equipment. The Dutch Cancer Foundation (grant # UVA 2006-3484, # UVA 2008-4019) and the Stichting Vanderes are acknowledged for financing personnel support

9. References

Auer, H.; Oehler, R., Lindner, R., Kowalski, H.; Sliutz, G., Orel, L., Kucera, E., Simon, M. M., & Glössl, J. (1997). Characterisation of genotoxic properties of 2',2'-difluorodeoxycytidine. *Mutation Research*, Vol.393, pp. 165-173.

Barendsen, G.W. (1982). Dose fractionation, dose rate and iso-effect relationships for normal tissue responses. *International Journal of Radiation Oncology, Biology and Physics*, Vol.8, pp. 1981-1997.

Barendsen, G.W. (1990). Mechanisms of cell reproductive death and shapes of radiation dose-survival curves of mammalian cells. *International Journal of Radiation Biology*, Vol.57, pp. 885-896.

Barendsen, G.W. (1994). The relationship between RBE and LET for different types of lethal damage in mammalian cells: biophysical and molecular mechanisms. *Radiation Research*, Vol.139, pp. 257-270.

Barendsen, G.W. (1997). Parameters of linear-quadratic radiation dose-effect relationships: dependence on LET and mechanisms of reproductive cell death. *International Journal of Radiation Biology*, Vol.71, pp. 649-655.

Barendsen, G.W.; van Bree, C. & Franken, N.A.P. (2001). Importance of cell proliferative state and potentially lethal damage repair on radiation effectiveness: implications for combined tumor treatments (review). *International Journal of Oncology*, Vol.19, pp257-256.

Begg, A.C.; Van der Kolk, P.J., Dewit, L. & Bartelink, H. (1986). Radiosensitization by cisplatin of RIF1 tumour cells in vitro. *International Journal of Radiation Biology*, Vol.50, pp. 871-884.

Bergman, A.M.; Giaccone, G.; Van Moorsel, C.J.A.; Mauritz, R.; Noordhuis, P.; Pinedo, H.M.; & Peters, G.J. (2000). Cross-resistance in the 2',2'-difluorodeoxycytidine (Gemcitabine)-resistant human ovarian cancer cell line AG6000 to standard and investigational drugs. *European Journal of Cancer*, Vol.36, p. 1974-1983.

Bergman, A.M.; Pinedo, H.M.; Jongsma, A.P.; Brouwer, M.; Ruiz van Haperen, V.W.T.; Veerman, G.; Leyva, A.; Eriksson, S. & Peters, G.J. (1999). Decreased resistance to gemcitabine (2',2'-difluorodeoxycitidine) of cytosine arabinoside-resistant myeloblastic murine and rat leukemia cell lines: role of altered activity and substrate specificity of deoxycytidine kinase. *Biochemical Pharmacology*; Vol.57, pp. 397-406.

Bergs, J.W.J.; Franken, N.A.P., ten Cate, R., Van Bree, C. & Haveman, J. (2006). Effects of cisplatin and gamma-irradiation on cell survival, the induction of chromosomal aberrations and apoptosis in SW-1573 cells. *Mutation Research*, Vol.594, pp. 148-154.

Bergs, J.W.J. (2007a). Hyperthermia, cisplatin and radiation trimodality treatment: In vitro studies on interaction mechanisms. PhD. Thesis, University of Amsterdam.

Bergs, J.W.J.; Haveman, J., ten Cate, R., Medema, J.P., Franken, N.A.P. & Van Bree, C. (2007b). Effect of 41 degrees C and 43 degrees C on cisplatin radiosensitization in two human carcinoma cell lines with different sensitivities for cisplatin. *Oncology Reports*, Vol.18, pp. 219-226.

Bergs, J.W.J.; Franken, N.A.P., Haveman, J., Geijsen, E.D., Crezee, J. & Van Bree, C. (2007c). Hyperthermia, cisplatin and radiation trimodality treatment: a promising cancer treatment? A review from preclinical studies to clinical application. *International Journal of Hyperthermia*, Vol.23, pp. 329-341. Review.

Bergs, J.W.J.; ten Cate, R., Haveman, J., Medema, J.P., Franken, N.A.P. & Van Bree, C. (2008). Chromosome fragments have the potential to predict hyperthermia-induced radio-sensitization in two different human tumor cell lines. *Journal of Radiation Research*, Vol.49, pp. 465-472.

Blackstock, A.W.; Lesser, G.J., Fletcher-Steede, J., Case, L.D.. Tucker, R.W., Russo, S.M., White. D.R. & Miller, A. (2001). Phase I Study of twice-weekly gemcitabine and concurrent thoracic radiation for patients with locally advanced son-small cell lung cancer. *International Journal of Radiation Oncology Biology Physics*, Vol.51, pp. 1281-1289.

Bouchard, V.J.; Rouleau, M. & Poirier, G.G. (2003). Parp1, a determinat of cell survival in response to DNA damage. *Experimental Hematology*, Vol.31, pp. 446-454.

Britten, R.A.; Peacock, J. & Warenius, H.M. (1992). Collateral resistance to photon and neutron irradiation is associated with acquired cis-platinum resistance in human ovarian tumour cells. *Radiotherapy and Oncology*, Vol.23, pp. 170-175.

Britten, R.A.;.Evans, A.J, Allalunis-Turner, M.J. & Pearcey, R.G. (1996). Effect of cisplatin on the clinically relevant radiosensitivity of human cervical carcinoma cell lines. *International Journal of Radiation Oncology, Biology, Physics*, Vol.34, pp. 367-374.

Castro Kreder, N.; Van Bree, C., Franken, N.A.P. & Haveman, J. (2003). Colour junctions as predictors of radiosensitivity: X-irradiation combined with gemcitabine in a lung carcinoma cell line. *Journal of Cancer Research and Clinical Oncology*, Vol. 129, pp. 597-603.

Castro Kreder, N.; Van Bree, C., Franken, N.A.P. & Havenman, J. (2004). Effects of gemcitabine on cell survival and chromosome aberrations after pulsed low dose-rate irradiation. *Journal of Radiation Research*, Vol.45, pp. 111-118.

Crezee J.; Barendsen, G.W., Westermann, A.M., Hulshof, M.C.C.M., Haveman, J., Stalpers, L.J.A., Geijsen, E.D. & Franken, N.A.P. (2009). Quantification of the contribution of hyperthermia to results of cervical cancer trials: In regard to Plataniotis an Dale (Int J Radiat Oncol Biol Phys 2009;73:1538-1544). *International Journal of Radiation Oncology, Biology, Physics*, Vol.75, p. 634.

Crul, M.; Van Waardenburg, R.C., Beijnen, J.H. & Schellens, J.H. (2002). DNA-based drug interactions of cisplatin. *Cancer treatment reviews*, 2002;28:291-303

De Silva, I.U.; McHugh, P.J., Clingen, P.H. & Hartley, J.A. (2002). Defects in interstrand cross-link uncoupling do not account for the extreme sensitivity of ERCC1 and XPF cells to cisplatin. *Nucleic Acids Research,* Vol.30, pp. 3848-3856.

Dewey, W.C.; Sapareto, S.A. & Betten, D.A. (1978). Hyperthermic radiosensitization of synchronous Chinese hamster cells: relationship between lethality and chromosomal aberrations. *Radiation research,* Vol.76, pp.48-59.

Dewhirst, M.W.; Vujaskovic, Z., Jones, E. & Thrall, D. (2005). Re-setting the biologic rationale for thermal therapy. *International Journal of Hyperthermia,* Vol.21, pp. 779-790.

Dewit, L. (1987). Combined treatment of radiation and cisdiamminedichloroplatinum (II): a review of experimental and clinical data. *International Journal of Radiation Oncology, Biology, Physics,* Vol.13, pp. 403-426.

Dolling, J.A., Boreham, D.R., Brown, D.L., Raaphorst, G.P. & Mitchel, R.E. (1999). Cisplatin-modification of DNA repair and ionizing radiation lethality in yeast, Saccharomyces cerevisiae. *Mutation Research* Vol.433, pp. 127-36.

Dronkert, M.L. & Kanaar, R. (2001). Repair of DNA interstrand cross-links, *Mutation Research* Vol.486, pp. 217-247.

Duenas-Gonzalez, A.; Cetina, L. & de la Mariscal, I.G.J. (2003). Modern management of locally advanced cervical carcinoma. *Cancer Treatment Reviews* Vol.29, pp. 389-399.

Fehlauer, F.; Barten-Van Rijbroek, A.D., Stalpers, L.J., Leenstra, S., Lindeman, J., Tjahja, I., Troost, D., Wolbers, J.G., van der Valk, P. & Sminia. P. (2000). Additive cytotoxic effect of cisplatin and X-irradiation on human glioma cell cultures derived from biopsy-tissue. *Journal of Cancer Research and Clinical Oncology,* Vol.126, pp. 711-716.

Fossella, F.V.; Lipmann, S.C., Shin DM, *et al.* (1997). Maximum-tolerated dose defined for single-agent gemcitabine: a Phase I dose-escalation study in chemotherapy-naïve patients with advanced non-small-cell lung cancer. *Journal of Clinical Oncology,* Vol.15, pp. 310-316.

Franken, N.A.P., Van Bree, C., Kipp, J.B.A. & Barendsen, G.W., (1997a). Modification of potentially lethal damage in irradiated chinese hamster V79 cells after incorporation of halogenated pyrimidines. *International Journal of Radiation Biology,* Vol.72, 101-109.

Franken, N.A.P.; Van Bree, C., Streefkerk, J.O., Kuper, I.M.J.A., Rodermond, H.M., Kipp, J.B.A. and Barendsen G.W. (1997b). Radiosensitization by iodo-deoxyuridine in cultured SW-1573 human lung tumor cells: Effects on α and β of the linear-quadratic model. *Oncology Reports,* Vol.4, pp. 1073-1076.

Franken, N.A.P.; Ruurs, P., Ludwików, G., Van Bree, C., Kipp, J.B.A., Darroudi, F. & Barendsen, G.W. (1999a). Correlation between cell reproductive death and chromosome aberrations assessed by FISH for low and high doses of radiation and sensitization by iododeoxyuridine in human SW-1573 cells. *International Journal of Radiation Biology,* Vol.75, pp. 293-299.

Franken, N.A.P.; van Bree, C., Veltmaat, M.A.T., Ludwików, G., Kipp, J.B.A. & Barendsen, G.W. (1999b) Increased chromosome frequencies in iodo-doxyuridine-sensitized human SW-1573 cells after γ-irradiation. *Oncology Reports* Vol.6, pp. 59-63.

Franken, N.A.P.; Van Bree, C., Veltmaat, M.A., Rodermond, H.M., Haveman, J. & Barendsen, G.W. (2001). Radiosensitization by bromodeoxyuridine and hyperthermia: analysis of linear and quadratic parameters of radiation survival curves of two human tumor cell lines. *Journal of Radiation Research*, Vol.42, pp. 179-190.

Franken N.A.P.; ten Cate, R., Van Bree, C. & Haveman, J. (2004). Induction of the early response protein EGR-1 in human tumour cells after ionizing radiation is correlated with a reduction of repair of lethal lesions and an increase of repair of sublethal lesions. *International Journal of Oncology*, Vol.24, pp. 1027-1031.

Franken, N.A.P.; Rodermond, H.M., Stap, J., Haveman, J. & Van Bree, C. (2006) Clonogenic assay of cells in vitro. *Nature Protocols*, Vol.1, pp. 2315-2319.

Froelich, J.J.; Schneller, F.R. & Zahn, R.K. (1999). The influence of radiation and chemotherapy-related DNA strand breaks on carcinogenesis: an evaluation. *Clinical Chemistry and Laboratory Medicine*, Vol.37, pp. 403-408.

Fujita, M.; Fujita, T., Kodama, T., Tsuchida, T. & Higashino, K. (2000). The inhibitory effect of cisplatin in combination with irradiation on lung tumor cell growth is due to induction of tumor cell apoptosis. *International Journal of Oncology*, Vol.17, pp. 393-397.

Gonzalez Gonzalez, D.; Van Dijk, J.D., & Blank, L.E. (1995). Radiotherapy and hyperthermia. *European Journal of Cancer*, Vol. 31A, pp.1351-1355.

Gorodetsky, R.; Levy-Agababa, F., Mou, X. & Vexler, AM. (1998). Combination of cisplatin and radiation in cell culture: effect of duration of exposure to drug and timing of irradiation. *International journal of Cancer*, Vol.75, pp. 635-642,.

Gregoire, V.; Hittelman, W.N., Rosier, J.F. & Milas, L. (1999). Chemo-radiotherapy: radiosensitizing nucleoside analogues. *Oncology Reports*, Vol.6, pp. 949-957.

Groen, H.J.M.; Sleijfer, S., Meijer, C, *et al.* (1995). Carboplatin- and cisplatin-induced potentiation of moderate dose radiation cytotoxicity in human lung cancer cell lines. *British Journal of Cancer*, Vol.72, pp. 1406-1411.

Guchelaar, H.J.; Vermes, I., Koopmans, R.P., Reutelingsperger, C.P. & Haanen, C. (1998). Apoptosis- and necrosis-inducing potential of cladribine, cytarabine, cisplatin, and 5-fluorouracil in vitro: a quantitative pharmacodynamic model, *Cancer Chemotherapy & Pharmacology*, Vol.42, pp. 77-83.

Hall, E.J. & Giaccia, A.J. (2006). Radiobiology for the Radiobiologist sixth Edition, Lippincott Williams & Wilkins. Chapter 28, Hyperthermia, pp. 469-490.

Harima, Y.; Nagata, K., Harima, K., Ostapenko, V.V., Tanaka, Y. & Sawada, S. (2001). A randomized clinical trial of radiation therapy versus thermoradiotherapy in stage IIIB cervical carcinoma. *International Journal of Hyperthermia*, Vol.17, pp. 97-105.

Hatzis, P.; Al-Madhoon, A.S., Jullig, M., Petrakis, T.G., Eriksson, S. & Talianidis, I. (1998). The intracellular localization of deoxycytidine kinase. *Journal of Biological Chemistry*, Vol.273, pp. 30239-30243.

Haveman, J.; Castro Kreder, N., Rodermond, H.M., Van Bree, C., Franken, N.A.P., Stalpers L.J., Zdzienicka, M.Z., & Peters. G.J. (2004). Cellular response of X-ray sensitive hamster mutant cell lines to gemcitabine, cisplatin and 5-fluorouracil, *Oncology Reports* Vol.12 pp. 187-192.

Haveman, J.; Rietbroek, R.C., Geerdink, A., Rijn van, J. & Bakker, P.J.M. (1995). Effect of hyperthermia on cytotoxicity of 2′,2′-difluorodeoxycytidine (gemcitabine) in cultured SW1573 cells. *International Journal of Cancer*, Vol.62, pp. 627-630

Heinemann, V.; Xu, Y.Z., Chubb S, Sen, A., Hertel, L.W. & Grindey, G.B. & Plunkett, W. (1992). Cellular elimination of 2',2'-difluorodeoxycytidine 5'-triphosphate: a mechanism of self-potentiation. *Cancer Research*, Vol.52, pp. 533-539.

Hildebrandt, B.; Wust, P., Ahlers, O., Dieing, A., Sreenivasa, G., Kerner, T., Felix, R. & Riess, H. (2002). The cellular and molecular basis of hyperthermia. *Critical reviews in oncology/hematology*, Vol.43, pp. 33-56.

Huang, H.; Huang, S.Y., Chen, T.T., Chen, J.C., Chiou, C.L. & Huang, T.M. (2004). Cisplatin restores p53 function and enhances the radiosensitivity in HPV16 E6 containing SiHa cells. *Journal of Cellular Biochemestry*, Vol.91, pp. 756-765.

Iliakis, G., Kurtzman, S., Pantelias, G. & Okayasu, R. (1989). Mechanism of radiosensitisation by halogenated pyrimidines: Effect of BrdU on radiation induction of DNA and chromosome damage and its correlation with cell killing. *Radiation Research*, Vol.119, pp. 286-304.

Iliakis, G., Wang, Y., Pantelias, G.E. & Metzger, L. (1992). Mechanism of radiosensitisation of halogenated pyrimidines. Effect of BrdU on repair of DNA breaks, interphase chromatin breaks and potentially lethal damage in plateau-phase CHO cells. *Radiation Research*, 1Vol.29, pp. 202-211.

Iliakis, G., Wright, E and Ngo, F.Q.H. (1987). Possible importance of PLD repair in the modulation of BrdUrd and IdUrd-mediated radiosensitisation in plateau-phase C3H10T1/2 mouse embryo cells. *International Journal of Radiation Biology*, Vol.51, pp. 541-548.

Iliakis, G.; Pantelias, G & Kurtzman, S., (1991). Mechanism of radiosensitisation by halogenated pyrimidines: Effect of BrdU on cell killing and interphase chromosome breakage in radiation sensitive cells. *Radiation Research*, Vol.25, pp. 56-64.

Joiner, M. & Van der Kogel, A. (2006). Basic Clinical Biology. Fourth Edition.

Jones, G.D.D., Ward, J.F., Limoli,C.L., Moyer, D.J. & Aguilera, J.A. (1995). Mechanisms of radiosensitization in iododeoxyuridine-substituted cells. *International Journal of Radiation Biology*, Vol.76, pp. 647-653.

Kampinga, H.H. & Dikomey, E. (2001). Hyperthermic radiosensitization: mode of action and clinical relevance. *International journal of radiation biology*, Vol.77, pp. 399-408.

Kinsella, T.J., Mitchell, J.R., Russo, A., Morstyn, G. & Glatstein,E. (1984). The use of halogenated thymidine analoge as clinical radiosensitisers: rationale, current status*and* future prospects: nonhypoxic cell sensitisers. *International Journal of Radiation Oncology, Biology and Physics*, Vol.10, pp. 1399-1406.

Krawczyk, P.M., Eppink, B., Essers, J., Stap, J., Rodermond, H.M., Odijk, H., Zelensky, A., Van Bree, C., Stalpers, L.J., Buist, M.R., Soullié, T., Rens, J., Verhagen, H.J. M., O'Connor, M., Franken, N.A.P., ten Hagen, T.L.M., Kanaar, R. & Aten, J.A. (2011) Temperature-controlled induction of BRCA2 degradation and homologous recombination deficiency sensitizes cancer cells to PARP-1 inhibition. *Proceedings of the National Academy of Science*, Vol. 108, pp. 9851-9856.

Krishnaswamy, G. & Dewey, W.C. (1993). Cisplatin induced cell killing and chromosomal aberrations in CHO cells: treated during G1 or S phase. *Mutation Research*, Vol.293, pp. 161-172.

Kumala, S.; Niemiec, P., Widel, M., Hancock, R. & Rzeszowska-Wolny, J. (2003). Apoptosis and clonogenic survival in three tumour cell lines exposed to gamma rays or chemical genotoxic agents. *Cellular and Molecular Biology Letters*, Vol.8, pp. 655-665.

Larsson, C. & Ng, C.E. (2003). p21+/+ (CDKN1A+/+) and p21-/- (CDKN1A-/-) human colorectal carcinoma cells display equivalent amounts of thermal radiosensitization. *Radiation research,* Vol.160, pp. 205-259.

Latz, D.; Fleckenstein, K., Eble, M., Blatter, J., Wannenmacher, M. & Weber, K.J. (1998). Radiosensitizing potential of gemcitabine (2',2'-difluoro-2'-deoxycytidine) within the cell cycle in vitro. *International Journal of Radiation Oncology, Biolology, Physics,* Vol.41, pp. 875-882.

Lawrence, T.S.; Blackstock, A.W. & McGinn, C. (2003). The mechanism of action of radiosensitization of conventional chemotherapeutic agents. *Seminars in radiation oncology,* Vol.13, pp. 13-21.

Lawrence, T.S.; Davis, M.A. & Normolle D.P. (1995). Effect of bromodeoxyuridine on radiation-induced DNA damage and repair based on DNA fragment size using pulsed-field gel electrophoresis. *Radiation Research,* Vol.144, pp. 282-287.

Loprevite, M.; Favoni, R.E., de Cupis, A., Pirani, P., Pietra, G., Bruno, S., Grossi, F., Scolaro, T. & Ardizzoni, A. (2001). Interaction between novel anticancer agents and radiation in non-small cell lung cancer cell lines. *Lung Cancer* Vol.33, pp. 27-39.

Löser, D.A.; Shibata, A., Shibata, A.K., Woodbine, L.J. Jeggo, P.A. & Chalmers, A.J. (2010). Sensitization to radiation and alkylating agents by inhibitors of poly(ADP-ribose) polymerase is enhanced in cells deficient in DNA double strand break repair. *Molecular Cancer Therapeutics,* Vol.9, pp. 1775-1787.

Mackey, M.A. & Ianzini, F. (2000). Enhancement of radiation-induced mitotic catastrophe by moderate hyperthermia. *International Journal of Radiation Biology,* Vol.76, pp. 273-280.

Manegold, C.; Zatloukal, P., Krejcy, K. & Blatter, J. (2000). Gemcitabine in non-small lung cancer (NSCLC). *Invesigationalt New Drugs,* Vol.18, pp. 29-42.

Meyn, R.E.; Meistrich, M.L. & White, R.A. (1980). Cycle-dependent anticancer drug cytotoxicity in mammalian cells synchronized by centrifugal elutriation. *Journal of the National Cancer Institute,* Vol.64, pp. 1215-1219.

Milas, L.; Fujii, T., Hunter, N., Elshaikh, M., Mason, K., Plunkett, W., Ang, K.K. & Hittelman, W. (1999). Enhancement of tumor radioresponse in vivo by gemcitabine. *Cancer Research,* Vol.59, pp. 107-114.

Miller, E.M.; Fowler, J.F. & Kinsella, T.J. (1992a) Linear-quadratic analysis of radiosensitisation by halogenated pyrimidines, I. Radiosensitisation of human colon cancer cells by iododeoxyuridine. *Radiation Research,* Vol.131, pp. 81-89.

Miller, E.M.; Fowler, J.F. & Kinsella, T.J. (1992b). Linear-quadratic analysis of radiosensitisation by halogenated pyrimidines. II. Radiosensitisation of human colon cancer cells by bromodeoxyuridine. *Radiation Research,* Vol.131, pp. 90-97.

Monk, B.J.; Burger. R.A., Parker, R., Radany, E.H., Redpath, L. & Fruehauf, J.P. (2002). Development of an in vitro chemo-radiation response assay for cervical carcinoma. *Gynecologic Oncology,* Vol.87, pp. 193-199.

Muggia, F.M. & Fojo, T. (2004). Platinums: extending their therapeutic spectrum. *Journal of Chemotherapy,* Vol.16 Suppl 4, pp. 77-82.

Munch-Petersen, B.; Cloos, L., Tyrsted, G. & Eriksson, S. (1991). Diverging substrate specificity of pure human thymidine kinases 1 and 2 against antiviral dideoxynucleosides. *Journal of Biological Chemistry,* Vol.266, pp. 9032-9038.

Murthy, A.K.; Harris, J.R. & Belli, J.A. (1977). Hyperthermia and radiation response of plateau phase cells. Potentiation and radiation damage repair. *Radiation Research*, Vol.70, pp. 241-247.

Myerson, R.J.; Roti Roti, J.L., Moros, E.G., Straube, W.L. & Xu, M. (2004). Modelling heat-induced radiosensitization: clinical implications. *International Journal of Hyperthermia*, Vol.20, pp. 201-12.

Myint, W.K., Ng, C. & Raaphorst, P. (2002). Examining the non-homologous repair process following cisplatin and radiation treatments, *International Journal of Radiation Biology* Vol.78, pp. 417-424.

Nakamoto, S.; Mitsuhashi, N., Takahashi, T., Sakurai, H. & Niibe, H. (1996). An interaction of cisplatin and radiation in two rat yolk sac tumour cell lines with different radiosensitivities in vitro. *International journal of radiation biology*, Vol.70, pp. 747-53.

Oleson, J.R. & Robertson, E. (1995). Special Lecture. Hyperthermia from the clinic to the laboratory: a hypothesis. *International Journal of Hyperthermia*, Vol.11, pp. 315-322.

Ostruszka, L.J. & Shewach, D.S. (2000). The role of cell cycle progression in radiosensitization by 2',2'-difluoro-2'-deoxycytidine. *Cancer Research*, Vol.60, pp. 6080-6088.

Peters, G.J.; Ruiz van Haperen, V.W., Bergman, A.M., Veerman, G., Smitskamp-Wilms, E., Van Moorsel, C.J.A., Kuiper, C.M. & Braakhuis, B.J.M. (1996). Preclinical combination therapy with gemcitabine and mechanisms of resistance. *Seminars in Oncology*, Vol.23 (Suppl. 10), pp. 16-24.

Plunkett, W., Huang, P &, Gandhi, V. (1995). Preclinical characteristics of gemcitabine. *Anti-Cancer Drugs*, Vol.6 (Suppl. 6), pp. 7-13.

Raaphorst, G.P.; Azzam, E.I. & Feeley, M. (1988). Potentially lethal radiation damage repair and its inhibition by hyperthermia in normal hamster cells, mouse cells, and transformed mouse cells. *Radiation research*, Vol.13,pp. 171-182.

Raaphorst, G.P.; Feeley, M.M., Danjoux, C.E., DaSilva, V. & Gerig, L.H. (1991). Hyperthermia enhancement of radiation response and inhibition of recovery from radiation damage in human glioma cells. *International Journal of Hyperthermia*, Vol.7, pp. 629-641.

Raaphorst, G.P.; Heller, D.P., Bussey, A. & Ng, C.E. (1994). Thermal radiosensitization by 41 degrees C hyperthermia during low dose-rate irradiation in human normal and tumour cell lines. *International Journal of Hyperthermia*, Vol.10, pp. 263-270.

Raaphorst, G.P.; Wang, G., Stewart, D. & Ng, C.E. (1996). Concomitant low dose-rate irradiation and cisplatin treatment in ovarian carcinoma cell lines sensitive and resistant to cisplatin treatment. *International Journal of Radiation Biology*, Vol.69, pp. 623-631.

Raaphorst, G.P.; Yang, D.P. & Niedbala, G. (2004). Is DNA polymerase beta important in thermal radiosensitization? *International Journal of Hyperthermia*, Vol.20, pp. 140-103.

Rabik, C.A. & Dolan, M.E. (2007). Molecular mechanisms of resistance and toxicity associated with platinating agents. *Cancer treatment reviews*, Vol.33, pp. 9-23.

Rockwell, S. & Grindey, G.B. (1992). Effect of 2',2'-difluorodeoxycytidine on the viability and radiosensitivity of EMT6 cells in vitro. *Oncology Research*, Vol.4, pp. 151-155.

Roti Roti, J.L.(2004). Introduction: radiosensitization by hyperthermia. *International Journal of Hyperthermia*, Vol.20, pp. 109-114.

Roti Roti, J.L.; Kampinga, H.H., Malyapa, R.S., Wright, W.D., Vander Waal, R.P. & Xu M. (1998). Nuclear matrix as a target for hyperthermic killing of cancer cells. *Cell stress & chaperones*, Vol.3, pp. 245-255.

Rots, M.G.; Willey, J.C., Jansen, G., Van Zantwijk, C. H., Noordhuis, P., DeMuth, J. P., Kuiper, E., Veerman, A. J., Pieters, R. & Peters, G. J. (2000). mRNA expression levels of methotrexate resistance-related proteins in childhood leukemia as determined by a standardized competitive template based RT-PCR method. *Leukemia*, Vol.4, pp. 2166-2175.

Rouleau, M.; Patel, A., Hendzel, M.J., Kaufmann' S.H. & Poirier, G.C. (2010). PARP inhibition: PARP1 and beyond. *Nature Reviews Cancer*, Vol.10, pp. 293-301.

Ruiz van Haperen, V.W.; Veerman, G., Eriksson, S., Boven, E., Stegmann, A.P., Hermsen, M., Vermorken, J.B., Pinedo, H.M. &, Peters, G.J. (1994). Development and molecular characterization of a 2',2'-difluorodeoxycytidine-resistant variant of the human ovarian carcinoma cell line A2780. *Cancer Research*, Vol.54, pp. 4138-4143.

Ryu, S.; Brown, S.L., Kim, S.H., Khil, M.S. & Kim J.H. (1996). Preferential radiosensitization of human prostatic carcinoma cells by mild hyperthermia. *International Journal of Radiation Oncology, Biology, Physics*, Vol.34, pp. 133-138.

Scalliet, P., Goor, C., Galdermans D, *et al*. (1998). Gemzar (Gemcitabine) with thoracic radiotherapy – a phase II pilot study in chemo-naïve patients with advanced non-small-cell lung cancer (NSCLC) (Abstract). *Proc ASCO*, Vol.17, pp.499a.

Shewach, D.S.; Hahn, T.M., Chang, E., Hertel, L.W., & Lawrence, T.S. (1994). Metabolism of 2',2'-difluoro-2'-deoxycytidine and radiation sensitization of human coloncarcinoma cells. *Cancer Research*, Vol.54, pp. 3218-3223.

Song, C.W.; Shakil, A., Osborn, J.L. & Iwata, K. (1996). Tumour oxygenation is increased by hyperthermia at mild temperatures. *International Journal of Hyperthermia*, Vol.12, pp. 367-373.

Spasokoukotskaja, T.; Arner, E.S.J., Brösjo, O., Gunvén, P., Julisson, G., Liliemark, J. & Eriksson, S. (1995). Expression of deoxycytidine kinase and phosphorylation of 2-chlorodeoxyadenosine in human normal and tumor cells and tissues. *European Journal of Cancer*, Vol.31, pp. 202-208.

Storniolo, A.M.; Enas, A.H., Brown, C.A., Voi, M., Rothenberg, M.L. & Schilsky, R. (1999). An investigational new drug treatment program for patients with gemcitabine-results for over 3000 patients with pancreatic carcinoma. *Cancer*, Vol.85, pp.1261-1268.

Turchi, J.J.; Henkels, K.M. & Zhou Y. (2000). Cisplatin-DNA adducts inhibit translocation of the Ku subunits of DNA-PK. *Nucleic Acids Research*, Vol.28, pp. 4634-4641.

Twentyman, P.R.; Wright, K.A. & Rhodes, T. (1991). Radiation response of human lung cancer cells with inherent and acquired resistance to cisplatin. *International Journal of Radiation Oncology, Biology, Physics*, Vol.20, pp. 217-220.

Van Bree, C.: Van Der Maat, B., Ceha, H.M., Franken, N.A.P., Haveman, J. & Bakker, P.J. (1992). Inactivation of p53 and of pRb protects human colorectal carcinoma cells against hyperthermia-induced cytotoxicity and apoptosis. *Journal of Cancer Research and Clinical Oncology*, Vol,125, pp. 549-55.

Van Bree, C.; Savonije, J.H., Franken, N.A., Haveman, J. & Bakker, P.J. (2000). The effect of p53-function on the sensitivity to paclitaxel with or without hyperthermia in

Radiosensitization with Hyperthermia and Chemotherapeutic Agents: Effects on Linear-Quadratic Parameters
of Radiation Cell Survival Curves

163

human colorectal carcinoma cells. *International Journal of Oncology*, Vol.16, pp. 739-744.

Van Bree, C.; Castro Kreder, N., Loves, W.J., Franken, N.A.P., Peters, G.J. & Haveman, J. (2002). Sensitivity to ionizing radiation and chemotherapeutic agents in gemcitabine-resistant human tumor cell lines. *International Journal of Radiation Oncology, Biology, Physics*, Vol.54, pp. 237-44.

Van Bree, C.; Franken, N.A.P., Bakker, P.J., Klomp-Tukker, L.J., Barendsen, G.W. & Kipp, J.B. (1997). Hyperthermia and incorporation of halogenated pyrimidines: radiosensitization in cultured rodent and human tumor cells. *International Journal of Radiation Oncology, Biology, Physics*, Vol.39, pp. 489-496.

Van Der Zee, J.; Gonzalez Gonzalez, D., Van Rhoon, G.C., Van Dijk, J.D., Van Putten, W.L. & Hart, A.A. (2000). Comparison of radiotherapy alone with radiotherapy plus hyperthermia in locally advanced pelvic tumours: a prospective, randomised, multicentre trial. Dutch Deep Hyperthermia Group. *Lancet*, Vol.355, pp. 1119-1125.

Van Der Zee, J.; Treurniet-Donker, A.D., The, S.K., Helle, P.A., Seldenrath. J.J., Meerwaldt, J.H., Wijnmalen, A.L., Van den Berg, A.P., Van Rhoon, G.C., Broekmeyer-Reurink, M.P. & Reinhold, H.S. (1988). Low dose reirradiation in combination with hyperthermia: a palliative treatment for patients with breast cancer recurring in previously irradiated areas. *International Journal of Radiation Oncology, Biology, Physics*, Vol.15, pp. 1407-1413.

Van Der Zee, J. & González, G.D. (2002). The Dutch Deep Hyperthermia Trial: results in cervical cancer. *International Journal of Hyperthermia*. Vol.18, pp. 1-12. Erratum in: *International Journal of Hyperthermia* (2003) Vol.19, p. 213.

Van Putten, J.W.G.; Groen, H.J.M., Smid, K., Peters, G. J. & Kampinga, H. H. (2001). Endjoining deficiency and radiosensitization induced by gemcitabine. *Cancer Research*, Vol.61, pp. 1585-1591.

Vujaskovic, Z. & Song, C.W. (2004). Physiological mechanisms underlying heat-induced radiosensitization. *International Journal of Hyperthermia*, Vol.20, pp. 163-174.

Wang, Y.; Pantelias, G.E. & Iliakis, G., 1994, Mechanism of radiosensitization by halogen ated pyrimidines: the contribution of excess DNA and chromosome damage in BrdU radiosensitization may be minimal in plateau cells. *Internat ional Journal of Radiation Biology*, Vol.**66**, 133± 142.

Warters, R.L. & Axtell, J. (1992). Repair of DNA strand breaks at hyperthermic temperatures in Chinese hamster ovary cells. *International Journal of Radiation Biology*, Vol.61, pp. 43-48.

Webb, C.F.; Jones, G.D.D., Ward, J.F., Moyer, D.J., Aguilera, J.A. & Ling, L.L. (1993). Mechanisms of radiosensitisation in bromodeoxyuridine-substituted cells. *International Journal of Radiation Biology*, Vol.64, pp. 695-705.

Wilkins, D,E.; Ng, C.E. & Raaphorst, G.P. (1996). Cisplatin and low dose rate irradiation in cisplatin resistant and sensitive human glioma cells. *International Journal of Radiation Oncology, Biology, Physics*, Vol.36, pp. 105-111.

Wilkins, D.E.; Heller, D.P. & Raaphorst, G.P. (1993). Inhibition of potentially lethal damage recovery by cisplatin in a brain tumor cell line. *Anticancer Research*, Vol.13, pp. 2137-2142.

Xu, M.; Myerson, R.J., Xia, Y., Whitehead, T., Moros, E.G., Straube WL, Roti Roti, J.L. (2007). The effects of 41 degrees C hyperthermia on the DNA repair protein, MRE11, correlate with radiosensitization in four human tumor cell lines. *International Journal of Hyperthermia,* Vol.23, pp. 343-351.

Xu, M.; Wright, W.D., Higashikubo, R., Wang, L.L. & Roti Roti, J.L. (1999). Thermal radiosensitization of human tumour cell lines with different sensitivities to 41.1 degrees C. *International Journal of Hyperthermia,* Vol.15, pp. 279-290.

Xu, M.; Myerson, R.J., Straube, W.L., Moros, E.G., Lagroye, I., Wang, L.L., Lee, J.T. & Roti Roti, J.L.(2002). Radiosensitization of heat resistant human tumour cells by 1 hour at 41.1 degrees C and its effect on DNA repair. *International Journal of Hyperthermia,* Vol.18, pp. 385-403.

7

The Effects of Antioxidants on Radiation-Induced Chromosomal Damage in Cancer and Normal Cells Under Radiation Therapy Conditions

Maria Konopacka[1], Jacek Rogoliński[1] and Krzysztof Ślosarek[2]
[1]Center for Translational Research and Molecular Biology of Cancer,
[2]Department of Radiotherapy and Brachytherapy Treatment Planning,
Maria Sklodowska-Curie Memorial Cancer Center and
Institute of Oncology, Gliwice,
Poland

1. Introduction

Radiotherapy is the major form of treatment for many human cancer. During the course of treatment, the ionising radiation produces many biological effects not only in cancer but also in normal cells. Due to the risk of toxicity to normal cells, the radioprotectors are needed to reduce the normal tissue injury during the irradiation of tumours without influence on effectiveness on cancer treatment. Most of the compounds that showed the radioprotective capacity in laboratory studies failed because of their toxicity to the normal cells. The good radioprotector should be non-toxic and should selectively eliminate the cancer cells. A number of natural dietary ingredients show capacity to protect cells from damage induced by ionising radiation (Arora et al., 2008).

It is well known that antioxidant vitamins such as ascorbic acid and vitamin E protect cellular DNA and membranes from radiation-induced damage (Noroozi et al., 1998; Konopacka & Rzeszowska-Wolny, 2001; Kumar et al., 2002; Jagetia, 2007). Recently, several flavonoids, polyphenols and phenolic acids have become more popular as diet compounds due to their beneficial impact on human health. One of them is ferulic acid. It renders preferential radioprotection to normal tissue, but not to tumour cells under both ex vivo and in vivo conditions (Maurya et al., 2005; Maurya & Nair, 2006). Ferulic acid is present in many plant products such as giant fennel, green tea, coffee beans and grains. It is monophenolic phenylpropanoid that acts as an antioxidant against peroxyl radicals-induced oxidation in neuronal culture and in synaptosomal membranes (Kanski et al., 2002). It was found that ferulic acid reduced the number of radiation-induced DNA strand breaks and enhanced the DNA repair processes in peripheral blood lymphocytes but not influenced the level of radiation-induced damage in fibroblastoma tumour cells in mice (Maurya et al., 2005).

It has been also showed that the α-tocopherole can preferentially reduce the level of radiation-induced chromosomal damage in normal human cells, but in cancer cells it actually increases the level of damage (Jha et al., 1999; Kumar et al., 2002). Although there is much evidence about the modulating effects of antioxidants in cells directly irradiated the effects of low dose of scattered radiation are poorly studied.

Our previous study showed that the cells placed outside the irradiation field were exposed to very low doses of scattered radiation that induced the micronuclei (Konopacka et al., 2009) and decreased the cell viability (Rogolinski et al., 2009). The extent of micronuclei formation and apoptosis as well as the decrease of viability of cells exposed to scattered radiation was higher that could be predicted by dosimetric methodology based on the linear non-treshold model (LNT). In this situation, the radioprotection of normal tissues placed outside the irradiation field during radiotherapy of cancer seems to be very crucial. In present study, we compared the influence of vitamin C, vitamin E and ferulic acid on micronuclei formation in human cancer cells directly irradiated and in normal human cells placed outside the irradiation field during exposure. This study was performed to answer the question whether the antioxidants selected by us could preferentially protect normal cells but not cancer cells during exposure to radiation in conditions mimicking the radiotherapy. We tested the modulating effects of vitamin C, vitamin E and ferulic acid on micronuclei formation in directly irradiated cancer A549 cells as well as in normal BEAS-2B cells exposed outside of the radiation field.

2. Experimental procedures

2.1 Cell culture
Human lung carcinoma cells (A549 line) and normal human bronchial epithelial cells (BEAS-2B) were grown on DMEM/F12 medium supplemented with 10% fetal bovine serum (Immuniq) in a humidified atmosphere of 5% CO_2 at 37°C. Before irradiation the cells were trypsinized and 20 µl of cell suspension, containing approximately 10^3 cells, were transferred to an Eppendorf`s tubes and then the tubes were filled with medium up 0.5 ml so the cells were irradiated without presence of air.

2.2 Preparation of antioxidants
Vitamin C (Serva, Germany) was dissolved in culture medium and filter sterilized. Vitamin E (α- tocopherole, Sigma) and ferulic acid (Linegal) were dissolved in ethanol. Antioxidants were added to cultures 1 h before irradiation at final concentration from 1 to 100 µg/ml. Experiments included control cultures treated with ethanol alone at the same volume as volume of vitamin E or ferulic acid added to cultures.

2.3 Irradiation of cells
The cancer cells in Eppendorf`s tubes were placed in a special stand on 3 cm of depth in a water phantom and expose to 5 Gy. The normal cells in Eppendorf`s tubes were placed in a water phantom at distance of 4 cm outside the radiation field. These cells were exposed to scattered radiation at dose of 0.2 Gy. The tubes were placed horizontally in the water phantom in such a way that the A549 cancer cells were within the radiation beam field, whereas the normal BEAS-2B cells were placed 4 cm outside the beam field, as is presented in Fig.1.

Experiments were performed for electron (22 MeV) radiation generated in a linear accelerator Clinac series Varian Medical system, for 300 Mu/min accelerator mode and dose

The Effects of Antioxidants on Radiation-Induced Chromosomal Damage in Cancer and Normal Cells Under
Radiation Therapy Conditions

167

of 5 Gy in build-up (3 cm) depth in a water environment. After irradiation the cells were transferred into plastic dishes (50 mm diameter) and supplemented with up 5 ml of the culture medium.

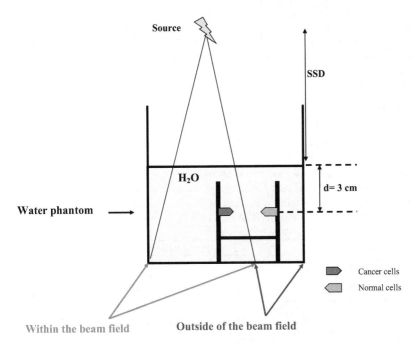

Fig. 1. The scheme of the irradiation set-up

2.4 Cytokinesis-block micronucleus test

The micronucleus test was performed according to the standard procedure (Fenech & Morley, 1985). After irradiation and transferring of cells to culture dishes, the cytochalasin B was added to medium to a final concentration 2 µg/ml and cells were incubated for 48h prior fixation. The cells were fixed *in situ* with a cold solution of 1 % glutaraldehyde (Sigma) in phosphate buffer (pH=7.5) and stained by Feulgen reaction. At least 500 binucleate cells were examined for the presence of micronuclei (MN-BN cells) under microscope.

2.5 Statistical analysis

Experiments were repeated three times. Means ± SD were calculated from experimental data and the Student`s t-test was used to determine the statistical significance of differences in the number of micronuclei between cells cultured and irradiated in the presence or absence of antioxidants.

3. Results

The effect of ferulic acid, vitamin C and vitamin E on the micronuclei formation in the normal human BEAS-2B cells is presented in Tab.1.

Concentration	Frequency of MN-BN cells (%)		
(µg/ml)	Ferulic acid	Vitamin C	Vitamin E
0	3.83 ± 0.44		
1	3.50 ± 0.32	3.44 ± 0.28	3.75 ± 0.43
0	3.33 ± 0.40	3.50 ± 0.42	3.80 ± 0.53
50	4.00 ± 0.62	3.85 ± 0.49	3.66 ± 0.44
100	4.98* ± 0.66	4.00 ± 0.55	4.10 ± 0.58
Ethanol	3.90 ± 0.42		

*- significantly different from control at $p < 0.01$.

Table 1. Influence of ferulic acid, vitamin C and vitamin E on the micronucleus formation in BEAS-2B cells. Means ±SD of three experiments are shown.

The background level of micronuclei was 3.83 ± 0.44 and none of the antioxidants at their concentration below 50 µg/ml caused any changes above this value. When the concentration of antioxidants was increased to 100 µg/ml the frequency of micronucleated cells was higher in comparison with background value but this difference was significant only for ferulic acid. Ethanol used as a solvent of ferulic acid and vitamin E did not change the spontaneous level of micronuclei in BEAS-2B cells.

Concentration	Frequency of MN-BN cells (%)		
(µg/ml)	Ferulic acid	Vitamin C	Vitamin E
0	2.10 ± 0.32		
1	2.18 ± 0.35	2.21 ± 0.33	2.25 ± 0.33
10	6.62* ± 0.55	2.26 ± 0.36	2.27 ± 0.20
50	9.16** ± 0.74	2.16 ± 0.29	2.20 ± 0.29
100	10.20** ± 0.98	2.23 ± 0.25	2.34 ± 0.42
Ethanol	2.16 ± 0.47		

Significantly different from control at: * $p < 0.01$, **$p < 0.001$.

Table 2. Influence of ferulic acid, vitamin C and vitamin E on the micronucleus formation in A 549 cells. Means ±SD of three experiments are shown.

Tab.2 presents the effect of antioxidants on the micronuclei formation in tumour A549 cells. Addition of vitamin C or vitamin E at concentrations ranging from 1 up to 100 µg/ml did not cause any measurable changes in the spontaneous level of micronuclei (2.10 ± 0.32). Ferulic acid at the concentrations 10 µg/ml and higher increased significantly the number of micronucleated cells in comparison with the control cells incubated without antioxidants. Ethanol did not cause any changes above the background in A549 cells.

In the next experiments we tested the effect of antioxidants on the level of radiation-induced micronuclei in cells irradiated in a water phantom (see Material and Methods). Antioxidants were tested at concentrations 1, 10 and 25 µg/ml. The results of this experiment in normal BEAS-2B cells exposed to radiation outside the field are showed in Fig.2.

Vitamin C at the concentrations of 1 and 10 µg/ml protected cells from radiation-induced DNA damage but at concentration of 25 µg/ml it was not effective in reducing this damage. In contrast to ascorbic acid, vitamin E was effective as a radioprotector at concentration above 10 µg/ml. Ferulic acid showed the best protective effect against radiation-induced micronuclei in normal cells. It inhibited significantly the micronuclei formation in dose-dependent manner.

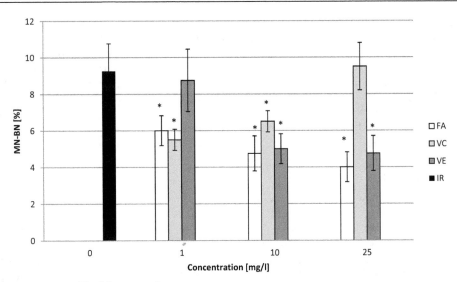

Values are means ±SD of three experiments.
* p < 0.01 refers to differences between irradiated only and preincubated with antioxidants cells
(Student's t – test).

Fig. 2. Effect of ferulic acid, vitamin C and vitamin E on the micronuclei formation in BEAS-2B cells exposed to radiation outside the field.

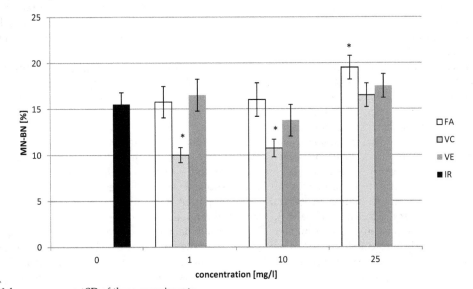

Values are means ±SD of three experiments.
* p < 0.01 refers to differences between irradiated only and preincubated with antioxidants cells
(Student's t – test).

Fig. 3. Effect of ferulic acid, vitamin C and vitamin E on the micronuclei formation in A-549 cells irradiated in a beam axis.

Fig.3 shows the radioprotective capacity of antioxidants in cancer cells irradiated in beam of axle. Vitamin C at low concentration (below 10 μg/ml) diminished the level of radiation-induced micronuclei. At the concentration of 25 μg/ml we observed no effect of this vitamin on formation of micronuclei in irradiated cells in comparison with the cells irradiated without vitamin treatment. Vitamin E at all concentrations did not influence the level of radiation-induced chromosomal damage. In cells irradiated in the presence of ferulic acid at concentrations 1 or 10 μg/ml we did not observe any protective effect in reducing the level of radiation-induced micronuclei but at the highest concentration (25 μg/ml) ferulic acid significantly increased the number of radiation induced micronuclei.

4. Discussion

Radioprotectors should selectively protect the normal tissues during radiotherapy of cancer without inhibition of damaging of cancer cells. In present work we tested the radioprotective activity of known antioxidants, namely, ferulic acid, vitamin C and vitamin E, in normal cells versus cancer cells. To answer the question whether these antioxidants can preferentially protect the normal cells during radiotherapy, we performed experiments under conditions like those that would be utilized in radiation therapy treatment procedures: tumour cells were irradiated directly (in a beam axle), whereas the normal cells were placed outside the radiation field during exposure. Our results indicate that the radioprotective effect of vitamin C is concentration-dependent and similar in normal and cancer cells; vitamin C at low concentration diminished the radiation-induced micronuclei whereas at high concentration it enhanced the level of damage, that was stronger in cancer than in normal cells. It is known that high concentration of ascorbic acid can potentate the production of hydroxyl radicals from hydrogen peroxide via Fenton reaction, which enhances the level of radiation-induced DNA damage (Halliwell & Gutteridge, 1985). This action of vitamin C varied among cell types due to the differences in the intracellular concentration of the ascorbic acid ranging from 10 μM in serum blood cells to 700 μM in bone marrow cells (Umegaki et al., 1995). It has been showed that vitamin C supplemented with vitamin K3 was effective in killing of cancer cells via activation of DNase, which degrades tumour cell DNA and induces cell death (Jamison et al., 2004).

Our results indicate that ferulic acid and vitamin E are potentially very good radioprotectors of normal cells during radiotherapy because they reduced the number of radiation-induced micronuclei in normal cells and simultaneously they did not influence the damaging effect of radiation in cancer cells and at high concentration they enhanced the damaging effect of radiation in cancer cells. Moreover ferulic acid showed selectively clastogenicity expressed as a micronuclei formation in tumour cells. The number of micronucleated cancer cells increased over the background value in concentration-dependent manner.

Our observations are in agreement with published data that indicated that ferulic acid preferentially protected normal mouse bone marrow and blood cells but not cancer fibroblastoma cells in mice exposed to 4 Gy of γ-radiation (Maurya & Nair, 2006). It was also shown that it protected human lymphocytes against radiation-induced chromosomal damage (Prasad et al., 2006). Moreover, it was found that vitamin E selectively protected normal human fibroblasts but not human cervical cancer and ovarian carcinoma cells against radiation-induced chromosomal damage (Kumar et al., 2002) and cell cycle inhibition (Jha et al., 1999). It has been suggested that the preferential protection of normal cells against radiation can be connected with alteration of genes encoding the elements of

The Effects of Antioxidants on Radiation-Induced Chromosomal Damage in Cancer and Normal Cells Under Radiation Therapy Conditions

171

the cell signalling pathways such as transcriptional factor E2F (Turley et al., 1997) and repair processes (Maurya & Nadir, 2006).

5. Conclusions

The results presented in this paper indicate that antioxidants such as ferulic acid and vitamin E protect to normal cells exposed to low dose of scattered radiation present outside the radiation field. Recently, there is increasing attention for the low dose radiation exposure including non-target phenomena such as bystander effect and low dose hypersensitivity and genetic instability (Morgan & Sowa, 2007) that can be responsible for the induction of secondary cancer. The protection of normal cells against these distant effects appears to be important element in radiotherapy and out to be taken into consideration in clinical practice. Ferulic acid and vitamin E seem to be the promising protectors of normal cells during radiotherapy.

6. Acknowledgment

This study was supported by a grant from Polish Ministry of Science and Higher Education, No. NN 402 4447 33.

7. References

Arora, R.; Kumar, R.; Sharma, A. & Tripathi, R.P. (2008). Radiomodulatory compounds of herbal origin for new frontiers in medicine, homeland security, management radiological incidents and space application. In: Herbal Radiomodulators, applications in medicine, homeland defence and space, R. Arora, (Ed.), 1 – 22, CABI North American Office, Cambridge

Fenech, M. & Morley, A.A. (1985). Measurement of micronuclei in lymphocytes. Mutation Research 147, 29-36

Halliwell, B. & Gutteridge, J.M.C. (1985). Free Radicals in Biology and Medicine, Clarendon Press, Oxford, pp 58-88

Jagetia, G.C. (2007). Radioprotective potential of plants and herbs against the effects of ionizing radiation. Journal of Clinical Biochemistry and Nutrition 40, 74-81

Jamison, J.M.; Gilloteaux, J.; Nassiri, M.R.; Venugopal, M.; Neal, D.R. & Summers, J.L. (2004). Cell cycle arrest and autoschizis in a human bladder carcinoma cell line following Vitamin C and Vitamin K3 treatment. Biochemical Pharmacology 15, 337-351

Jha, M.N.; Bedford, J.S.; Cole, W.C.; Edward-Prasad, J. & Prasad, K.N. (1999). Vitamin E (d-alpha-tocopheryl succinate) decreases mitotic accumulation in gamma-irradiated human tumor, but not in normal cells. Nutrition and Cancer 35, 189-194

Kanski, J.; Aksenova, M.; Stoyanova, A. & Butterfield, D.A. (2002). Ferulic acid antioxidant protection against hydroxyl and peroxyl radical oxidation in synaptosomal and neuronal cell culture systems in vitro: structure-activity studies. Journal of Nutritional Biochemistry 13, 273-281

Konopacka, M. & Rzeszowska-Wolny, J. (2001). Antioxidant vitamins C, E and beta-carotene reduce DNA damage before as well as after gamma-ray irradiation of human lymphocytes in vitro. Mutation Research 49, 1-7

Konopacka, M.; Rogoliński, J. & Ślosarek, K. (2009). Comparison of dose distribution of ionizing radiation in a water phantom with frequency of cytogenetic damage in a human bronchial cells. In: *World Congress on Medical Physics and Biomedical Engineering. IFMBE Proceedings,* Dossel Schlegel (Ed.), Munich, 7-12 September 2009, 25/3, pp 379-382

Kumar, B.; Jha, M.N.; Cole, W.C.; Bedford, J.S. & Prasad, K.N. (2002). D-alpha-tocopheryl succinate (vitamin E) enhances radiation-induced chromosomal damage levels in human cancer cells, but reduces it in normal cells. *Journal of American College of Nutrition* 21, 339-343

Maurya, D.K.; Salvi, V.P. & Nair, C.K. (2005). Radiation protection of DNA by ferulic acid under in vitro and in vivo conditions. *Molecular and Cellular Biochemistry* 280, 209-217

Maurya, D.K. & Nair, C.K. (2006). Preferential radioprotection to DNA of normal tissues by ferulic acid under ex vivo and in vivo conditions in tumor bearing mice. *Molecular and Cellular Biochemistry* 285, 181-190

Morgan, W.F. & Sowa, M.B. (2007). Non-targeted bystander effects induced by ionizing radiation. *Mutation Research* 616, 159-164

Noroozi, M.; Angerson, W.J. & Lean, M.E. (1998). Effects of flavonoids and vitamin C on oxidative DNA damage to human lymphocytes. *American Journal of Clinical Nutrition* 67, 1210-1218

Prasad, N.R.; Srinivasan, M.; Pugalendi, K.V. & Menon, V.P. (2006). Protective effect of ferulic acid on gamma-radiation-induced micronuclei, dicentric aberration and lipid peroxidation in human lymphocytes. *Mutation Research* 28, 129-134

Rogolinski, J.; Konopacka, M.; Sochanik, A. & Ślosarek, K. (2009). Scattering medium depth and cell monolayer positioning with respect to beam field affect cell viability. In: *World Congress on Medical Physics and Biomedical Engineering. IFMBE Proceedings,* Dossel Schlegel (Ed.), Munich, 7-12 September 2009, 25/3: pp 403-405

Turley, J.M.; Ruscetti, F.W.; Kim, S.J.; Fu, T.: Gou, F.V. & Birchenall-Roberts, M.C. (1997). Vitamin E succinate inhibits proliferation of BT-20 human breast cancer cells: increased binding of cyclin A negatively regulates E2F transactivation activity. *Cancer Research* 57, 2668-2675

Umegaki, K.; Aoki, S. & Esashi, T. (1995). Whole body X-ray irradiation to mice decreases ascorbic acid concentration in bone marrow: comparison between ascorbic acid and vitamin E. *Free Radicals in Biology and Medicine* 19, 493-497

8

Melatonin for Protection
Against Ionizing Radiation

M. A. El-Missiry[1], A. I. Othman[1] and M. A. Alabdan[2]
[1]Department of Zoology, Faculty of Sciences, Mansoura University, Mansoura,
[2]Department of Zoology, Faculty of Science,
Princess Nora Bint AbdulRahman University, Riyadh,
[1]Egypt
[2]Kingdom of Saudi Arabia

1. Introduction

Radiations exist ubiquitously in the environment since the Earth's creation in soil, water and plants. Radiation exposure is a concern in the health industry and other occupations in the world. Apart from diagnostic, therapeutic and industrial purposes, humans also are exposed to ionizing radiations during air and space travel and exploration, background radiation, nuclear accidents, and nuclear terror attacks. Elevated radiation levels have been detected following Chernobyl on April 1986 at Ukraine, and recently Fukushima Daiichi Nuclear Power Plants on March 2011 at Japan. This raised the need for finding out efficient and reliable radioprotectors especially when a whole nation is exposed at high or even low levels for a prolonged period. The fallout and radioactivity cause concern during the weeks and months after the accidents. In addition, radiations are commonly used in a number of medical and industrial situations; however, their pro-oxidative effects limit their applications. Therefore, it is essential to protect humans from ionizing radiations by efficient pharmacological intervention. A valid approach to halt normal tissue radiotoxicity is the use of radioprotectors that when present prior to radiation exposure protect normal tissues from radiation effects. This view has also been used as a successful preventative measure for possible nuclear/radiological situation. From a practical point of view radioprotectors should perfectly have several criteria that relate to the ability of the agent to improve the therapeutic outcome. Ionizing radiation causes oxidative damage to tissues within an extremely short period, and possible protection against it would require the rapid transfer of smart antioxidants to the sensitive sites in cells. At this point, melatonin (N-acetyl-5-methoxytryptamine; MW= 232), an innate antioxidant produced mainly by the pineal gland, seems unique among antioxidants because of its multiple properties and reactions which reviewed and documented in several publications and summarised herein.

While ionizing radiation exposures, due to free radical generation, present an enormous challenge for biological and medical safety, melatonin is a potent radioprpotector. In several investigations, melatonin has been recognized for successful amelioration of oxidative injury and illness due to direct and indirect effects of ionizing radiation and against oxidative stress in several experimental and clinical settings. Furthermore, numerous studies have established that melatonin is a highly efficient free radical scavenger, broad antioxidant and stimulator of several antioxidants in biological systems. Because of its unique characteristics; melatonin has effects not only at the cell level but also within subcellular organelles and structures. The antioxidant and prophylatic properties of melatonin allow the use of radiation during radiotherapy to get better therapeutic outcomes. Several published articles documented that melatonin's anticancer and oncostatic effects make melatonin an excellent candidate and good choice to be used in routine radiotherapies, space travel and following nuclear accidents occupational settings where accidental exposure may occur. This article will review antioxidant features that put melatonin on top of potentially efficient pharmacological radioprotectors.

2. Ionizing radiation, free radicals & oxidative stress

Ionizing radiations are types of particle radiation (such as neutron, alpha particles, beta particles and cosmic ray) or electromagnetic (such as ultraviolet, X-rays and gamma rays) with sufficient energy to ionize atoms or molecules by detaching electrons from their valence orbitals. The degree and nature of such ionization depends on the energy of the individual particles or on frequency of electromagnetic wave. It is well known that exposure to ionizing radiation at sufficiently high doses results in various types of adverse biological effects. The biological effect of radiation involves direct and indirect actions. Both actions produce molecular changes that mostly need enzymatic repair. Indirect effect involves the production of reactive free radicals which produce oxidative mutilation on the key molecules. The environmental sources of oxidative attack include, in particularly, specific exposures of the organism to ionizing radiations like X-, γ- or cosmic rays and α-particles from radon decay as well as UVA and UVB solar light. Ionizing radiations prevalent in space, involve a broad range of radiation types and energies from cosmic and unpredictable solar sources, representing a very diverse range of ionization qualities and biological effectiveness. Linear energy transfer (LET) is a measure of the energy transferred to tissue or cells as an ionizing particle travels through it. The LET of the potential radiations can cover several orders of magnitude from <1.0 keV μm^{-1} to > several 100 keV μm^{-1} (Blakely and Chang 2007) Low LET radiation causes damage through reactive oxygen species (ROS) production mainly by the radiolysis of water present in living system.

From a chemical point of view, reactive oxygen species (ROS) and reactive nitrogen species (RNS) are oxygen and nitrogen containing molecules constitute the main category of free radicals which may be defined as any chemical moiety generated with an unpaired number of electrons in valancy orbital. ROS include oxygen-based free radicals, the superoxide anion ($O_2^{\bullet-}$), hydroxyl ($\bullet OH$), alkoxyl ($RO\bullet$), peroxyl ($ROO\bullet$), and hydroperoxyl ($HOO\bullet$). RNS include peroxynitrite ($ONOO-$), nitric oxide ($NO\bullet$), and nitrogen dioxide ($NO2\bullet$). ROS may be radical, such as, $O_2^{\bullet-}$ and $\bullet OH$, or non-radical, such as, hydrogen peroxide (H_2O_2) and singlet oxygen ($^1O_2$). At high concentrations, free radicals can be harmful to living organisms. Some ROS damage biomolecules indirectly. For example, H_2O_2 and $O_2^{\bullet-}$ initiate DNA and lipids damage by interaction with transition metal ion, in particular iron and copper, in the metal-catalysed Haber–Weiss reaction, producing $\bullet OH$. It is the most electrophilic and reactive of the ROS, with a half-life of~10^{-9} s (Draganic and Draganic 1971). $\bullet OH$ can be produced by ultraviolet and ionizing radiations (Von Sonntag 1987). This radical is considered the most frequently damaging species. It has been estimated that the $\bullet OH$ is responsible for 60–70% of the tissue damage caused by ionizing radiations (Galano et al. 2011). Moreover, •OH has great ability to react with almost any molecule in the vicinity of where it is generated (Reiter et al. 2010). Chemical nature and reactivity of free radicals in biological systems has been recently reviewed (Galano et al. 2011). Once formed ROS and RNS can produce a chain reaction. The transfer of the free radical to a biological molecule can be sufficiently damaging to cause bond breakage or inactivation of key functions. The organic ROO• can transfer the radical from molecule to molecule causing damage at each encounter. Thus, a cumulative effect can occur, greater than a single ionization or broken bond.

A variety of external events, in particular, exposure to ionizing or ultraviolet radiation, can lead to an increase in the generation of ROS in comparison with available antioxidants leading to oxidative stress. Oxidative stress is caused by the presence of excessive amount of ROS which the cell is unable to counterbalance. This implies that the steady state balance of pro-oxidant/anti-oxidant systems in intact cells is shifted to the former. When excessive oxidative events occur, the pro-oxidants outbalance the anti-oxidant systems. Moreover, oxidative stress may result by overwhelming of antioxidant and DNA repair mechanisms in the cell by ROS. In radiation sickness oxidative stress is a factor as either cause or effect. The result is oxidation of critical cellular macromolecules including DNA, RNA, proteins and lipids eventually leading to cell death in severe oxidative stress. On the other hand, moderate oxidative stress may lead to activation of cytoplasmic/nuclear signal transduction pathways, modulation of gene and protein expression and alteration of DNA polymerase activity, affect the endogenous anti-oxidant systems by down-regulating proteins that participate in these systems, and by depleting cellular reserves of anti-oxidants (Acharya et al. 2010, Cadet et al. 2010, Little 2000).

3. Molecular biology effects of ionizing radiation due to free radical generation

As a matter of fact, ionizing radiation penetrating living tissue and can damage all important cellular components both through direct ionization and through generating ROS due to water radiolysis and induce oxidative damage. Radiation-induced oxidative stress was evaluated by three independent approaches; DNA damage, lipid peroxidation and protein oxidation.

3.1 DNA damage
Cells and their genomic constituent of the living organisms are continually exposed to oxidative attacks. Acute exposure to ionizing radiation can create oxidative stress in a cell and chronic exposure to this stress can result in permanent changes in the genome (Cooke et al. 2003). The main target of ionizing radiation has long since been indicated to be DNA which shows wide range of lesions. The oxidatively DNA damage commonly are apurinic/apyrimidinic (abasic) DNA sites, oxidized purines and pyrimidines, single strand (SSBs) and double strand (DSB) DNA breaks and non-DSB (Kryston et al. 2011). Other initial chemical events induced in DNA by ionizing radiation include cross-links, oxidative base modification (Hutchinson 1985) and clustered base damage (Goodhead 1994), sugar moiety modifications, and deaminated and adducted bases (Cooke et al. 2003, Sedelnikova et al. 2010, Sutherland et al. 2000, Ward 1994). The numbers of DNA lesions per cell that are detected immediately after a radiation dose of 1 Gy have been estimated to be approximately greater than 1000 base damage, 1000 SSBs, 40 DSBs, 20 DNA–DNA cross-

links, 150 DNA-protein cross-links and 160-320 non-DSB clustered DNA damage and defective DNA mismatch repair proteins (MMP) (Martin et al. 2010). Recently, it is suggested that radiation dose and the type of DNA damage induced may dictate the involvement of the MMP system in the cellular response to ionizing radiation. In particular, the literature supports a role for the MMP system in DNA damage recognition, cell cycle arrest, DNA repair and apoptosis (Martin et al. 2010). In addition, The DNA oxidation products are a direct risk to genome stability, and of particular importance are oxidative clustered DNA lesions, defined as two or more oxidative lesions present within 10 bp of each other (Sedelnikova et al. 2010).

The most common cellular DNA base modifications are 8-oxo-7,8-dihydroguanine (8-oxoGua) and 2,6- diamino-4-hydroxy-5-formamidopyrimidine. Both originate from the addition of the •OH to the C8 position of the guanine ring producing a 8-hydroxy-7,8-dihydroguanyl radical which can be either oxidized to 8-oxoGua or reduced to give the ring-opened FapyGua (Altieri et al. 2008, Kryston et al. 2011). The •OH interact with pyrimidines (thymine and cytosine) at positions 5 or 6 of the ring, and yield several base lesions. The most abundant and well known products, 5,6-dihydroxy-5,6-dihydrothymine (thymine glycol) and 5,6-dihydroxy-5,6-dihydrocytosine (cytosine glycol). It is generally accepted that 8-oxodG and thymine glycol are reliable biomarkers of high levels of oxidative stress and damage in the human body. These lesions ultimately are not lethal to the cell, but are considered to be highly mutagenic. Oxidized bases in DNA are potentially mutagenic and so are implicated in the process of carcinogenesis. It has been reported that X-radiation induced a significant increase in 8-OHdG concentration in mammary gland DNA (Haegele et al. 1998). High levels of 8-OHdG have been observed in normal human epidermis or purified DNA exposed to ultraviolet radiation (Wei et al. 1997). Thus, genome stability is crucial for maintaining cellular and individual homeostasis, but it is subject to many changes due to free radicals attack induced by the exposure to ionizing radiation. DNA breaks and fragments resulted from chromosomal damage appear as micronuclei in rapidly proliferating cells micronuclei frequency was markedly enhanced in bone marrow cells of mouse exposed to 5Gy radiation (Verma et al. 2010).

Comprehensive reviews have been appeared to give structural and mechanistic information on the radiation-induced damage to DNA (Cadet et al. 2010, Cadet et al. 2005). However, these authors showed that there is still a dearth of precise data on the formation of radiation-induced base injury to DNA in cells and tissues. This is because the determination of the radiation-induced base damage within DNA is achieved indirectly by methods utilizing hydrolysis of the biopolymer then followed by analysis of the free fragments. The situation is even more difficult for cellular DNA since highly sensitive assays are required to monitor the formation of very low amounts of injury, typically within the range of one modified base per 10^6 normal nucleotides.

Direct damage to DNA caused by ionizing radiation has been considered as a significant initiator of mutation and cancer. However, some reports suggest that extracellular and extranuclear targets may contribute to the genotoxic effects of radiation (Little 2000, Morgan 2003). In addition, it has been shown that irradiation of the cytoplasm produces gene mutations in the nucleus of the hit cells and that this process is mediated by free radicals (Wu et al. 1999). Recently, it is proposed that a possible extracellular signal-related kinase pathway involving ROS/RNS and COX-2 in the cytoplasmic irradiation-induced genotoxicity effect (Hong et al. 2010). Furthermore, it has been demonstrated that nitric oxide synthase (NOS) produces sustained high concentrations of nitric oxide (NO) in various mammalian cells after exposure to radiation (Matsumoto et al. 2001). In cytoplasmic-irradiated cells, 3-nitrotyrosine, a nitrosated protein product used as a marker of ONOO-, was significantly elevated and dramatically inhibited by L-NMMA, (NO inhibitor) implicating a critical role of RNS in the mutagenicity induced by cytoplasmic irradiation (Hong et al. 2010).

3.2 Lipid peroxidation

The direct and indirect destructive effects of ionizing radiation lead to peroxidation of macromolecules, especially those present in lipid-rich membrane structures, lipoproteins and chromatin lipids. Phospholipids in membranes and triglycerides in LDL are highly susceptible to free radical attacks. Once the process of lipid peroxidation is started, it proceeds as a free radical-mediated chain reaction involving initiation, propagation, and termination (Gago-Dominguez et al. 2005). The first step (initiation) in the lipid peroxidation process is the abstraction of a hydrogen atom, from a methylene group next to a double bond in polyunsaturated fatty acids. This produces a carbon centered radical which undergo rearrangement of the double bond to form a stable conjugated diene. In propagation step, carbon centered radicals react with oxygen to form new ROO• that react further with another neighboring lipid molecule forming a hydroperoxy group and a new carbon centered radicals. The lipid hydroperoxide will react further to form cyclic peroxide, cyclic endoperoxide, and finally aldehydes. The propagation phase can repeat many times until it is terminated by chain breaking antioxidants (Halliwell 2009, Reed 2011).

During lipid peroxidative pathway, several end products are formed such as malondialdehyde (MDA) and 4-hydroxy-2-nonenal (4-HNE), pentane and ethane, 2,3 transconjugated dienes, isoprostanes and cholesteroloxides (Catala 2009, Tuma 2002). These aldehydes are highly reactive and bind with DNA and proteins and form adducts which inhibit proteins functions and can disrupt nuclear events. The aldehydes are more diffusible than free radicals, thus injury can occur in distant locations. The aldehyde can be found in measurable concentrations in biological fluids and analytical methods used are sometimes complex and require sample preparation involving extraction and purification steps. Isoprostanes, prostaglandin like compounds, are generated from the free radical-initiated peroxidation of arachidonic acid. F2-isoprostanes are the most specific markers of lipid peroxidation and the most difficult to measure (Comporti et al. 2008). The mostly frequently lipid peroxidation markers used in free radical research are MDA and 4-HNE. Lipid peroxidation was assessed as thiobarbituric acid reactive substances (TBARS) in biological materials using thiobarbituric acid reaction method (Esterbauer and Zollner 1989, Moller and Loft 2010).

Several studies have examined the radiation-induced free radical damage evidenced by the elevation of lipid peroxidation levels. Lipid peroxidation-derived products have been implicated in the pathogenesis of oxidative stress-associated radiation sickness and diseases. Aldehydes showed significant increase with increasing doses of ionizing radiation in several organs (Bhatia and Manda 2004, Sener et al. 2003) and mitochondrial membranes (Kamat et al. 2000). A significant increase in DNA strand breaks and TBARS concentrations was found in rat brain exposed to 10Gy ionizing radiation (Undeger et al. 2004). Exposure to 5 Gy irradiation led to considerable elevation of MDA level in thymus, brain, jejunum liver and kidney of total body irradiated mice (Taysi et al. 2003, Verma et al. 2010) after 24 hrs of irradiation continued up to 48 hrs. Lipid peroxidation due to •OH attack were found to be in a radiation dose-dependent manner but no significant differences between radiation resistant and radiation sensitive rats were detectable after whole-body-irradiation with x-rays at 2, 4, and 6 Gy. Among the subcellular organelles mitochondria are one of the key components of the cell injured by radiation-induced oxidative stress. In an interesting study, mitochondria from rat brain and liver was isolated then exposed to 450Gy gamma radiation. In this study there was considerable increase in lipid hydroperoxide (LOOH) and MDA in rat liver and brain mitochondria (Lakshmi et al. 2005). Recent research showed that increased free radicals due to radiation exposure damage membrane lipids, which results in cell lysis due to altered membrane fluidity (Gulbahar et al. 2009).

The ideas about the leading role of lipid peroxidation in radiation damage of cells and tissues arose from the damaging of cell membrane structures. Peroxidation of lipids can greatly alter the physicochemical properties of membrane lipid bilayers, resulting in severe cellular dysfunction. It causes the change in structure, fluidity and permeability of membranes and inactivates several membrane associated enzymes and protein receptors. In biological membranes, lipid peroxidation is also usually accompanied by oxidation of membrane proteins. In consequence, peroxidation of lipids may change the agreement of proteins in bilayers and by that interfere with their physiological role on the membrane function.

3.3 Protein oxidation

As defined earlier, ionizing radiation can interact and modify all cellular components both through direct ionization and through induction of ROS resulting in a variety of subtle and profound biological effects. Radiation-induced oxidative protein damage can be started by even quite low doses of radiation and can produce an alteration of the cellular redox balance, which lasts for substantial time after exposure and may contribute to changes in cell survival, proliferation, and differentiation(Shuryak and Brenner 2009). Several damages to the peptide chain or to the side-chains of amino acid residues have been identified, and some of their mechanisms of formation has been described (Griffiths et al. 2002).

Available data from various studies raveled that the most sensitive amino acids, cysteine, tryptophan, tyrosine and methionine, bear aromatic or sulphur-containing side-chains. Furthermore, protein oxidation can lead to hydroxylation of aromatic groups and aliphatic amino acid side chains, nitration of aromatic amino acid residues, nitrosylation of sulfhydryl groups, sulfoxidation of methionine residues, chlorination of aromatic groups and primary amino groups, and to conversion of some amino acid residues to carbonyl derivatives (Catala 2007). The fundamental mechanisms involved in the oxidation of proteins by ROS were described by studies in which amino acids, peptides, and proteins were exposed to ionizing radiations under conditions where $\bullet OH$ or a mixture of $\bullet OH$ and $O_2\bullet -$ are formed (Stadtman 2004). It has been demonstrated that the attack by $\bullet OH$ leads to an abstraction of a hydrogen atom from the protein polypeptide backbone and form a carbon-centered radical (Klaunig et al. 2011, Stadtman 2004). Oxidation due to radiation exposure can lead also to cleavage of the polypeptide chain and formation of cross-linked protein aggregates. Because the generation of carbonyl derivatives occurs by many different mechanisms, the level of carbonyl groups in proteins is widely used as a marker of oxidative protein damage (Guajardo et al. 2006). Studies performed with various tissues have revealed that radiation increases protein oxidation and carbonyl levels as well as produces structural and functional changes (Gulbahar et al. 2009).

In most reports describing the in vivo experiments on radiation sickness, the carbonyl levels were determined in tissue homogenates or soluble cytoplasmic proteins. Moreover, it is very important to consider the carbonyl levels in different subcellular fractions, since they may show different susceptibility to oxidative damage, probably due to differences in their protein composition or activities of their antioxidant defenses, and, therefore, make different contribution to the impairment of cell functioning with radiation responses.

Numerous investigations showed that protein carbonyls, a marker of primary protein damage indicated a higher magnitude of damage in irradiated mice brains exposed to 1.5 Gy high-LET 56Fe beams (500 MeV/nucleon, 1.5 Gy). This effect was associated with impaired cognitive behavior of mice at day 30 post-exposure as well as apoptotic and necrotic cell death of granule cells and Purkinje cells (Manda et al. 2008). γ-irradiation of rats at a dose of 10 Gy caused increases in protein carbonyl groups in mitochondria and cytoplasm both in liver and spleen. Similar results have been obtained for homogenates of different tissues isolated from γ-irradiated gerbils and rats (Sohal et al. 1995). Post-irradiation accumulation of oxidized proteins in subcellular fractions, especially if occurring in nuclei, might probably affect not only the catalytic properties of enzymes but also the regulation of radiation-induced gene expression by interfering with the activation of transcription factors (Whisler et al. 1997). Among the nuclear proteins, histones are likely most susceptible to oxidative modification, due to high contents of lysine and arginine residues in their molecules. Information on the formation of radicals on peptides and proteins and how radical damage may be propagated and transferred within protein structures have been reviewed (Hawkins and Davies 2001).

4. Defenses against free radicals

Human and all of the aerobic organisms have a very efficient defense network of antioxidants against oxidative stress. An antioxidant can be defined as a molecule or an element that, when present at low concentrations compared to those of the oxidizable substrate, significantly combat, delays and inhibit oxidation of that substrate, thus, prevent free radicals from damaging healthy cells (Halliwell 1997, 2009). Under normal condition, cells have well coordinated and efficient endogenous antioxidant defense systems, which protect against the injurious effects of oxidants.

From the viewpoint of mechanistic functions, antioxidant defense mechanisms can be classified into the following five lines of defenses: preventing antioxidants, scavenging antioxidants, repair and de novo antioxidants, adaptive antioxidants, and finally cellular signaling messenger (Halliwell 1997, Niki 2010). The first line of defense is the preventing antioxidants which act by suppressing the formation of ROS and RNS by reducing H_2O_2 and lipid hydroperoxides that are generated during lipid peroxidation, to water and lipid hydroxides, respectively, or sequestering pro-oxidant metal ions such as iron and copper by some binding of proteins (e.g., transferrin, metallothionein). The second line of defense can be described as the scavenging antioxidants which exist to intercept, or scavenge free radicals and remove active species rapidly before attacking biologically essential molecules. For example; superoxide dismutase (SOD) converts $O_2 \bullet^-$ to H_2O_2, while α-tocopherol and carotenoids are efficient scavenger of $^1O_2$ (Inoue et al. 2011). Many phenolic compounds and aromatic amines act as a free radical-scavenging antioxidant. There is a general agreement that electron transfer and hydrogen transfer are the main mechanisms involved in the reactions of melatonin with free radicals. The third line of defenses is various enzymes which function by repairing damages, clearing the wastes, and reconstituting the lost function. The adaptation mechanism is considered the fourth line of defense, in which appropriate antioxidants are released at the right time and transported to the right site in right concentration. Some antioxidants constitute the fifth line of defense by functioning as a cellular signaling messenger to control the level of antioxidant compounds and enzymes (Niki and Noguchi 2000, Noguchi and Niki 2000).

5. Radioprotectors and mitigators of radiation induced injury

Generally, any chemical/biological agents given before to or at the time of irradiation to prevent or ameliorate damage to normal tissues are termed radioprotectors. While mitigators of normal tissue injury are agents delivered at the end of irradiation, or after irradiation is complete, but prior to the manifestation of normal tissue toxicity. The estimated time scale to use mitigators efficiently ranges from seconds to hours after radiation exposure. Agents delivered to improve established normal tissue toxicity are considered treatments which can be monitored over weeks to years after radiation exposure (Citrin et al. 2010). Since radiotherapy, occupational, accidental exposure to radiation or space travel and exploration can produce unwanted side effects, it is important to prevent such effects by the use of radioprotectors or mitigators. Ideally, radioprotective and mitigative agent should fulfill several characteristics that relate to the ability of the agent to improve the therapeutic results. First, the agent should have protective effects on the majority of organs and tissues. Second, the agent must reach all cells and organelles and can easily penetrate cellular membranes. Third, it must have an acceptable route of administration (preferably oral or alternatively intramuscular) and with minimal toxicity. Fourth, to be useful in the radiotherapy settings, radioprotectant should be selective in protecting normal tissues from radiotherapy without protecting tumor tissue. Finally, to a large extent radioprotectors should be compatible with the wide range of other drugs that will be prescribe to patients. Moreover, because free radicals are responsible for injury caused by ionizing radiation, therefore, for an agent to protect cells from primary free radical damage, the agent needs to be present at the time of radiation and in sufficient concentration to compete with radicals produced through radical-scavenging mechanisms (Citrin et al. 2010, Hosseinimehr 2007, Shirazi et al. 2007).

A large body of literature describes radioprotection or mitigation with a variety of agents after total body or localized exposures. A complete and comprehensive review of these agents is outside the scope of this chapter. Herein, we briefly highlight melatonin that have been described as radiation protectors and mitigators, and attempt to focus on it with demonstrated or anticipated usefulness for therapeutic radiation exposures. As defined above, an ideal radioprotectors need to have radical-scavenging properties and can also exert broad antioxidant activity. Whereas all antioxidants cannot afford full radioprotection, melatonin verify most of the criteria needed for efficient radioprotector , mitigators and treatment agent with antioxidant potential, radical scavenging characteristics and stimulator of intrinsic antioxidants.

6. Melatonin

6.1 Synthesis, distribution, and metabolism
Melatonin synthesis in the pineal gland has been reviewed in significant detail (Reiter 2003). In summary, pinealocytes take up L-tryprophane from blood. Via several enzymatic steps including tryptophan 5-hydroxylation, decarboxylation, N-acetylation and O-methylation,

In that sequence, N-acetyl-5-methoxytryptamine (melatonin) is synthesized. It is secreted upon biosynthesis into the extracellular fluid to the general circulation from which it easily crosses various cellular membranes. It is secreted by the pineal gland and its levels have diurnal variation and also fluctuate with sleep stages. They are higher during night (Luboshitzky et al. 1999). The diurnal/nocturnal levels of blood melatonin can range between 8 ± 2 pg/mL (light phase) and 81 ± 11 pg/mL (dark phase). The synthesis and presence of melatonin have also been demonstrated in non-pineal tissues such as retina, Harderian gland, gastrointestinal track, testes, and human lymphocytes. Furthermore, the distribution of melatonin in the human being is very broad (Reiter 2003). Once synthesized, the majority of melatonin diffuses directly towards the cerebrospinal fluid of the brain's third ventricle, while another fraction is released into the blood stream where it is distributed to all tissues and body fluids (Cheung et al. 2006). It is found in serum, saliva (Cutando et al. 2011, Novakova et al. 2011), cerebrospinal fluids (Rousseau et al. 1999), and aqueous humor of the eye (Chiquet et al. 2006), ovarian follicular fluid, hepato-gastrointestinal tissues (Messner et al. 2001). Melatonin in the milk of lactating mothers exhibits a marked daily rhythm, with high levels during the night and undetectable levels during the day (Illnerova et al. 1993, Sanchez-Barcelo et al. 2011). Moreover, melatonin production is not confined exclusively to the pineal gland, but other tissues including retina, Harderian glands, gut, ovary, testes, bone marrow and lens also produce it (Esposito and Cuzzocrea 2010).

Melatonin has two important functional groups which determine its specificity and amphiphilicity; the 5-methoxy group and the N-acetyl side chain. In liver melatonin is metabolized by P- 450 hepatic enzymes, which hydroxylate this hormone at the 6- carbon position to yield 6- hydroxymelatonin which conjugated with sulfuric or glucuronic acid, to produce the principal urinary metabolite, 6-sulfatoxymelatonin. In the final stage, conjugated melatonin and minute quantities of unmetabolized melatonin are excreted through the kidney. In addition to hepatic metabolism, oxidative pyrrole-ring cleavage appears to be the major metabolic pathway in other tissues, including the central nervous system (Esposito and Cuzzocrea 2010).

A plethora of evidence suggests that melatonin mediates a variety of physiological responses through membrane and nuclear binding sites. In mammals, the mechanisms of action of melatonin include the involvement of high affinity G protein-coupled membrane receptors (MT1, MT2), cytosolic binding sites (MT3 and calmodulin), and nuclear receptors of the RZR/ROR family. Melatonin also has receptor-independent activity and can directly scavenge free radicals. A disulfide bond between Cys 113 and Cys 190 is essential for high-affinity melatonin binding to MT2 and possibly to MT1 receptors (Dubocovich and Markowska 2005). RZR/ROR family is expressed in a variety of organs. It presumably regulates the immune system and circadian cycles via the nuclear receptor and these also may be involved in its regulation of antioxidative enzymes (Cutando et al. 2011).

6.2 Melatonin and factors that determine antioxidant capacity

A number of criteria that characterize an ideal free radical scavenging antioxidant can be identified. First, because free radicals are highly reactive with very short half life time, therefore, an efficient antioxidant should be ubiquitous and present in adequate amounts in tissues and cells. Furthermore, the biological systems are heterogeneous in nature, which affects the action and efficacy of antioxidants. Both hydrophilic and lipophilic antioxidants act at respective site. Some antioxidants are present in free form, but others as metabolite or in bound form (Niki 2010). In contrast to other antioxidants that are either hydrophilic or lipophilic, melatonin is an amphiphilic small size molecule (Giacomo and Antonio 2007). These features of melatonin allow it to cross all morphophysiological barriers and to interact with toxic molecules throughout the cell and its organelles, thereby reducing oxidative damage to molecules in both the lipid and aqueous environments of cells. Numerous articles documented that melatonin is widely distributed and found in all body fluids, organs, cells and organelles. Recently, melatonin levels in the cell membrane, cytosol, nucleus, and mitochondrion was found to vary over a 24-hr cycle, although these variations do not exhibit circadian rhythms. The cell membrane has the highest concentration of melatonin followed by mitochondria, nucleus, and cytosol (Venegas et al. 2011).

Second, an efficient antioxidant should react with most of free radicals because as it is well known that free radicals are variable in their biological, chemical and physical properties that involved in the oxidative stress. In functional terms, have reported that melatonin exerts a host of antioxidant effects that can be described as a broad spectrum antioxidant (Karbownik and Reiter 2000). Initially, (Hardeland et al. 1993a, Hardeland et al. 1993b, Tan et al. 1993) are the first who documented that melatonin is a remarkably potent scavenger of the particularly reactive, destructive, mutagenic and carcinogenic •OH. It is documented that, melatonin is a more efficient •OH scavenger than either glutathione or mannitol, and that melatonin reacts at a diffusion-controlled rate with the •OH. Thus, melatonin is probably an important endogenously produced antioxidant. Extensive studies have established that melatonin is much more specific than its structural analogs in undergoing reactions which lead to the termination of the radical reaction chain and in avoiding prooxidant, carbon or oxygen centered intermediates (Hardeland et al. 2011, Poeggeler et al. 2002, Tan et al. 1993). Besides the •OH, melatonin in cell-free systems has been shown to directly scavenge H_2O_2, $^1O_2$ and HOCl with little ability to scavenge the O_2•-. Furthermore, melatonin scavenges nitric oxide (NO•) and suppresses the activity of its rate limiting enzyme, nitric oxide synthase (NOS), thereby inhibiting the formation of the ONOO-. In addition, melatonin scavenges a number of RNS including ONOO- and peroxynitrous acid (ONOOH). Moreover, melatonin has proven to scavenge alkoxyl, peroxyl radicals. The peroxyl radical (POO•), which is formed during the complex process of lipid peroxidation, is highly destructive to cells, because, once formed, it can propagate the process of lipid peroxidation. Therefore, agents that neutralize the POO• radical are generally known as chain breaking antioxidants, is important to maintaining the optimal function of not only cell membranes, but of cells themselves. It is estimated that each molecule of melatonin scavenged four POO• molecules (Pieri et al. 1994, Pieri et al. 1995), which would make it twice as effective as vitamin E, the principal well known chain-breaking lipid antioxidant and POO• scavenger. The most important products of melatonin's interaction with H_2O_2, N1-acetyl-N2-formyl-5-methoxykynuramine (AFMK), N-acetyl-5-methoxykynuramine(AMK), and 6-hydroxymelatonin, are also a highly efficient radical scavenger (Maharaj et al. 2003). The cascade of reactions where the secondary metabolites are also effective scavenges is believed to contribute to melatonin's high efficacy in reducing oxidative damage.

Very relevant to the development of this chapter is an alternate concept proposed to explain the protective effects of melatonin at the level of radical generation rather than detoxification of radicals already formed. It has been suggested that if melatonin is capable of decreasing the processes leading to enhanced radical formation, this might be achieved by low concentrations of the indole (Hardeland et al. 2011). Therefore, the main sources of free radicals should be investigated with regard to their modulation by melatonin and their sensitivity to the stressor. In addition to free radicals generated by leukocytes, mitochondria should be mentioned as main sources. In support of this suggestion, melatonin may also reduce free radical generation in mitochondria by improving oxidative phosphorylation, thereby lowering electron leakage and increasing ATP generation (Acuna-Castroviejo et al. 2001).

In addition to its direct free radical scavenging actions, melatonin influences both functional integrity of other antioxidative enzymes and cellular mRNA levels for these enzymes. Many studies documented the influence of melatonin on the activity and expression of the antioxidants both under physiological and under conditions of elevated oxidative stress. As an indirect antioxidant, melatonin stimulates gene expression and activity of SOD, thereby inducing the rapid conversion of $O_2 \bullet -$ to H_2O_2. The removal of $O_2 \bullet -$ by SOD also leads to a reduced formation of the highly reactive and damaging ONOO-. Catalase (CAT) and glutathione peroxidase (GSH-Px), enzymes that metabolizes H_2O_2 to H_2O, have also been shown to be stimulated by melatonin (Karbownik and Reiter 2000). In fact, it was demonstrated by several investigator that melatonin stimulates the rate-limiting enzyme, γ-glutamylcysteine synthase thereby increasing the level of an important endogenous antioxidant, glutathione (GSH) (El-Missiry et al. 2007, Othman et al. 2008, Urata et al. 1999) which, besides being a radical scavenger, is used by GSH-Px as a substrate to metabolize H_2O_2. In this process, GSH is converted to oxidized glutathione (GSSG). To maintain high levels of GSH, melatonin promotes the activity of glutathione reductase, which converts GSSG back to GSH. The possible intracellular mechanisms and pathways by which melatonin regulates antioxidant enzymes were reviewed (Rodriguez et al. 2004).

An added value of melatonin is that its metabolite N1-acetyl-N2-formyl-5-methoxykynuramine (AFMK) also has remarkable antioxidant properties and redox potential (Tan et al. 2001). It is formed when melatonin interacts with ROS, in particular, $^1O_2$ and H_2O_2. AFMK can be then enzymatically converted, by CAT, to N1-acetyl-5-methoxykynuramine (AMK). Cyclic 3-hydroxymelatonin (C3-OHM) is another product formed from melatonin by its interactions with free radicals, (Tan et al. 1998), which can be further metabolized by free radicals to AFMK (Tan et al. 2003). All these findings indicate that AFMK is a central metabolite of melatonin oxidation especially in nonhepatic tissues.

Interestingly, melatonin was shown to prevent the loss of important dietary antioxidants including Vitamins C and E (Susa et al. 1997), bind iron and participate in maintaining iron pool at appropriate level resulting in control of iron haemostasis, thereby providing tissue protection (Othman et al. 2008). Furthermore, melatonin enhances antioxidant action of a-tocopherol and ascorbate against NADPH- and iron-dependent lipid peroxidation in human placental mitochondria (Milczarek et al. 2010).

The bioavailability is the main factor that determines the capacity of antioxidants in vivo. The antioxidants should be absorbed, transported, distributed, and retained properly in the biological fluids, cells and tissues (Cheeseman and Slater 1993, Niki 2010). Recently, it has been suggested that melatonin present in edible plants may improve human health, by virtue of its biological activities and its good bioavailability (Iriti et al. 2010). This could add a new factor to the of health benefits for patients associated to radiotherapy. Melatonin's interactions with other drugs that influence its bioavailability were summarized (Tan et al. 2007). Furthermore, melatonin can be administered by virtually any route, including orally, via submucosal or transdermal patches, sublingually, intranasally, intravenously (Reiter 2003).

6.3 Radioprotective effect of melatonin and its metabolites

The radiosensitivity of cells to ionizing radiation depends on several factors including the efficiency of the endogenous antioxidative defense systems to prevent oxidative stress. A number of natural and synthetic compounds have been investigated for their antioxidative as well as radioprotective potential. Most of the effective compounds were found to have some inconvenient side effects such as hypotention, hypocalcemia, nausea, vomiting and hot flashes. Furthermore, most of compounds must be given intravenously or subcutaneously which restrict their clinical application outside of controlled clinical situations. Thus, a need still exists for identifying a non-toxic, effective, and convenient compound to protect humans against radiation damage in accidental, occupational, clinical settings and space-travel. Melatonin received much attention for its unique antioxidative potential at a very low concentration compared with other antioxidants.

A number of in vitro and in vivo studies have reported that exogenously administered melatonin provides profound protection against radiation induced lipid peroxidation and oxidative stress (El-Missiry et al. 2007, Taysi et al. 2003). As indicated above, radiation-induced lipid peroxidation is a three steps free radical process. Melatonin inhibits lipid peroxidation by preventing the initiation phase of lipid peroxidation and interrupting the chain reaction. This is mainly due to melatonin ability to quench •OH and several other ROS and RNS. It has been reported to scavenge the several ROS, in particular •OH and the ONOO-•. When melatonin interacts with •OH it becomes indolyl (melatonyl) radical which has very low toxicity. After some molecular rearrangements, the melatonyl radical scavenges a second •OH to form cyclic 3-hydroxymelatonin (3-OHM). Thus, this reaction pathway suggests that 3-OHM is the footprint product of the interaction between melatonin with •OH. 3-OHM was also detected in the urine of both rats and humans. This provides direct evidence that melatonin, under physiological conditions, functions as an antioxidant to detoxify the most reactive and cytotoxic endogenous •OH (Tan et al., 1999). When rats were exposed to ionizing radiation which results in •OH generation, urinary 3-OHM increased significantly compared to that of controls (Tan et al. 1998). This provided direct evidence that radiation induced oxidative stress increases melatonin consumption in rats. The rapid decrease in circulating melatonin under conditions of excessive stress can be considered a protective mechanism for organisms against highly damaging free radicals; in this sense, melatonin can be categorized as a first line of defensive molecule (Tan et al. 2007). Along this line, the melatonin's metabolite AFMK protected against space radiation induced impairment of memory and hippocampal neurogenesis in adult C57BL mice (Manda et al. 2008). This study demonstrated that radiation exposure (2.0 Gy of 500 MeV/nucleon [56]Fe beams, a ground-based model of space radiation) significantly reduced the spatial memory of mice without affecting their motor activity. It is also reported that AFMK pretreatment significantly ameliorated radiation induced neurobehavioral ailments and reduced the loss of doublecortin and cell proliferation. Radiation exposure dramatically augmented the level of 8-OHdG in serum as well as DNA migration in the comet tail were impaired by AFMK pretreatment. In addition, radiation-induced augmentation of protein carbonyl content and 4-HNE + MDA and reduced the level of brain sulfhydryl contents was ameliorated by AFMK pretreatment (Manda et al. 2008). The ameliorating action of AFMK against radiation induced lipid peroxidation was attributed to free-radical scavenging property of AFMK. In vitro, AFMK showed a very high level of •OH scavenging potential which was measured by an electron spin resonance spin study of the 2-hydroxy-5,5-dimethyl-1-pyrrolineN-oxide (DMPO-OH) adduct. In this experiment, 10 Gy of X-ray for the radiolysis of water with different concentration of AFMK was used and intensity of spin adduct (DMPO-OH) were measured by ESR (Manda et al. 2007).

Extensive studies have established that pretreatment with melatonin at physiological dose 5 mg/kg or pharmacological dose 10 mg/kg bw significantly decreased MDA and NO• levels. The data documented that melatonin reduces tissue damage inflicted by irradiation when given prior to the exposure to ionizing radiation (Babicova et al. 2011, Taysi et al. 2003, Verma et al. 2010). These authors explained that NO• is formed in higher amounts from L-arginine by inducible nitric oxide synthase (iNOS) during early response to ionizing radiation presumably as a part of signal transduction pathways (Babicova et al. 2011). Its cytotoxicity is primarily due to the production of ONOO-, a toxic oxidant, generated when the NO• couples with O_2•- (El-Sokkary et al. 2002). The processes triggered by ONOO- include initiation of lipid peroxidation, inhibition of mitochondrial respiration, inhibition of membrane pumps, depletion of GSH, and damage to DNA. Melatonin is reported to scavenge ONOO- both in vitro and in vivo (El-Sokkary et al. 1999, Gilad et al. 1997) and to inhibit iNOS activity thereby reducing excessive NO• generation.

It should emphasized that ionizing radiation causes oxidative damage to tissues within an extremely short period, and possible protection against it would require the rapid transfer of antioxidants to the sensitive sites in cells. At this point, melatonin seems unique among cellular antioxidants because of its physical and chemical properties; it can easily cross biological membranes and reach the cytosol, nucleus, and cellular compartments (Menendez-Pelaez and Reiter 1993). The effect of melatonin in maintaining normal hepatic and renal functions may be related to its ability to localize mainly in a superficial position in the lipid bilayer near the polar heads of membrane phospholipids (El-Sokkary et al. 2002). The protective action of melatonin against lipid and protein oxidation as a factor modifying membrane organization may also be related to melatonin's ability to scavenge the oxidation-initiating agents, which are produced during the oxidation of proteins and lipids. Since membrane functions and structure are influenced by proteins in membranes and radiation is known to damage thiol proteins (Biaglow et al. 2003), it is possible that the protective action of melatonin against membrane damage may be related partially to the ability of melatonin to prevent lipid and protein oxidative damage (Le Maire et al. 1990). Changes in membrane structure and fluidity due to ROS reactivity after irradiation are also attributed to graded alterations in the lipid bilayer environment (Karbownik and Reiter 2000). It has been suggested that the protective role of melatonin in preserving optimal levels of fluidity of the biological membranes may be related to its ability to reduce lipid peroxidation (Garcia et al. 1997, Garcia et al. 2001).

Moreover, melatonin prevents inflammation and MDA caused by abdominopelvic and total body irradiation of rat (Taysi et al. 2003). Thus, the radioprotective effect of melatonin is likely achieved by its ability to function as a scavenger for free radicals generated by ionizing radiation. Furthermore, these findings may suggest that melatonin may enable the use of higher doses of radiation during therapy and may therefore allow higher dose rates in some patients with cancer. In another study, pretreatment with melatonin (10mg/kg bw) for 4 days before acute γ-irradiation significantly abolished radiation induced elevations in MDA and protein carbonyl levels in the liver and significantly prevented the decrease in hepatic GSH content, GST, and CAT activities. Moreover, preirradiation treatment with melatonin showed significantly higher hepatic DNA and RNA contents than irradiated rats. The levels of total lipids, cholesterol, triglycerides (TG), high density lipoproteins (HDL), low density lipoproteins (LDL), total proteins, albumin, total globulins, creatinine, and urea, as well as the activities of AST, ALT, and GGT in serum were significantly ameliorated when melatonin was injected before irradiation. The protection evidenced by normalization of the clinical parameters was associated with and attributed to decreased lipid peroxidation in the presence of melatonin. This data indicate that melatonin has a radioprotective impact against ionizing-radiation induced oxidative stress and organ injury (El-Missiry et al. 2007). At sufficiently high radiation doses, GSH becomes depleted, leaving highly reactive ROS, beyond the immediate and normal needs of the cell, to react with critical cellular biomolecules and cause tissue damage. The concentration of intracellular GSH, therefore, is the key determinant of the extent of radiation-induced tissue injury. Thus, interest has been focused on melatonin, which acts as an antioxidant and is capable of stimulating GSH synthesis. It has been shown that melatonin enhances intracellular GSH levels by stimulating the rate-limiting enzyme in its synthesis, γ-glutamylcysteine synthase, which inhibits the peroxidative enzymes NOS and lipoxygenase. Experimentally, melatonin is demonstrated to increase hepatic GSH content, (El-Missiry et al. 2007, Sener et al. 2003), and hence to inhibit formation of extracellular and intracellular ROS (Reiter et al. 2004). It is also proposed that prevention of GSH depletion is the most efficient method of direct protection against radiation-induced oxidative toxicity. A significant decrease in hepatic GST activity and increase in serum GGT activity was recorded after exposure to 2&4 Gy (El-Missiry et al. 2007, Sridharan and Shyamaladevi 2002), and 10 Gy gamma irradiation (Samanta et al. 2004). Melatonin treatment at a dose level of 10mg/Kg bw before irradiation significantly countered radiation-induced decrease in the activities of these enzymes in the liver and serum. Furthermore, melatonin increases the activity of GSH-Px (Barlow-Walden et al. 1995) and superoxide dismutase (Antolin et al. 1996). These findings support the conclusion that melatonin affords radioprotection by modulating antioxidative enzyme activities in the body (El-Missiry et al. 2007).

Al the level of clinical markers, like cholesterol, TG, LDL, and free radicals are also risk factors that tend to damage arteries, leading to cardiovascular diseases. Pre-irradiation treatment with melatonin is found to reduce serum cholesterol, TG, and LDL levels in serum, indicating modulation of lipid metabolism in cells. It is well reported that antioxidants reduce oxidation susceptibility of HDL (Schnell et al. 2001) and control hyperlipidaemia (Mary et al. 2002). This might potentiate antiatherogenic effects of antioxidants including melatonin. Therefore, it is suggested that preirradiation treatment with melatonin appears to be hypolipidemic which might potentate its beneficial use in occupational, clinical, and space settings (El-Missiry et al. 2007).

Ionizing irradiation is among reproductive harmful agent and is widely identified to affect testicular function, morphology and spermatogenesis. Irradiation of the testes can produce reversible or permanent sterility in males. In an experiment, rats were subjected to sublethal irradiation dose of 8 Gy, either to the total body or abdominopelvic region using a ^{60}Co source. In this experiment, melatonin pretreatment resulted in less apoptosis as indicated by a considerable decrease in caspace-3 immunoreactivity. Electron microscopic examination showed that all spermatogenic cells, especially primary spermatocytes, displayed considerably inhibited of degeneration in the groups treated with melatonin before total body and abdominopelvic irradiation (Take et al. 2009).

An extensive literature implicates cellular DNA as the primary target for the biological and lethal effects of ionizing radiation. Melatonin has the ability to protect the DNA of hematopoietic cells of mice from the damaging effects of acute whole-body irradiation (Vijayalaxmi et al. 1999). The radioprotective ability of melatonin was investigated in the

Indian tropical rodent, Funambulus pennanti during its reproductively inactive phase when peripheral melatonin is high and the animal is under the influence of environmental stresses. Exogenous melatonin with its anti-apoptotic and antioxidant properties additively increased the immunity of the squirrels, by protecting their hematopoietic system and lymphoid organs against X-ray radiation induced cellular toxicity (Sharma et al. 2008). Human keratinocytes is the main target cells in epidermal photodamage. Protection against UVR-induced skin damage was manifested by suppression of UV-induced erythema by topical pretreatment with melatonin with / without combination of vitamins C and E (Bangha et al. 1997, Dreher et al. 1998). Melatonin increased cell survival of HaCaT keratinocytes and ensured keratinocyte colony growth under UV-induced stress and showed decrease of UV-induced DNA fragmentation. Also, transcription of several classical target genes which are up-regulated after UV-exposure and play an important role in the execution of skin photodamage were down-regulated in HaCaT keratinocytes by melatonin pretreatment. It has been previously reported that melatonin reduces UVB-induced cell damage and polyamine levels in human skin fibroblasts (Lee et al. 2003). Furthermore, it was reported that melatonin increases survival of HaCaT keratinocytes by suppressing UV-induced apoptosis (Fischer et al. 2006). The molecular mechanisms underlying protective effects of melatonin on human keratinocytes and human fibroblasts upon UVB induced-apoptotic cell death was investigated (Cho et al. 2007). In this study, cDNA microarray analysis was perform from HaCaT keratinocytes, exposed to 100 mJ /cm2 and pretreated with melatonin for 30 min. Data showed that melatonin inhibits the expression of apoptosis related protein-3, apoptotic chromatin condensation inducer in the nucleus, and glutathione peroxidase 1 in UVB-irradiated HaCaT cells. The inhibitory effect of melatonin upon UVB irradiation is likely to be associated with antioxidant capacity of melatonin, thereby suggesting that melatonin may be used as a sunscreen agent to reduce cell death of keratinocytes after excessive UVB irradiation.

Radiotherapy plays an important role as part of the multimodality treatment for a number of malignancies in children. In young children, significant growth arrest was demonstrated with fractionated doses of 15 Gy and above as well as, in children less than one year of age, with doses as low as 10 Gy (Robertson et al. 1991). Proliferating chondrocytes, distal metaphyseal vessels, and epiphyseal vasculature are main targets for radiation-induced injury of bone growth plate (Kember 1967). Proliferating chondrocytes show marked cytological changes and cell death with a single fraction of 5 Gy. Melatonin with its antioxidant capacity protected against the hypoxic conditions of chondrocytes and promoted continued proliferation despite exposure to radiation. Moreover, in vitro studies showed that melatonin is capable of promoting osteoblast proliferation directly and stimulating these cells to produce increased amounts of several bone matrix proteins such as bone sialoprotein, alkaline phosphatase, osteocalcin (Roth et al. 1999) and procollagen type I c-peptide (Nakade et al. 1999), responsible for bone formation. Osteoprotegerin, an osteoblastic protein that inhibits the differentiation of osteoclasts is also increased by melatonin in vitro (Koyama et al. 2002). This data may support the radioprotective effect of melatonin on bone growth. The effects of fractionated radiotherapy combined with radioprotection by melatonin compared with fractionated radiotherapy alone in preserving the integrity and function of the epiphyseal growth plate from radiation damage in a weanling rat model was investigated. Data revealed that melatonin is more protective for bone growth protection than amifostine and a potential exists to implement the use of melatonin in an effort to maximize the radiotherapeutic management of children with less long-term morbidity than previous clinical experience (Yavuz et al. 2003).

Recent studies have documented that radiation in general and radiotherapy in particular has effects on brain function, such as thinking, memory and learning ability (Hsiao et al. 2010). Because cognitive health of an organism is maintained by the ability of hippocampal precursors to proliferate and differentiate, radiation exposures have been shown to inhibit neurogenesis and are associated with the onset of cognitive impairments. In recent investigation, on the protection by melatonin against the delayed effects of cranial irradiation on hippocampal neurogenesis melatonin maintained adult hippocampal neurogenesis and cognitive functions after irradiation (Manda and Reiter 2010). In this study the pretreatment with melatonin showed a significantly higher count of microtubule binding protein doublecortin and the proliferation marker Ki-67 positive cells compared with irradiated only animals. The protection was achieved by a single intraperitoneal injection of 10 mg melatonin/kg bw prior to irradiation. These protective effects were accompanied by significant control of oxidative stress indicated by reduction in the count of immunohistochemical localization of DNA damage and lipid peroxidation using the anti-8-hydroxy-2-deoxyguanosin the anti-hydroxynonenal. This indicated that melatonin minimize cell death.

6.4 Melatonin modulates apoptosis in radiotherapy and space radiation

Recent studies have showed the exposure to heavy ions such as ^{56}Fe or ^{12}C particle can induce detrimental physiological and histological changes in the brain, which lead to behavioral changes, spatial learning, and memory deficits. During space travel, astronauts are exposed to high-LET galactic cosmic rays at higher radiation doses and dose rates than humans received on Earth is one of the acknowledged showstoppers for long duration manned interplanetary missions. Hadrontherapy is an innovative modality of high precision tool for radiotherapy which consists in using hadrons (mainly protons or carbon ions) to irradiate tumors. This technique is used in certain cases to treat patients whose tumors are resistant to conventional X-ray radiotherapy. Given cancer therapy and space radiation protection, there is a demand for reliable agent for the protection of the brain against oxidative stress induced by heavy-ion radiation. It is well known that ionizing radiations can induce apoptosis. It is well known that oxidative stress is a mediator of apoptosis by compromising the fine balance between intracellular oxidant and their defense systems to produce abnormally high levels of ROS. Melatonin supplementation at 1,3&10mg/Kg bw reduced irradiation-induced oxidative damage, and stimulated the activities of SOD & CAT together with total antioxidant capacity in brain of rats exposed to heavy-ion radiation. Furthermore, pretreatment with melatonin significantly elevated the expression of Nrf2 which regulates redox balance and stress. In addition, pre-irradiation treatment with melatonin mitigated apoptotic rate, maintained mitochondrial membrane potential, decreased cytochrome C release from mitochondria, down-regulated Bax/Bcl-2 ratio and caspase-3 levels, and consequently inhibited the important steps of irradiation-induced activation of mitochondrial pathway of apoptosis (Liu et al. 2011). Studies performed by other investigators documented that melatonin pretreatment inhibited the cerebellum cell apoptosis after mice received 2 Gy ^{56}Fe particle irradiation (Manda and Reiter 2010) and that decreased apoptosis by melatonin was associated with a reduction in Bax/Bcl-2 ratio in mice splenocytes exposed to 2 Gy X-ray irradiation (Jang et al. 2009). Along this line, melatonin supplementation suppresses NO-induced apoptosis via induction of Bcl-2 expression in immortalized pineal PGT-β cells (Yim et al. 2002). A similar pathway of inhibitory effects of melatonin on apoptosis induced by ischemic neuronal injury has been determined (Sun et al. 2002).

Relevant to this context, it has been recently proposed that ERK MAPK plays a central role to determine whether cells will live or die in response to apoptotic stimuli. It is well documented that the apoptotic signaling activated during UVB stress mainly converges at the mitochondrial level into intrinsic pathway and supporting evidence consider this pathway might be the principle target of melatonin to prevent apoptosis in human leukocytes (Radogna et al. 2008) as well as in other tumor cell lines and in vivo models (Acuna-Castroviejo et al. 2007). An in vitro stress model for the cell protection and antiapoptotic functions of melatonin was studied using U937 cells exposed to UVB radiation (Luchetti et al. 2006). Melatonin sustained the activation of the survival-promoting pathway ERK MAPK (extracellular signal-regulated kinase) which controls the balance between survival and death-promoting genes throughout the MAPK pathway, and is required to antagonize UVB induced apoptosis of U937 cells. This kinase was found to modulate the antioxidant and mitochondrial protection effects of melatonin that may find therapeutic applications in inflammatory and immune diseases associated with leukocyte oxidative stress and accelerated apoptosis (Luchetti et al. 2009, Luchetti et al. 2006). Recently, it is reported that redox-sensitive components are included in the cell protection signaling of melatonin and in the resulting transcriptional response that involves the control of NF-κB, AP-1, and Nrf2. Through these pathways, melatonin stimulates the expression of antioxidant and detoxification genes, acting in turn as a glutathione system promoter (Luchetti et al. 2010).

It is suggested that ionizing radiation produces oxidative stress due formation of mitochondrial ROS resulting in calcium influx into the mitochondria with opening of the mitochondrial permeability transition pore (MPTP) (Andrabi et al. 2004, Halestrap 2006) and depolarization of the mitochondrial membrane potential as the end result of radiation-induced mitochondrial damage and cell apoptosis. The different regulatory mechanisms of apoptosis and their modification by treatment with melatonin were tested in different cells after irradiation. It is found that the mitochondrial pathway was strongly influenced by melatonin by reducing mitochondrial ROS generation and calcium release as well as inhibiting the opening of the MPTP as shown in rat brain astrocytes (Jou et al. 2004), mouse striatal neurons (Andrabi et al. 2004) and rat cerebellar granule neurons (Han et al. 2006). Moreover, the prevented decreases in the mitochondrial membrane potential resulted from irradiation suggests that melatonin, due to its physiochmeical characters crosses the blood–brain barrier and biological membranes to easily reach mitochondria, stabilizes oxidative stress-mediated dysfunction and integrity of mitochondria by preserving its membrane potential and increasing the efficiency of mitochondrial electron transfer chain and ATP synthesis (Acuna-Castroviejo et al. 2001). When melatonin treated cultured keratinocytes were irradiated with UVB radiation (50 mJ/cm2), there were less cell leaky, more uniform shape and less nuclear condensation as compared to irradiated, nonmelatonin-treated controls (Fischer et al. 2006). Exogenous melatonin with its anti-apoptotic and antioxidant properties additively increased the immunity of the animals, by protecting their hematopoietic system and lymphoid organs against X-ray radiation induced cellular toxicity (Sharma et al. 2008). These findings strongly highlight melatonin as a potential antiaopoptotic neuroprotective and mitigative agent against the space radiation hazards and the side effect risk in hadrontherapy. Consistent with all the overwhelming experimental evidence described above it may be concluded that melatonin can efficiently protect against and mitigate radiation induced oxidative stress. The majority of the published works on its ROS scavenging action coincide on the conclusion that melatonin is excellent for this task and make melatonin efficient pharmacological radioprotectors.

7. Toxicity and biosafety of melatonin

The melatonin doses chosen in several studies were between 5 and 15 mg/kg bw, which are rather minimal effective doses as reported in animal studies. Whereas, pharmacological studies in rats of up to 250 mg/kg doses did not indicate any adverse effects. In addition, human volunteers who ingested a single oral dose of 1–300 mg and 1 g of melatonin daily for 30 days did not report any adverse side effects. In a study, none of the 15 weanling rats administered with 5–15 mg/kg of melatonin died during the 6-wk observation period (Yavuz et al. 2003). In addition, ip treatment with melatonin for 45 days did not show abnormal singes (El-Missiry et al. 2007). All of these observations provide support for the non-toxic nature of melatonin (Cheung et al. 2006).

8. Conclusion

Apart from nuclear accidents, radiation has been used increasingly in medicine and industry to help with diagnosis, treatment, and technology. However, radiation hazards present an enormous challenge for the biological and medical safety. The deleterious effects of ionizing radiation in biological systems are mainly mediated through the generation ROS in cells as a result of water radiolysis leading to oxidative stress. •OH considered the most damaging of all free radicals generated in organisms, are often responsible for biomolecular damage caused by ionizing radiation. Oxidative stress greatly contributes to radiation-induced cytotoxicity and to metabolic and morphologic changes in animals and humans during occupational settings, radiotherapy, and space flight. Melatonin is an indoleamine hormone synthesized from tryptophan in pinealocytes. It is distributed ubiquitously in organisms and in all cellular compartments, and it easily passes through all biological membranes. Several studies have indicated that melatonin may act as a scavenger of ROS such as hydroxyl radical, alkoxyl radical, hypochlorous acid and singlet oxygen. A number of in vitro and in vivo studies have reported that exogenously administered melatonin provides protection against radiation induced oxidative stress in different species. Its ability to reduce DNA damage, lipid peroxidation, and protein damage may originate from its function as a preventive antioxidant (scavenging initiating radicals directly or indirectly). Furthermore, this indoleamine manifests its antioxidative properties by upregulation of endogenous antioxidant defense mechanisms, increases the efficiency of the electron transport chain thereby limiting electron leakage and free radical generation, protects the integrity of the mitochondria and promotes ATP synthesis. Furthermore, several metabolites that are formed when melatonin neutralizes damaging reactants are themselves scavengers suggesting scavenging cascade reaction that greatly increase the efficacy of melatonin in preventing oxidative damage. Several observations provide support for the non-toxic nature of melatonin. The radioprotective and mitigative effects of melatonin against cellular damage caused by oxidative stress and its low toxicity make this innate antioxidant a potential drug in situations where the effects of ionizing radiation are to be controlled.

9. References

Acharya MM, Lan ML, Kan VH, Patel NH, Giedzinski E, Tseng BP, Limoli CL. 2010. Consequences of ionizing radiation-induced damage in human neural stem cells. Free Radic Biol Med 49: 1846-1855.

Acuna-Castroviejo D, Escames G, Rodriguez MI, Lopez LC. 2007. Melatonin role in the mitochondrial function. Front Biosci 12: 947-963.

Acuna-Castroviejo D, Martin M, Macias M, Escames G, Leon J, Khaldy H, Reiter RJ. 2001. Melatonin, mitochondria, and cellular bioenergetics. J Pineal Res 30: 65-74.

Altieri F, Grillo C, Maceroni M, Chichiarelli S. 2008. DNA damage and repair: from molecular mechanisms to health implications. Antioxid Redox Signal 10: 891-937.

Andrabi SA, Sayeed I, Siemen D, Wolf G, Horn TF. 2004. Direct inhibition of the mitochondrial permeability transition pore: a possible mechanism responsible for anti-apoptotic effects of melatonin. FASEB J 18: 869-871.

Antolin I, Rodriguez C, Sainz RM, Mayo JC, Uria H, Kotler ML, Rodriguez-Colunga MJ, Tolivia D, Menendez-Pelaez A. 1996. Neurohormone melatonin prevents cell damage: effect on gene expression for antioxidant enzymes. FASEB J 10: 882-890.

Babicova A, Havlinova Z, Pejchal J, Tichy A, Rezacova M, Vavrova J, Chladek J. 2011. Early changes in L-arginine-nitric oxide metabolic pathways in response to the whole-body gamma irradiation of rats. Int J Radiat Biol.

Bangha E, Elsner P, Kistler GS. 1997. Suppression of UV-induced erythema by topical treatment with melatonin (N-acetyl-5-methoxytryptamine). Influence of the application time point. Dermatology 195: 248-252.

Barlow-Walden LR, Reiter RJ, Abe M, Pablos M, Menendez-Pelaez A, Chen LD, Poeggeler B. 1995. Melatonin stimulates brain glutathione peroxidase activity. Neurochemistry International 26: 497-502.

Bhatia AL, Manda K. 2004. Study on pre-treatment of melatonin against radiation-induced oxidative stress in mice. Environ Toxicol Pharmacol 18: 13-20.

Biaglow JE, Ayene IS, Koch CJ, Donahue J, Stamato TD, Mieyal JJ, Tuttle SW. 2003. Radiation response of cells during altered protein thiol redox. Radiat Res 159: 484-494.

Blakely EA, Chang PY. 2007. A review of ground-based heavy ion radiobiology relevant to space radiation risk assessment: Cataracts and CNS effects. Advances in Space Research 40: 1307-1319.

Cadet J, Douki T, Ravanat JL. 2010. Oxidatively generated base damage to cellular DNA. Free Radic Biol Med 49: 9-21.

Cadet J, Douki T, Gasparutto D, Ravanat JL. 2005. Radiation-induced damage to cellular DNA: measurement and biological role. Radiation Physics and Chemistry 72: 293-299.

Catala A. 2007. The ability of melatonin to counteract lipid peroxidation in biological membranes. Curr Mol Med 7: 638-649.

Catala A. 2009. Lipid peroxidation of membrane phospholipids generates hydroxy-alkenals and oxidized phospholipids active in physiological and/or pathological conditions. Chem Phys Lipids 157: 1-11.

Cheeseman KH, Slater TF. 1993. An introduction to free radical biochemistry. Br Med Bull 49: 481-493.

Cheung RT, Tipoe GL, Tam S, Ma ES, Zou LY, Chan PS. 2006. Preclinical evaluation of pharmacokinetics and safety of melatonin in propylene glycol for intravenous administration. J Pineal Res 41: 337-343.

Chiquet C, Claustrat B, Thuret G, Brun J, Cooper HM, Denis P. 2006. Melatonin concentrations in aqueous humor of glaucoma patients. Am J Ophthalmol 142: 325-327 e321.

Cho JW, Kim CW, Lee KS. 2007. Modification of gene expression by melatonin in UVB-irradiated HaCaT keratinocyte cell lines using a cDNA microarray. Oncol Rep 17: 573-577.

Citrin D, Cotrim AP, Hyodo F, Baum BJ, Krishna MC, Mitchell JB. 2010. Radioprotectors and mitigators of radiation-induced normal tissue injury. Oncologist 15: 360-371.

Comporti M, Signorini C, Arezzini B, Vecchio D, Monaco B, Gardi C. 2008. F2-isoprostanes are not just markers of oxidative stress. Free Radic Biol Med 44: 247-256.

Cooke MS, Evans MD, Dizdaroglu M, Lunec J. 2003. Oxidative DNA damage: mechanisms, mutation, and disease. FASEB J 17: 1195-1214.

Cutando A, Aneiros-Fernandez J, Lopez-Valverde A, Arias-Santiago S, Aneiros-Cachaza J, Reiter RJ. 2011. A new perspective in Oral health: Potential importance and actions of melatonin receptors MT1, MT2, MT3, and RZR/ROR in the oral cavity. Arch Oral Biol.

Draganic IG, Draganic ZD. 1971. The Radiation Chemistry of Water. Academic Press, New York.

Dreher F, Gabard B, Schwindt DA, Maibach HI. 1998. Topical melatonin in combination with vitamins E and C protects skin from ultraviolet-induced erythema: a human study in vivo. Br J Dermatol 139: 332-339.

Dubocovich ML, Markowska M. 2005. Functional MT1 and MT2 melatonin receptors in mammals. Endocrine 27: 101-110.

El-Missiry MA, Fayed TA, El-Sawy MR, El-Sayed AA. 2007. Ameliorative effect of melatonin against gamma-irradiation-induced oxidative stress and tissue injury. Ecotoxicol Environ Saf 66: 278-286.

El-Sokkary GH, Omar HM, Hassanein AF, Cuzzocrea S, Reiter RJ. 2002. Melatonin reduces oxidative damage and increases survival of mice infected with Schistosoma mansoni. Free Radic Biol Med 32: 319-332.

El-Sokkary GH, Reiter RJ, Cuzzocrea S, Caputi AP, Hassanein AF, Tan DX. 1999. Role of melatonin in reduction of lipid peroxidation and peroxynitrite formation in non-septic shock induced by zymosan. Shock 12: 402-408.

Esposito E, Cuzzocrea S. 2010. Antiinflammatory activity of melatonin in central nervous system. Curr Neuropharmacol 8: 228-242.

Esterbauer H, Zollner H. 1989. Methods for determination of aldehydic lipid peroxidation products. Free Radic Biol Med 7: 197-203.

Fischer TW, Zbytek B, Sayre RM, Apostolov EO, Basnakian AG, Sweatman TW, Wortsman J, Elsner P, Slominski A. 2006. Melatonin increases survival of HaCaT keratinocytes by suppressing UV-induced apoptosis. J Pineal Res 40: 18-26.

Gago-Dominguez M, Castelao JE, Pike MC, Sevanian A, Haile RW. 2005. Role of lipid peroxidation in the epidemiology and prevention of breast cancer. Cancer Epidemiol Biomarkers Prev 14: 2829-2839.

Galano A, Tan DX, Reiter RJ. 2011. Melatonin as a natural ally against oxidative stress: a physicochemical examination. Journal of Pineal Research 51: 1-16.

Garcia JJ, Reiter RJ, Guerrero JM, Escames G, Yu BP, Oh CS, Munoz-Hoyos A. 1997. Melatonin prevents changes in microsomal membrane fluidity during induced lipid peroxidation. FEBS Lett 408: 297-300.

Garcia JJ, Reiter RJ, Karbownik M, Calvo JR, Ortiz GG, Tan DX, Martinez-Ballarin E, Acuna-Castroviejo D. 2001. N-acetylserotonin suppresses hepatic microsomal membrane rigidity associated with lipid peroxidation. Eur J Pharmacol 428: 169-175.

Giacomo CG, Antonio M. 2007. Melatonin in cardiac ischemia/reperfusion-induced mitochondrial adaptive changes. Cardiovasc Hematol Disord Drug Targets 7: 163-169.

Gilad E, Cuzzocrea S, Zingarelli B, Salzman AL, Szabo C. 1997. Melatonin is a scavenger of peroxynitrite. Life Sciences 60: PL169-174.

Goodhead DT. 1994. Initial events in the cellular effects of ionizing radiations: clustered damage in DNA. Int J Radiat Biol 65: 7-17.

Griffiths HR, et al. 2002. Biomarkers. Mol Aspects Med 23: 101-208.

Guajardo MH, Terrasa AM, Catala A. 2006. Lipid-protein modifications during ascorbate-Fe2+ peroxidation of photoreceptor membranes: protective effect of melatonin. J Pineal Res 41: 201-210.

Gulbahar O, Aricioglu A, Akmansu M, Turkozer Z. 2009. Effects of radiation on protein oxidation and lipid peroxidation in the brain tissue. Transplant Proc 41: 4394-4396.

Haegele AD, Wolfe P, Thompson HJ. 1998. X-radiation induces 8-hydroxy-2'-deoxyguanosine formation in vivo in rat mammary gland DNA. Carcinogenesis 19: 1319-1321.

Halestrap AP. 2006. Calcium, mitochondria and reperfusion injury: a pore way to die. Biochem Soc Trans 34: 232-237.

Halliwell B. 1997. Antioxidants: the basics--what they are and how to evaluate them. Adv Pharmacol 38: 3-20.

Halliwell B. 2009. The wanderings of a free radical. Free Radic Biol Med 46: 531-542.

Han YX, Zhang SH, Wang XM, Wu JB. 2006. Inhibition of mitochondria responsible for the anti-apoptotic effects of melatonin during ischemia-reperfusion. J Zhejiang Univ Sci B 7: 142-147.

Hardeland R, Poeggeler B, Balzer I, Behrmann G. 1993a. A hypothesis on the evolutionary origins of photoperiodism based on circadian rhythmicity of melatonin in phylogenetically distant organisms. Gutenbrunner, C., Hildebrandt, G., Moog, R. (Eds.), Chronobiology & Chronomedicine. 1: 113-120.

Hardeland R, Reiter RJ, Poeggeler B, Tan DX. 1993b. The significance of the metabolism of the neurohormone melatonin: antioxidative protection and formation of bioactive substances. Neurosci Biobehav Rev 17: 347-357.

Hardeland R, Cardinali DP, Srinivasan V, Spence DW, Brown GM, Pandi-Perumal SR. 2011. Melatonin--a pleiotropic, orchestrating regulator molecule. Prog Neurobiol 93: 350-384.

Hawkins CL, Davies MJ. 2001. Generation and propagation of radical reactions on proteins. Biochim Biophys Acta 1504: 196-219.

Hong M, Xu A, Zhou H, Wu L, Randers-Pehrson G, Santella RM, Yu Z, Hei TK. 2010. Mechanism of genotoxicity induced by targeted cytoplasmic irradiation. Br J Cancer 103: 1263-1268.

Hosseinimehr SJ. 2007. Trends in the development of radioprotective agents. Drug Discov Today 12: 794-805.

Hsiao KY, Yeh SA, Chang CC, Tsai PC, Wu JM, Gau JS. 2010. Cognitive Function before and after Intensity-Modulated Radiation Therapy in Patients with Nasopharyngeal Carcinoma: A Prospective Study. International Journal of Radiation Oncology Biology Physics 77: 722-726.

Hutchinson F. 1985. Chemical changes induced in DNA by ionizing radiation. Prog Nucleic Acid Res Mol Biol 32: 115-154.

Illnerova H, Buresova M, Presl J. 1993. Melatonin rhythm in human milk. J Clin Endocrinol Metab 77: 838-841.

Inoue S, Ejima K, Iwai E, Hayashi H, Appel J, Tyystjarvi E, Murata N, Nishiyama Y. 2011. Protection by alpha-tocopherol of the repair of photosystem II during photoinhibition in Synechocystis sp. PCC 6803. Biochim Biophys Acta 1807: 236-241.

Iriti M, Varoni EM, Vitalini S. 2010. Melatonin in traditional Mediterranean diets. J Pineal Res 49: 101-105.

Jang SS, Kim WD, Park WY. 2009. Melatonin exerts differential actions on X-ray radiation-induced apoptosis in normal mice splenocytes and Jurkat leukemia cells. Journal of Pineal Research 47: 147-155.

Jou MJ, Peng TI, Reiter RJ, Jou SB, Wu HY, Wen ST. 2004. Visualization of the antioxidative effects of melatonin at the mitochondrial level during oxidative stress-induced apoptosis of rat brain astrocytes. J Pineal Res 37: 55-70.

Kamat JP, Devasagayam TP, Priyadarsini KI, Mohan H. 2000. Reactive oxygen species mediated membrane damage induced by fullerene derivatives and its possible biological implications. Toxicology 155: 55-61.

Karbownik M, Reiter RJ. 2000. Antioxidative effects of melatonin in protection against cellular damage caused by ionizing radiation. Proc Soc Exp Biol Med 225: 9-22.

Kember NF. 1967. Cell survival and radiation damage in growth cartilage. Br J Radiol 40: 496-505.

Klaunig JE, Wang Z, Pu X, Zhou S. 2011. Oxidative stress and oxidative damage in chemical carcinogenesis. Toxicol Appl Pharmacol 254: 86-99.

Koyama H, Nakade O, Takada Y, Kaku T, Lau KH. 2002. Melatonin at pharmacologic doses increases bone mass by suppressing resorption through down-regulation of the RANKL-mediated osteoclast formation and activation. Journal of Bone and Mineral Research 17: 1219-1229.

Kryston TB, Georgiev AB, Pissis P, Georgakilas AG. 2011. Role of oxidative stress and DNA damage in human carcinogenesis. Mutat Res 711: 193-201.

Lakshmi B, Tilak JC, Adhikari S, A. DTP, andJanardhanan KK. 2005. Inhibition of lipid peroxidationinduced by g-radiation and AAPH inrat liver and brain mitochondria by mushrooms. CURRENT SCIENCE 88: 484-488.

Le Maire M, Thauvette L, Deforesta B, Viel A, Beauregard G, Potier M. 1990. Effects of Ionizing-Radiations on Proteins - Evidence of Nonrandom Fragmentations and a Caution in the Use of the Method for Determination of Molecular Mass. Biochemical Journal 267: 431-439.

Lee KS, Lee WS, Suh SI, Kim SP, Lee SR, Ryoo YW, Kim BC. 2003. Melatonin reduces ultraviolet-B induced cell damages and polyamine levels in human skin fibroblasts in culture. Exp Mol Med 35: 263-268.

ittle JB. 2000. Radiation carcinogenesis. Carcinogenesis 21: 397-404.

iu Y, Zhang L, Zhang H, Liu B, Wu Z, Zhao W, Wang Z. 2011. Exogenous melatonin modulates apoptosis in the mouse brain induced by high-LET carbon ion irradiation. J Pineal Res.

uboshitzky R, Lavi S, Lavie P. 1999. The association between melatonin and sleep stages in normal adults and hypogonadal men. Sleep 22: 867-874.

uchetti F, Betti M, Canonico B, Arcangeletti M, Ferri P, Galli F, Papa S. 2009. ERK MAPK activation mediates the antiapoptotic signaling of melatonin in UVB-stressed U937 cells. Free Radic Biol Med 46: 339-351.

uchetti F, Canonico B, Curci R, Battistelli M, Mannello F, Papa S, Tarzia G, Falcieri E. 2006. Melatonin prevents apoptosis induced by UV-B treatment in U937 cell line. J Pineal Res 40: 158-167.

uchetti F, Canonico B, Betti M, Arcangeletti M, Pilolli F, Piroddi M, Canesi L, Papa S, Galli F. 2010. Melatonin signaling and cell protection function. FASEB J 24: 3603-3624.

Maharaj DS, Limson JL, Daya S. 2003. 6-Hydroxymelatonin converts Fe (III) to Fe (II) and reduces iron-induced lipid peroxidation. Life Sciences 72: 1367-1375.

Manda K, Reiter RJ. 2010. Melatonin maintains adult hippocampal neurogenesis and cognitive functions after irradiation. Prog Neurobiol 90: 60-68.

Manda K, Ueno M, Anzai K. 2007. AFMK, a melatonin metabolite, attenuates X-ray-induced oxidative damage to DNA, proteins and lipids in mice. J Pineal Res 42: 386-393.

Manda, K.Ueno, M. Anzai, K. 2008. Space radiation-induced inhibition of neurogenesis in the hippocampal dentate gyrus and memory impairment in mice: ameliorative potential of the melatonin metabolite, AFMK. J Pineal Res 45: 430-438.

Martin LM, Marples B, Coffey M, Lawler M, Lynch TH, Hollywood D, Marignol L. 2010. DNA mismatch repair and the DNA damage response to ionizing radiation: making sense of apparently conflicting data. Cancer Treat Rev 36: 518-527.

Mary NK, Shylesh BS, Babu BH, Padikkala J. 2002. Antioxidant and hypolipidaemic activity of a herbal formulation--liposem. Indian J Exp Biol 40: 901-904.

Matsumoto H, Hayashi S, Hatashita M, Ohnishi K, Shioura H, Ohtsubo T, Kitai R, Ohnishi T, Kano E. 2001. Induction of radioresistance by a nitric oxide-mediated bystander effect. Radiat Res 155: 387-396.

Menendez-Pelaez A, Reiter RJ. 1993. Distribution of melatonin in mammalian tissues: the relative importance of nuclear versus cytosolic localization. J Pineal Res 15: 59-69.

Messner M, Huether G, Lorf T, Ramadori G, Schworer H. 2001. Presence of melatonin in the human hepatobiliary-gastrointestinal tract. Life Sciences 69: 543-551.

Milczarek R, Hallmann A, Sokolowska E, Kaletha K, Klimek J. 2010. Melatonin enhances antioxidant action of alpha-tocopherol and ascorbate against NADPH- and iron-dependent lipid peroxidation in human placental mitochondria. J Pineal Res 49: 149-155.

Moller P, Loft S. 2010. Oxidative damage to DNA and lipids as biomarkers of exposure to air pollution. Environ Health Perspect 118: 1126-1136.

Morgan WF. 2003. Is there a common mechanism underlying genomic instability, bystander effects and other nontargeted effects of exposure to ionizing radiation? Oncogene 22: 7094-7099.

Nakade O, Koyama H, Ariji H, Yajima A, Kaku T. 1999. Melatonin stimulates proliferation and type I collagen synthesis in human bone cells in vitro. J Pineal Res 27: 106-110.

Niki E. 2010. Assessment of antioxidant capacity in vitro and in vivo. Free Radic Biol Med 49: 503-515.

Niki E, Noguchi N. 2000. Evaluation of antioxidant capacity. What capacity is being measured by which method? IUBMB Life 50: 323-329.

Noguchi N, Niki E. 2000. Phenolic antioxidants: a rationale for design and evaluation of novel antioxidant drug for atherosclerosis. Free Radic Biol Med 28: 1538-1546.

Novakova M, Paclt I, Ptacek R, Kuzelova H, Hajek I, Sumova A. 2011. Salivary Melatonin Rhythm as a Marker of the Circadian System in Healthy Children and Those With Attention-Deficit/Hyperactivity Disorder. Chronobiol Int 28: 630-637.

Othman AI, El-Missiry MA, Amer MA, Arafa M. 2008. Melatonin controls oxidative stress and modulates iron, ferritin, and transferrin levels in adriamycin treated rats. Life Sciences 83: 563-568.

Pieri C, Marra M, Moroni F, Recchioni R, Marcheselli F. 1994. Melatonin: a peroxyl radical scavenger more effective than vitamin E. Life Sciences 55: PL271-276.

Pieri C, Moroni F, Marra M, Marcheselli F, Recchioni R. 1995. Melatonin is an efficient antioxidant. Arch Gerontol Geriatr 20: 159-165.

Poeggeler B, Thuermann S, Dose A, Schoenke M, Burkhardt S, Hardeland R. 2002. Melatonin's unique radical scavenging properties - roles of its functional substituents as revealed by a comparison with its structural analogs. J Pineal Res 33: 20-30.

Radogna F, Cristofanon S, Paternoster L, D'Alessio M, De Nicola M, Cerella C, Dicato M, Diederich M, Ghibelli L. 2008. Melatonin antagonizes the intrinsic pathway of apoptosis via mitochondrial targeting of Bcl-2. J Pineal Res 44: 316-325.

Reed TT. 2011. Lipid peroxidation and neurodegenerative disease. Free Radic Biol Med 51: 1302-1319.

Reiter RJ. 2003. Melatonin: clinical relevance. Best Pract Res Clin Endocrinol Metab 17: 273-285.

Reiter RJ, Manchester LC, Tan DX. 2010. Neurotoxins: free radical mechanisms and melatonin protection. Curr Neuropharmacol 8: 194-210.

Reiter RJ, Tan DX, Gitto E, Sainz RM, Mayo JC, Leon J, Manchester LC, Vijayalaxmi, Kilic E, Kilic U. 2004. Pharmacological utility of melatonin in reducing oxidative cellular and molecular damage. Polish Journal of Pharmacology 56: 159-170.

Robertson WW, Butler MS, Dangio GJ, Rate WR. 1991. Leg Length Discrepancy Following Irradiation for Childhood Tumors. Journal of Pediatric Orthopaedics 11: 284-287.

Rodriguez C, Mayo JC, Sainz RM, Antolin I, Herrera F, Martin V, Reiter RJ. 2004. Regulation of antioxidant enzymes: a significant role for melatonin. J Pineal Res 36: 1-9.

Roth JA, Kim BG, Song F, Lin WL, Cho MI. 1999. Melatonin promotes osteoblast differentiation and bone formation (vol 274, pg 22041, 1999). Journal of Biological Chemistry 274: 32528-32528.

Rousseau A, Petren S, Plannthin J, Eklundh T, Nordin C. 1999. Serum and cerebrospinal fluid concentrations of melatonin: a pilot study in healthy male volunteers. J Neural Transm 106: 883-888.

Samanta N, Kannan K, Bala M, Goel HC. 2004. Radioprotective mechanism of Podophyllum hexandrum during spermatogenesis. Molecular and Cellular Biochemistry 267: 167-176.

Sanchez-Barcelo EJ, Mediavilla MD, Reiter RJ. 2011. Clinical uses of melatonin in pediatrics. Int J Pediatr 2011: 892624.

Schnell JW, Anderson RA, Stegner JE, Schindler SP, Weinberg RB. 2001. Effects of a high polyunsaturated fat diet and vitamin E supplementation on high-density lipoprotein oxidation in humans. Atherosclerosis 159: 459-466.

Sedelnikova OA, Redon CE, Dickey JS, Nakamura AJ, Georgakilas AG, Bonner WM. 2010. Role of oxidatively induced DNA lesions in human pathogenesis. Mutat Res 704: 152-159.

Sener G, Jahovic N, Tosun O, Atasoy BM, Yegen BC. 2003. Melatonin ameliorates ionizing radiation-induced oxidative organ damage in rats. Life Sciences 74: 563-572.

Sharma S, Haldar C, Chaube SK. 2008. Effect of exogenous melatonin on X-ray induced cellular toxicity in lymphatic tissue of Indian tropical male squirrel, Funambulus pennanti. Int J Radiat Biol 84: 363-374.

Shirazi A, Ghobadi G, Ghazi-Khansari M. 2007. A radiobiological review on melatonin: a novel radioprotector. J Radiat Res (Tokyo) 48: 263-272.

Shuryak I, Brenner DJ. 2009. A model of interactions between radiation-induced oxidative stress, protein and DNA damage in Deinococcus radiodurans. J Theor Biol 261: 305-317.

Sohal RS, Agarwal S, Sohal BH. 1995. Oxidative stress and aging in the Mongolian gerbil (Meriones unguiculatus). Mech Ageing Dev 81: 15-25.

Sridharan S, Shyamaladevi CS. 2002. Protective effect of N-acetylcysteine against gamma ray induced damages in rats--biochemical evaluations. Indian J Exp Biol 40: 181-186.

Stadtman ER. 2004. Role of oxidant species in aging. Curr Med Chem 11: 1105-1112.

Sun FY, Lin X, Mao LZ, Ge WH, Zhang LM, Huang YL, Gu J. 2002. Neuroprotection by melatonin against ischemic neuronal injury associated with modulation of DNA damage and repair in the rat following a transient cerebral ischemia. Journal of Pineal Research 33: 48-56.

Susa N, Ueno S, Furukawa Y, Ueda J, Sugiyama M. 1997. Potent protective effect of melatonin on chromium(VI)-induced DNA single-strand breaks, cytotoxicity, and lipid peroxidation in primary cultures of rat hepatocytes. Toxicol Appl Pharmacol 144: 377-384.

Sutherland BM, Bennett PV, Sidorkina O, Laval J. 2000. Clustered DNA damages induced in isolated DNA and in human cells by low doses of ionizing radiation. Proc Natl Acad Sci U S A 97: 103-108.

Take G, Erdogan D, Helvacioglu F, Goktas G, Ozbey G, Uluoglu C, Yucel B, Guney Y, Hicsonmez A, Ozkan S. 2009. Effect of melatonin and time of administration on irradiation-induced damage to rat testes. Braz J Med Biol Res 42: 621-628.

Tan DX, Manchester LC, Terron MP, Flores LJ, Reiter RJ. 2007. One molecule, many derivatives: a never-ending interaction of melatonin with reactive oxygen and nitrogen species? J Pineal Res 42: 28-42.

Tan DX, Poeggeler B, Reiter RJ, Chen LD, Chen S, Manchester LC, Barlow-Walden LR. 1993. The pineal hormone melatonin inhibits DNA-adduct formation induced by the chemical carcinogen safrole in vivo. Cancer Lett 70: 65-71.

Tan DX, Manchester LC, Reiter RJ, Plummer BF, Hardies LJ, Weintraub ST, Vijayalaxmi, Shepherd AM. 1998. A novel melatonin metabolite, cyclic 3-hydroxymelatonin: a biomarker of in vivo hydroxyl radical generation. Biochem Biophys Res Commun 253: 614-620.

Tan DX, Hardeland R, Manchester LC, Poeggeler B, Lopez-Burillo S, Mayo JC, Sainz RM, Reiter RJ. 2003. Mechanistic and comparative studies of melatonin and classic antioxidants in terms of their interactions with the ABTS cation radical. J Pineal Res 34: 249-259.

Tan DX, Manchester LC, Burkhardt S, Sainz RM, Mayo JC, Kohen R, Shohami E, Huo YS, Hardeland R, Reiter RJ. 2001. N1-acetyl-N2-formyl-5-methoxykynuramine, a biogenic amine and melatonin metabolite, functions as a potent antioxidant. FASEB J 15: 2294-2296.

Taysi S, Koc M, Buyukokuroglu ME, Altinkaynak K, Sahin YN. 2003. Melatonin reduces lipid peroxidation and nitric oxide during irradiation-induced oxidative injury in the rat liver. J Pineal Res 34: 173-177.

Tuma DJ. 2002. Role of malondialdehyde-acetaldehyde adducts in liver injury. Free Radic Biol Med 32: 303-308.

Undeger U, Giray B, Zorlu AF, Oge K, Bacaran N. 2004. Protective effects of melatonin on the ionizing radiation induced DNA damage in the rat brain. Experimental and Toxicologic Pathology 55: 379-384.

Urata Y, Honma S, Goto S, Todoroki S, Iida T, Cho S, Honma K, Kondo T. 1999. Melatonin induces gamma-glutamylcysteine synthetase mediated by activator protein-1 in human vascular endothelial cells. Free Radic Biol Med 27: 838-847.

Venegas C, Garcia JA, Escames G, Ortiz F, Lopez A, Doerrier C, Garcia-Corzo L, Lopez LC, Reiter RJ, Acuna-Castroviejo D. 2011. Extrapineal melatonin: analysis of its subcellular distribution and daily fluctuations. J Pineal Res.

Verma S, Gupta ML, Dutta A, Sankhwar S, Shukla SK, Flora SJ. 2010. Modulation of ionizing radiation induced oxidative imbalance by semi-fractionated extract of Piper betle: an in vitro and in vivo assessment. Oxid Med Cell Longev 3: 44-52.

Vijayalaxmi, Meltz ML, Reiter RJ, Herman TS, Kumar KS. 1999. Melatonin and protection from whole-body irradiation: survival studies in mice. Mutat Res 425: 21-27.

Von Sonntag C. 1987. The Chemical Basis of Radiation Biology. Taylor & Francis, London.

Ward JF. 1994. The complexity of DNA damage: relevance to biological consequences. Int J Radiat Biol 66: 427-432.

Wei H, Cai Q, Rahn R, Zhang X. 1997. Singlet oxygen involvement in ultraviolet (254 nm) radiation-induced formation of 8-hydroxy-deoxyguanosine in DNA. Free Radic Biol Med 23: 148-154.

Whisler RL, Chen M, Beiqing L, Carle KW. 1997. Impaired induction of c-fos/c-jun genes and of transcriptional regulatory proteins binding distinct c-fos/c-jun promoter elements in activated human T cells during aging. Cell Immunol 175: 41-50.

Wu LJ, Randers-Pehrson G, Xu A, Waldren CA, Geard CR, Yu Z, Hei TK. 1999. Targeted cytoplasmic irradiation with alpha particles induces mutations in mammalian cells. Proc Natl Acad Sci U S A 96: 4959-4964.

Yavuz MN, Yavuz AA, Ulku C, Sener M, Yaris E, Kosucu P, Karslioglu I. 2003. Protective effect of melatonin against fractionated irradiation-induced epiphyseal injury in a weanling rat model. J Pineal Res 35: 288-294.

Yim SV, et al. 2002. Melatonin suppresses NO-induced apoptosis via induction of Bcl-2 expression in PGT-beta immortalized pineal cells. Journal of Pineal Research 33: 146-150.

9

Alternatives to Radiation Investigations in Orthodontics

Shazia Naser-Ud-Din
*School of Dentistry, University of Queensland, Brisbane,
Australia*

1. Introduction

There has been a massive paradigm shift in investigatory tools at present. A greater surge of interest in soft tissue analysis, real time imaging and functional recordings to enhance our understanding of the complexity of the biological system is underway; with high emphasis on non-invasiveness and cost effectiveness. Current studies are looking at the possibilities of using non-ionizing alternatives to derive the required parameters to diagnose, treatment plan and evaluate the orthodontic case. This is in stark contrast to some decades ago where static hard tissue investigations were considered gold standard.

This chapter will explore the different investigations traditionally used for Orthodontic treatment planning and evaluation of treatment objectives. Since the introduction of lateral cephalometrics in 1931 by Broadbent and Hoffrath this 2 dimensional investigation has been considered the ultimate. However, it has several key limitations – radiation exposure particularly for growing children, 2 dimensional representation of a 3 dimensional living body with static image output. The need to provide a more holistic view to the case assessment with new investigatory tools like 3D imaging, ultrasonography, video recordings and EMG are perhaps the way for the future to assess growing living functioning body.

At the time of writing there is a wide array of imaging technology with innovative developments occurring at a phenomenal pace. Nearly every time one searches the world wide web something new is presented. The new hype- 3D video softwares are considered to be the next big thing [1]. The imaging industry finds wide application in defence, architecture, resource industries and general entertainment. In the medical field, it has immense application for diagnosis and potential to generate understanding not previously possible with 2D investigations. The classic examples are MRI and CT scans [2-6].

In orthodontics, the focus is more based on the external features and soft tissues and 3D reproductions are invaluable not only for diagnosis in general medical scenarios, but more so for treatment evaluation and long-term growth changes. However, the majority of published work tends to come from Japan followed by European and USA researchers [7-9].

In this chapter a brief historical background of the different imaging techniques namely Lateral cephalometrics, sEMG, Ultrasonography & 3D imaging as used in contemporary investigations along with important findings from recent studies from author's research involving such technology are discussed.

2. Lateral cephalometrics

Lateral cephalometric techniques have dominated in mainstream orthodontic, orthognathic and dentofacial growth studies for over seven decades. It is considered a versatile tool and has served the profession well since it was first introduced in 1931 by Broadbent [10]. Cephalograms have improved our understanding of growth through seminal research [11,12]. It is an important diagnostic tool utilized extensively in orthodontics. Recent literature search brought out 28,200 items related to "lateral cephalometrics in Orthodontics" while a similar search on PubMed found 808 documents on related topics. This highlights the volume of work done in the field.

There are two major limitations with cephalometrics; primarily, the additive radiation dosage of progressive films and, secondly, it is a two dimensional representation of the three dimensional craniofacial region. Moreover, the validity of cephalometric analyses has been questioned[13] and there have been concerns about the application of such simple analyses in diagnosis and treatment planning. Errors generally are attributed to orientation, geometry and association. A meta-analysis was conducted on six studies for random and systematic error along with repeatability and reproducibility [14] with good repeatability found for limited landmarks namely menton (Me) posterior nasal spine (PNS) anterior nasal spine (ANS) sella (S) pterygomaxillary fissure(Ptm) point A deepest point on the anterior maxillary margin (A) and point B deepest point on the anterior symphysis region of the mandible(B).

A number of computer software sites such as the Dolphin® (Los Angeles CA, USA), Mona Lisa® (Tidbinbilla, Canberra, Australia) (Fig.2) etc. are available to analyse the lateral cephalogram with multiple analyses at a click or a mouse scroll. However, the computer analyses have also been criticized for error through loss of contrast, related to pixel size and poor calibration. Overall, the trained eye can effectively trace and resolve finer details up to 0.1mm or smaller [15] as compared to 2.48mm accuracy recorded from Active Appearance Models (AAM) which is a computer program designed to locate landmarks [16]. Such technology needs refinement with mathematical model in order to provide a higher level of accuracy and reproducibility with work in progress by most of the softwares.

Furthermore, the lateral cephalogram has been supplemented with other radiographic conventional images such as panoramic radiography, periapicals, and hand wrist films. The later is now being replaced by the cervical maturation index (CMI) which can be derived from standard lateral cephalograms reinforcing the trend towards minimizing radiation exposure in growing individuals.

There is a galaxy of analyses to choose from. One needs to be mindful of the analyses being a composite of several unique dimensions often with a signature eponym. So it is imperative to choose specific dimensions of relevance to the study at hand. Prudent selection of analysis proves to be time efficient. Figure 1 depicts the basic linear, angular and proportional analysis for majority of diagnostic features commonly required for treatment planning in Orthodontics.

The common range and standard deviations for the different variables from a recent study are given in Table 1. These reflect the general spread for the variables elicited in Fig.1(part of the research project of Dr S Naser-ud-Din for the fulfillment of PhD University of Adelaide 2009)[18].

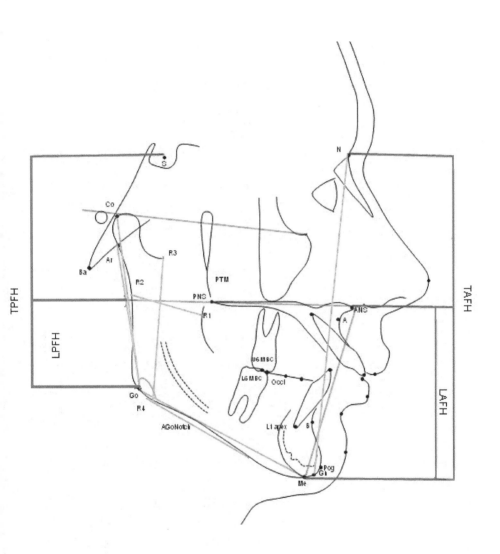

Fig. 1. **Lateral Cephaolmetric with Linear (Blue)-** 9 linear (Co-Go, Ar-Go, Go-Me, Masseter Length, R1-R2, R3-R4, N-Me, N-Gn, AGoNotch).**Angular (Green)** -6 angular (GoAngle, UGoAngle, LGoAngle, Max-Man, Frankfort Horizontal to Mandibular plane, ODI) & **Proportional (Red) analysis-**5 proportional (NGn-ArGo, LAFH/TAFH,LPFH/TPFH, ArGo-GoMe, Jarabak Ratio)[17]

	Variables	Abbreviations	Mean	SD	Range
Linear Variables 1	Condylion-Gonion	Co-Go	64.3mm	(6.0)	55.4-79.7
2	Ante-Gonial Notch	Ago-Notch	1.9mm	(0.6)	0.9-3.2
3	Articulare-Gonion	Ar-Go	45.8mm	(5.2)	35.3-55.4
4	Gonion-Menton	Go-Me	73.7mm	(4.6)	65.4-80.8
5	Nasion-Gnathion	N-Gn	107.7mm	(6.8)	96.9-121.3
6	Ricketts R1-R2	R1-R2	29.2mm	(3.9)	21.7-36.2
7	Ricketts R3-R4	R3-R4	48.0mm	(6.5)	39.6-58.0
8	Masseter Length	ML	56.3mm	(5.6)	44.6-65.8
9	Nasion-Menton	N-Me	110.4mm	(7.4)	99.2-126.6
Angular Variables 1	Maxillary - Mandibular Plane	Max-Man	21.3°	(6.2)	10.2-36.5
2	Frankfort Horizontal –Mandibular Plane	FH-Man	21.8°	(6.3)	13.2-36.1
3	Upper Gonial Angle	Go Angle U	49.9°	(5.5)	41.6-63.6
4	Lower Gonial Angle	Go Angle L	68.3°	(5.2)	60.1-82.1
5	Gonial Angle	Go Angle	118.2°	(7.8)	102.7-134.6
6	Overbite Depth Indicator	ODI	73°	(7.8)	56.5-92.3
Proportional Variables 1	Nasion-Gnathion/ Articulare Gonion	NGn/ArGo	2.2	(0.2)	1.8-2.7
2	Articulare Gonion / Gonion Menton	ArGo/GoMe	0.7	(0.1)	0.6-0.9
3	Lower Anterior Face Height	LAFH/TAFH	57.1%	(3.9)	47.1-63.3
4	Lower Posterior Face Height	LPFH/TPFH	73.4%	(6.4)	59.1-84.2
5	Jarabak Ratio	J Ratio	83.6%	(6.5)	71.0-95.7

Table 1. Common distribution for the variables from cephalometrics. (Part of the doctorate project Dr S Naser-ud-Din University of Adelaide Australia, 2009)

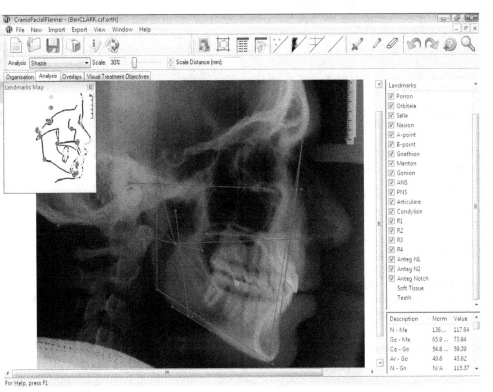

Fig. 2. MonaLisa ® Program for custom made cephalometric analysis. For efficient and standardized analysis computer based software packages are widely utilized.

3. sEMG (surface Electro Myo Graphy)

Our understanding of the neurophysiology of the muscles of mastication has increased substantially in the past 60 years. With the introduction of electrophysiological tools in the 1950′s [19], an array of experimental protocols have become available around the globe. Although, the majority of studies have used animal models due to the invasiveness of the procedures, results from such studies still provide valuable data that is extrapolated into humans. The earliest EMG studies reported on muscles of mastication during basic mandibular movements were from Carlsöö(1952), Göpfert & Göpfert (1955) in Moller(Moller 1966[20]. These preliminary studies improved our fundamental understanding of the complex masticatory system. Other work of historical interest is from Perry & Ahlgren[21,22]. EMG registers signals of muscle contractility through action potentials delivered by the motoneurones. Highly refined bipolar surface electrodes are sensitive to these electrical signals and, once amplified, are visible as the EMG recordings.

In the human model, indirect methods are employed and are preferably, non-invasive. Surface electromyography (sEMG) is a tool that is used extensively to explore the neural circuitry. Although, it has its limitations (discussed later) is overall considered user-friendly, non-invasive and has extensive applications in neurophysiology.

3.1 Development of sEMG devices

The first chewing apparatus was described by Moller in 1966 [20]. It was a landmark achievement in which bipolar recordings could be saved on photographic oscilloscopes rather than printed by inkwriters. Therefore, several experiments were designed to enhance our understanding of muscle activity during chewing, postural activities, full effort clenching, swallowing patterns and even facial morphology. This innovative apparatus only required an EMG recorder with electrodes. The electrical activity was recorded on three channels. Although, an advanced methodology for its time, Moller's apparatus had the serious limitation of un-standardized positioning of the subject. This produced a lack of reproducibility, albeit the reproducibility factor poses a challenge even with contemporary advancements. Other methodological issues arose from interference from neighbouring potentials, rejection of common voltage, distortion of amplitude and uneven distribution of electrical activity [20]. Over time, several global studies have been conducted which have improved designs and chewing devices [23,24] leading to an enhanced understanding of the reflexes in the oro-facial regions.

The periodontal-masseteric reflex was first described by Goldberg 1971 [25] highlighting the importance of not only PMR but also the role of the masseter muscle and central connections, due to the short latency. This study concluded that there was a central connection located in the mesencephalic nucleus that was responsible for the excitatory masseteric reflex evoked by PMR and gingival receptors. One can appreciate that, even with a very simplistic methodology and manual tapping of the teeth, Goldberg was able to deduce these important findings.

3.2 EMG studies and significance in orthodontics

Work in this field has resulted in few conclusive findings. A canonical correlation analysis between facial morphology, age, gender and EMG during rest and contraction has not found a statistically significant correlation [26]. EMG and bite force have been studied extensively[20,27,28] with consistent findings of reduced force levels in dolichofacial patterns. Different vertical facial types, both in adults and children, produce differences in EMG responses recorded over the course of a day [29]. Masseter muscle EMG activity was found to be consistently longer in short vertical dimension facial types as compared to high mandibular plane angle individuals. The variables of bite force, muscle efficiency and mechanical advantage in children with vertical growth patterns [30] were negatively correlated with muscle efficiency and vertical proportions.

Morimitsu et al. documented recordings of EMG from masseter muscles which showed positive correlations with linear cephalometric measurements; particularly, the muscle activity was significantly related to mandibular dimensions such as (Co-Go) and body length (Go- Me) [31]. In fact, the muscle activity increased with decrease in vertical proportions such as mandibular plane angle. Furthermore, they compared the muscle activities with anterior-posterior skeletal base relationships indicated by Sella Nasion Point A (SNA) and Sella Nasion Point B (SNB) variables. Morimitsu et al. strongly advocated the use of EMG for examining masticatory activity, but one could argue that the reproducibility with electrode placement at subsequent visits may prove to be a challenge [32].

A comprehensive study by Tuxen et al. [33] compared masseter muscle fibre types along with function from EMG and facial morphology. They concluded that even with intensive analysis of different dimensions, linear regression analyses failed to show any significant association highlighting, once again, the complexity of the craniofacial region.

Muscle spindle reflexes are stronger in short face height individuals[17,34] which may explain the phenomenon with bruxism and brachyfacial tendency. Moreover, this information may translate into better designs for splints and Orthodontic appliances.

A recent publication [35] has utilized EMG to analyse masseter muscle differences in individuals who brux in their sleep. The authors believe their system can help diagnose different types of bruxism and is aptly termed the innovative bruxism analysing system. Once again, the literature supports the use of sEMG as a useful mode of investigation.

3.3 Limitations & methodological issues
EMG has provided insight into the neuromuscular system but there are inherent limitations which must be minimized in order to derive meaningful data. This has been reviewed thoroughly [36]. EMG accuracy may be enhanced by the use of intramuscular electrodes compared with sEMG electrodes. However, sEMG electrodes with built-in amplifiers can assist in reducing cross-talk and movement artifact. Passive surface electrodes require skin preparation and monitoring of resistance below 10Ω for clear data recording. Contemporary sEMG involves the use of bipolar electrodes, which suppress noise. The amplification should counter distortion, preferably 10 x higher than the electrode-to-skin electrical resistance. The filter level is dependent upon skin thickness and the frequencies recorded. It should not eliminate any of the frequencies within the range of recording and the choice lies between upper (high-cut or low pass) and lower (low-cut or high pass).

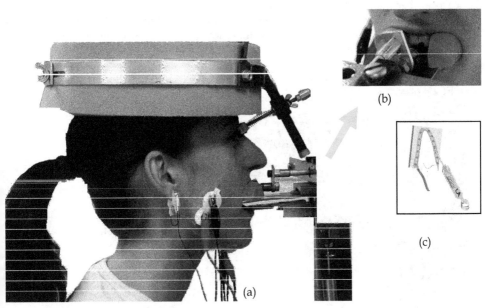

Fig. 3. Subject seated comfortably throughout the experiment session. sEMG recorded bilaterally from the masseter and the digastric muscles: (a) biting on a custom made impression; (b) left central incisor mechanically stimulated by an orthogonal probe;(c) local anaesthetic was infiltrated to eliminate periodontal mechanoreceptor input during the spindle trials. An ear clip provided the electrical ground.

In addition, the quality of data production is dependent on the recording and display devices, the sophistication of the computer programs to reliably identify the initiation of muscle activity, the type of ground used, and obviating the use of long leads which contribute to noise and, therefore, contamination of the data. Several different types of ground have been utilized in human studies such as forehead, wrist, elbow, ear lobe and lip clips. Currently the ear lobe is selected due to ease of placement and distance from the experiment site (Fig 3).

Minimizing artifacts is essential for clear data recording and minimizing movements from adjacent muscles is very helpful. In the study by author[34], the customized nose rest and the head halo provided stabilization during the chewing and static phases of the experiments (Fig. 3).

Cross-talk is a phenomenon commonly experienced in sEMG recordings where adjacent muscle activity leads to electrical volume added to the data. Cross-talk can be minimized by standardizing the experimental conditions. Moreover, double deferential techniques can also assist in eliminating cross-talk [37-43]. This is the prime cause of criticism in telemetry sEMG recording where cross-talk and noise can compromise the results [44].

Finally, correct processing of the data is essential and custom-made software programs are routinely used. IZZY® [45], in Fig. 4 provides an array of systematic offline analysis which allows the data from sEMG to be full wave rectified and further processed as CUSUM (Cumulative Sum).

Normalization (Fig.5) is a standardization process for data acquisition which allows comparison between different subjects and with the same subject data on different occasions. It means normalizing sEMG levels to the percentage of an individual subject's MVC (maximum voluntary contraction) for each muscle group, hence reducing the variability between records.

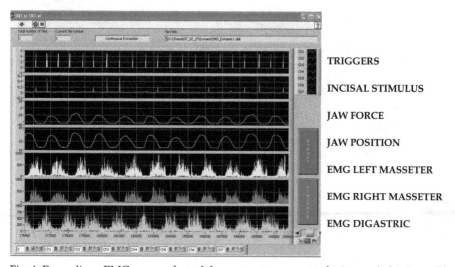

Fig. 4. Recording sEMG screen shot of the computer generated triggers (white), profile of stimulus (red), Jaw force (green), Jaw position in opening and closing (blue) and raw sEMG recorded from left masseter (yellow), right masseter (pink) and anterior digastric (orange). Total of seven channels were operating at the time of the experiments.

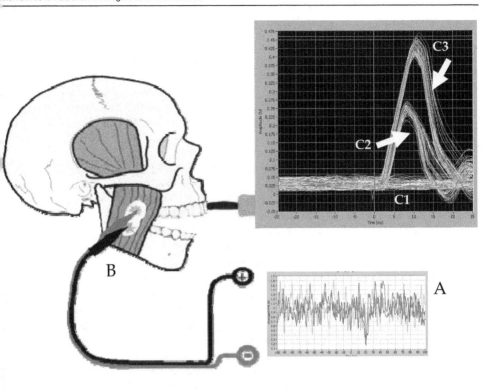

Fig. 5. sEMG (A) was recorded from the masseter (B) during experimental conditions with normalization of the raw scan in A to C1, C2 and C3 (control, with low tap and high force tap to the central incisors).

4. US (Ultra Sonography)

Ultrasonography is developing rapidly as the mainstream investigatory tool.

Ultrasonography (US), as the name implies, utilizes sound waves within the frequency of 2-18MHz that bounce off the tissue to provide the depth and density of the image conventionally captured on the screen as a real time image. It was developed for medical use in the 1940's with parallel developments across the Atlantic in the USA and Sweden. It has since been widely accepted as the safest investigating tool and is applied widely in obstetrics. However, it has applications across several medical disciplines including physiotherapy therapeutic outcomes for muscular-skeletal conditions. It is now easily available infact handheld devices which are easy to use and with higher resolution are making way into the clinical set ups and for field work. The 3D graphics along with real time color coding will indeed provide greater understanding and acceptance in the future to evaluate functioning muscles. There has been a major paradigm shift with realization that form and function are inextricably intertwined in Orthodontics. Our understanding of the hard tissue relationship has been comprehensive, with extensive clinical trials both cross sectional and longitudinal over the past several decades. The majority of studies have used standardized cephalograms that reflect our current concepts. However, the same cannot be stated for soft tissues and, more specifically, the muscles that envelope the skeletal bases. Hence, the interest with non-invasive soft tissue real time evaluation with US.

The hypothesis was that if such a simple non-invasive investigation can provide diagnostic data then it would obviate the need for radiation investigations, in particular for randomized controlled clinical trials (RCCT) that are currently considered the highest caliber of evidence based literature. Such study designs require several radiographs which are difficult to justify ethically. Hence, non-invasive US could prove vital for long term studies.

US is advancing from 2D to 3D imaging whereby a series of 2D images are collated and rendered into 3D. The main limitation is cost as it needs a specialized probe and the image acquisition speed is relatively slow, especially in moving tissues such as contracting muscles. Further advances with colour Doppler technology are of value for imaging tortuous vessels. Generally speaking, conventional US serves well and is good for imaging muscles and soft tissues, but has poor acoustic impedance for bone.

Muscles of mastication have been studied extensively with US which has been considered to be a valuable, precise technique for analysing muscle shapes [46-49].

Furthermore, US is superior to radiographs for soft tissue evaluation and definitely overcomes the radiation hazards. The surface topography for the masseter muscle is excellent due to its superficial anatomical position Fig. 6 &7); however, the volume and cross-section assessments have been difficult [50]. Hence, the present study sought to evaluate these dimensions with a simplified approach and using arithmetical formulas [51].

A comprehensive meta analysis was recently presented by Serra in 2008[50] which reported masseter muscle thickness in contraction to range from 5-14.1mm [52,53]. One could argue that this wide variation would be due to racial or ethnic diversity not to mention the time of recording be the muscle relaxed or tensed in contraction. US has recorded thickness variations in masseter muscle contraction and relaxation [49] during function which would otherwise have been impossible with snap exposure type investigations.

US has been extended to determine association between several variables such as the masticatory muscle thickness, TMD and bite forces [54]. The study found a positive correlation between masseter muscle thickness and posterior facial dimensions. The study concluded that muscle thickness is related to vertical facial dimensions and bite force. Previous studies have had similar findings in large samples where not only negative relations between muscle thickness and anterior facial heights and mandibular lengths were noted but positive relationships existed with intergonial width and bizygomatic facial width according to anthropometric measurements[55,56]. Likewise, a Swedish study consistently found a relationship between thin masseter and longer faces in females [57]. This study found US to be a reliable and accurate method for muscle assessment.

Hatch and associates have clearly indicated the importance of multivariate analyses of the masticatory system for comprehensive diagnosis in dentistry [58]. They went so far as to include blood glucose levels along with bite force and cross-sectional area of the masseter muscle.

Overall, US has been documented as a reproducible and reliable investigating tool for masseter muscle assessment [59,60] and could produce useful, non-radiological information to enhance orthodontic practice.

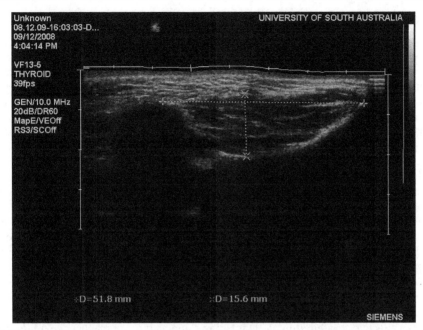

Fig. 6. US Scan of the left masseter showing (dotted lines) the length (horizontal line) and the thickness (vertical line)

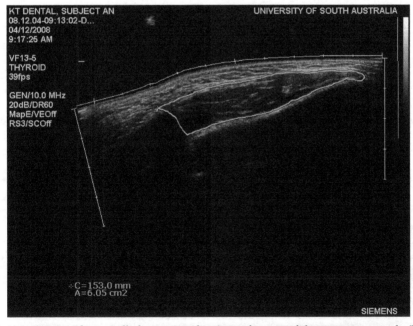

Fig. 7. Area measured manually by tracing the circumference of the masseter muscle (B).

5. 3D imaging

3D imaging has come a long way in the last 20 years, particularly since its introduction six decades ago [Thalmann-Degen 1944][61]. Primarily developed for application in industry [62] as an effective and non-invasive method, it quickly found its way into clinical applications, particularly in orthodontics [63] and oral surgery. 3D applications in restorative dentistry with the Cerec technique and CAD-CAM are already in main stream clinical application and boast high accuracy at the level of 70 microns [64]. Scanning of comparably smaller dimensions of hard surfaces with accurate reproduction has evolved over the past few years[7]. A similar level of precision is desired for facial soft tissue imaging. It's only a matter of time and refinement of technology until 3D imaging becomes a routine part of private orthodontic practice [65]. With innovative robotic wire bending, tailormade appliances both labial and lingual (Insignia, SureSmile[66] and Incognito-3M Unitek USA) 3 D imaging is gaining an integral component of how Orthodontics shall be delivered in 21st century.

In orthodontics, 3D imaging was tested for the first time in 1981 with optical surface scans [67-69]. A major change came with the use of a hand-held scanner making the scan portable (McCallum et al.; 1996 in Moss[70]) followed by a probe to record 3D coordinates [55]. Imaging is sensitive to the surface acquisition and sufficient data are required to appreciate the subtle changes that occur in soft tissues over a period of time.

There are various types of imaging techniques ranging from stereo photogrammetry, to 3D laser scanning, vision-based scans like Moiré tomography along with the latest, safest and most cost- effective structured light 3D imaging [9]. Innovative approaches are under way to integrate the conventional lateral cephalogram with stereo photogrammetry and digital fusion. It is claimed that such techniques will obliterate errors through image sensors and 3D orientation [71].

Structured light creates a superficial shell-like reproduction of the face enabling the digitized topology of the face to be displayed in 3D without any ionizing radiation [71]. It is a simple and cost effective method of generating three-dimensional images with minimal time required for exposure, usually within one second.

The structured light (SL) technique provides reasonable accuracy when following certain protocols such as the exposure should be frontal with deviations of up to +/-15° only because with increasing profile view there is proportionally reduced accuracy. Linear measurements can be erroneous up to 1mm and if that is acceptable to the operator and duly accounted for it should not be of much concern. Likewise, the smaller inter-distance between the two cameras creates a limited field of view leading to diminished accuracy in z-coordinate measurements [72].

Overall, SL is considered a simple, cost effective and readily applied 3D imaging system [73]. It is based on stereoimaging and triangulation to produce a 3D image. From a light grid or pattern which is usually horizontal Fig.8. The mean absolute error for linear measurements with SL 3D imaging in the current study was 0.53mm [18,74,75].

3D imaging is valuable in assessing the growth changes in soft tissues over time, because previous investigating methods were not able to adequately assess soft tissue changes related to growth. A large scale growth study with 3D optical surface scanning [76] found significant changes in the vertical facial dimension with increase in the cross sectional cohort age range 5-10 years old. Moreover, they found a significant increase in dimension in the masseter muscle mass across the age range. Hence, age-related changes can be appreciated and quantified with 3D imaging. Furthermore, a longitudinal study analysing facial

morphology changes also found similar results with an increase in the vertical dimension being gender dependent with greater significance in males [8].
Similarly, growth changes have been evaluated with 3D facial morphometry and Fourier analyses depicting changes in profile [77]. Another area where 3D imaging is used extensively concerns pre- and post-orthodontic treatment effects on soft tissues. The very controversial and anecdotal references often made regarding functional orthopaedic appliances has been backed with evidence [78-80]. Yet another highly contentious issue in orthodontics has been the extraction and non-extraction debate, where the treatment decision pendulum keeps swinging by the decade. A study in London [81] used 3D optical surface scans for patients with fixed appliances; one group with extractions and another without. The results were conclusive and laid the age-long controversy to rest with no significant facial soft tissue changes discernable in either study group.

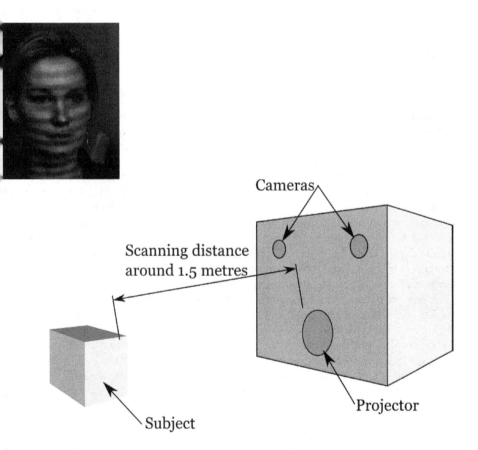

Fig. 8. Capturing an image with the 3D structured light projector placed 1.5m from the subject. The grid of light with horizontal pattern is flashed on the subject's face for 1sec. Triangulation helps to generate the 3D effect.

Work is underway to fabricate user-friendly analysis methods especially as 3D imaging is relatively contemporary and lacking well-established quantification and comparison techniques. 3D studies conducted with Procrustes analysis have determined the translation of soft tissues in monozygotic and dizygotic twins [70]. Another study has used the Procrustes registration method in conjunction with 3D CT scan and Virtual Reality Modelling Language (VRML) to assess the registration error associated with such techniques and found it to be within +/- 1.5mm in most parts [82]. A South Korean study [83] found negligible error of 0.37mm and less than 0.66% magnification with laser scanners. They claim that the soft tissue rendering was highly reliable and reproducible. However, one would argue about the safety of such imaging as the study was not conducted on humans but mannequins.

Unique to anthropometry and 3D imaging is the ability to measure and evaluate the transverse dimensions of the face. Although, a large list of measurements are presented from the frontal aspect in the Farkas textbook [55] for diagnostic purposes one should consider selected ones.

Two horizontal planes commonly used for facial widths are the bizygomatic diameter, inter-zygonion (zy-zy), and mandibular or lower face width, inter-gonion (go-go). Generally, such evaluations assist in picking up asymmetry greater than 2mm [84]. These dimensions form indices for facial proportions. Thus, the Facial Index was calculated with horizontal (zy-zy) as a percentage of vertical from nasion and gnathion (n-gn). Similarly, the (go-go) width is used in the Jaw Index with stomion to gnathion as the lower face proportion (Fig.9).

The midline landmarks were used for vertical proportions which were Index of lower jaw to facial height and Index of Jaw to facial height. These proportions are related to the lateral cephalogram and the correlations can be beneficial substitutes in progressive evaluation over the course of the treatment.

The curvilinear measurements were included because they are more biologically meaningful than straight line representations of complex 3D structures (Fig.10). For example, our findings show a significant difference ($p < 0.0001$) at 95% CI and with a standard error of 0.52 and SD = +/- 1.65 for the linear mandibular depth compared to the curvilinear mandibular depth. The mandibular curvilinear arc extended from left tragus to subnasale to right tragus (t-sn-t) and did not have an equivalent linear counterpart[75]. The use of curvilinear measures is gaining application in tooth morphology studies where curvilinear measurements are deemed superior to the conventional linear ones [85,86].

3D imaging was used for the superficial soft tissue measurements and transverse indices (Figs.9-11). Once again the findings were similar to past work. Currently, the imaging is still a single capture in time and presents limited insight into the true functional assessment. Even though there are several existing packages available for 3D imaging it has yet to become a routine diagnostic tool. There are several reasons for not having 3D imaging in mainstream clinical set ups. Primarily the cost, followed by *not completely realistic* imaging which still needs refinement for accurate and life-like reproduction of the face. Also, the data acquired from such analyses needs to be unique and provide information that cannot be generated by other more conventional means. Finally, the product needs to be user-friendly and easy to apply by the clinician or auxiliary staff. Above all, the suggestion that 3D imaging could be a surrogate for lateral cephalograms with predictive correlations obviating the need for serial radiography and radiation exposure, could make it an attractive choice.

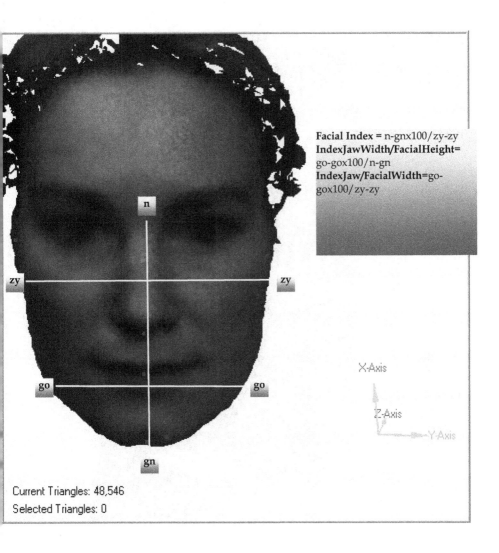

Fig. 9. Anthropometric Indices of Farkas[55] applied to 3D images for transverse and vertical assessments.

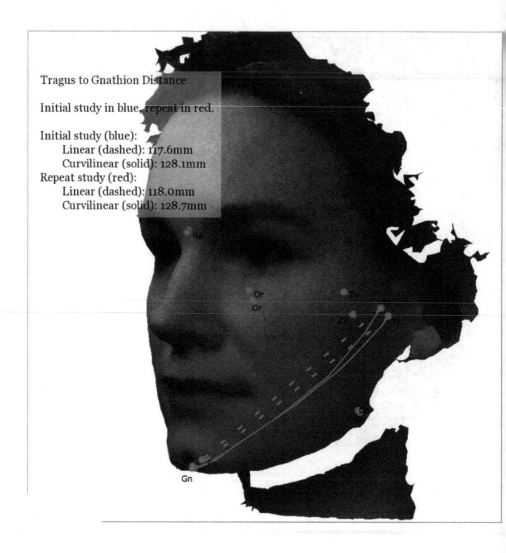

Tragus to Gnathion Distance

Initial study in blue, repeat in red.

Initial study (blue):
    Linear (dashed): 117.6mm
    Curvilinear (solid): 128.1mm
Repeat study (red):
    Linear (dashed): 118.0mm
    Curvilinear (solid): 128.7mm

Fig. 10. Curvilinear and linear mandibular measurements for the left side of the subject. There is a clear advantage with 3D images over photographs.

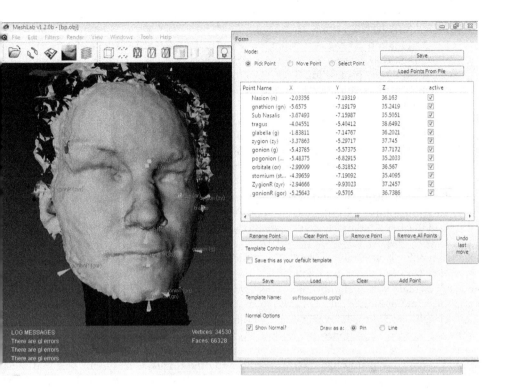

Fig. 11. Analysis of 3D scanned images using MeshLab® shows point selection for computing distances between the landmarks.

Author's study has found high correlation of certain lateral cephalometric variables, particularly the vertical ones, with 3D imaging indices. However, before it enters into mainstream clinical practice, the following criteria need to be fulfilled.

a. *Reliability* is considered a composite of repeatability and reproducibility [87] and different 3D imaging techniques have been scrutinized for accuracy and reliability. Kusnoto and Evans [88] used a surface laser scanner (Minolta Vivid 700) to assess various objects ranging from cylinders to cast analyses and facial models. They found accuracy in the range of 1.9 +/- 0.8mm for facial imaging which was not as high as for cast analyses 0.2 +/- 0.2mm. Hence, one can conclude that the larger the surface to scan and the greater the data acquisition, the higher the error rate with current 3D scans. Likewise, the suggestions for improvement and exploring more user-friendly options in 3D technology have been highlighted in a review article addressing these issues [89].

b. *Standardization* is equally essential for the translation of the predictive equations across the board with different imaging packages. This has been norm with many software cephalometric packages currently in use and could become an overall package.

c. *Cost Effectiveness* and ease of accessibility with perhaps even free downloads such as used in this study can enhance its wider applicability. Such off-line computer analyses can provide flexibility and effective access [90]. However, one has to be wary of the quality of such softwares; reproducibility, reliability, accuracy of reproduction and ease of manipulation. Perhaps custom-made software packages specifically tailored to orthodontic case assessment [85,86] with higher resolution and reproducibility would be welcomed by the profession. Overall such endeavour will aid in patient education and serve as a vital teaching tool to students to appreciate subtle changes during treatment and growth.

d. *Safety* Efforts are being made to assess mandibular growth with 3D MRI [91]. However, there is real concern about the justification of such expensive investigations and is often under fire for overuse in certain countries. Clearly, for the conventional orthodontic patient such scans would not be routinely recommended.

Multiple investigations used in the past such as CT, MRI and radiographs can be replaced with carefully selected diagnostic and evaluating tools which provide maximum information with minimal hazard along with the added bonus of being cost-effective. The current work has addressed this issue and future investigations should be geared to validate it.

Recent work has explored the correlations between anthropometric and cephalometric measurements of vertical profile [92]. The study found nasion and menton to be highly correlated between the two investigation, but landmarks such as subnasale, supradentale and infradentale varied significantly. This could be due to the overlying soft tissue and its variability among individuals. For this particular reason we chose landmarks that would be more closely related to the cephalometric counterparts such as glabella, nasion, gnathion, subnasale, gonion, zygonion, orbitale, tragus and stomion.

However, due to difficulties in lateral projections inherent to the Structured Light (SL) technique we consistently had difficulty in locating the tragus. Overall, our best and most highly repeatable landmarks were soft tissue subnasale (y-axis) with Dahlberg's statistic (DS) = 0.009, followed by zygonion (y-axis) DS=0.004, glabella (z-axis) with DS=0.005 and stomion (z-axis) DS=0.01. The above mentioned landmarks provide a reliable array of reference points to include in future 3D assessments. However, the majority of the 3D landmarks were within an acceptable range. Perhaps the repeatability could be enhanced with refined software with greater matrix size and pixel ratio [87]. This would further refine the accuracy of the prediction equations and improve overall reliability.

Currently, work is in progress in Liverpool where Brook *et al.* are evaluating 3D imaging for better assessment of surface contours of teeth [85,86,93]. Precision and accuracy with e-tools will improve the way we measure. Thus, more studies will be needed to explore landmark assessment and measurement.

6. Prediction equations

Predictive equations were developed from multiple linear regressions which were significant[18,75]. Unfortunately very few variables qualified as the projection from one medium of investigation to the other was diverse, highlighting that each has its own individual data set that is novel and there is very little overlap. However, those that showed somewhat strong correlations were developed with ß- weights which act as constants for a given equation.

Predictive Equation	R^2	MSE	SIG
Linear Variables			
R3-R4 = 25.44932 + 0.48688*t-go	0.33	5.46	*
Masseter = -37.9501 + 0.81574*FaceHt	0.88	2.19	*
N-Gn = 75.76050 + 0.20701*go-gn + 0.10290*ManDepthLinear	0.42	3.83	*
Masseter = -45.1784 + 0.67128*FaceHt + 0.20006*Mandiblewidth	0.92	1.98	*
Go-Me = 58.13533 + 0.06039*Mandibulararc	0.17	4.06	
Co-Go = 28.11536 + 0.76730*t-go	0.58	5.14	*
Ar-Go = -8.69762 + 2.50677*ManDepthLinear + -1.93239*ManDepthCurvilinear	0.67	4.16	*
Angular Variables			
GoangleU = 71.06816 + -0.36734*ManIncli	0.49	4.13	
Goangle = 144.2894 + -0.46133*ManIncli	0.47	5.33	
Proportional Variables			
AFH = 52.67724 + 0.11825*Indexjawwidthfacialheight	0.01	3.06	
AFH = 52.71398 + 0.18845*JawIndex	0.07	2.96	
AFH = 54.06144 + -0.03102*Indexjawwidthfacialheight + 0.19646*JawIndex	0.07	3.17	
JarabakR = 109.0283 + -0.20828*Indexjawfacialwidth	0.11	6.62	
JarabakR = 69.87241 + 0.24662*IndexJaw-FH	0.13	6.52	
JarabakR = 101.1132 + -0.27080*Indexjawfacialwidth + 0.30918*IndexJaw-FH	0.32	6.2	

Table 2. Predictive Tables for the various 3D imaging variables that can provide information pertaining to lateral cephalometric variables. Starred equations show significance at the level of $p < 0.05$. Linear measurements have the statistically significant predictive equations. Gray bars are the ß-weights.

We used simple methods to produce prediction equations rather than the more resource intensive Besaynian or Monte Carlo factor models with the aim of, decreasing the complexity and making it more clinically applicable. The concept of generating predictive equations from 3D images for lateral cephalometric variable estimation has been presented for the first time. Although, it can be debated that with a small sample size the value of such prediction is questionable, it does set the scene for future work[75].
Compared to the reported Eigenvalues from Benington et al [94] our values were lower. Also their study was the first to document the masseter volume measurements with high technology 3D US scanning. Likewise, the current study by author did measure masseter volume, albeit indirectly, utilizing a simple, portable US unit that could find application in clinical settings[75]. Similarly, when compared with Radsheer [95], our values were slightly lower but their study included, 121 adults and previous work with 329 subjects aged 7-22 from a Greek population [56]. Our result differences could be attributed to the sample size and also the population from which they are selected.
The method error indicated that there were certain landmarks that had reduced repeatability. This may be explained by background noise and lack of clear definition in some scans. Palpation is generally advantageous in precise positioning of soft tissue gonion, but it is considered a difficult point overall to determine even in anthropology [55]. Nasion (n)

and subnasale (sn) were relatively easily identified on the 3D images. For a better view, the image was rotated upwards by 30°. This is an advantage of the scans because repositioning the live patient can be difficult and sometimes embarrassing. The accuracy and repeatability of 3D Imaging has been documented as exceptional [96] and, as with most recordings, in order to eliminate inter-operator variability all measurements should be made by the same investigator. The angular measurements are dependent on head position and need to be standardized across the sample [55].

Baumrind eloquently stated that the "craniofacial complex system needs to be evaluated from different perspectives as no single view would do justice in its evaluation and future treatment planning" [97]. 3D imaging has extended not only to facial surface analyses but is extending to cephalometry as well [98]. Subtle soft tissue changes have been studied extensively in recent work particularly with pre- and post-treatment outcomes [8,64,80,89,91,99,100] . Similar interest with mandibular growth changes[91] and four-dimensional analyses for the TMD [100] are gaining momentum in diagnostic imaging. A common critique with new imaging techniques has been the reliability which is of concern and can be addressed with the development of self –calibrating measuring systems [64].

New directions in technology will soon be evident with the need to enhance the quality of 3D images with more real-life effect. This would assist in patient education and teaching sessions for dental students. Moreover, programs with substantial mouse manipulation create undue hand and wrist fatigue whereas touch screen options would be ergonomically valuable.

Clinical applications of 3D imaging are vast and numerous. From treatment planning to pre- and post-treatment evaluation, the images can be manipulated from any direction providing in-depth analysis of the case without patient recalls and inconvenience of anthropometry. It is becoming a vital tool in orthognathic surgery planning, patient education, "tele-orthodontics" [7] to manage global movement of patients, and where interdisciplinary treatments are sought. It is an effective recording method for various facial asymmetries [96] and severe craniofacial dysostoses, hemifacial microsomias [101], and where simple photographs have serious limitations. The non-invasive nature of 3-D imaging will make its application widely accepted for sequential evaluation in growing children and on controls for research [96].

However, the idea of obviating lateral cephalograms entirely may not yet be the case because radiographs still provide valuable skeletal and dental information vital for initial treatment planning. Moreover, we do acknowledge the limitations of prediction because as with any prognostic approach, particularly in a biological system, there will be variations. However, a method that can assist the clinician in reducing the number of radiographic investigations would surely be welcomed by the public and profession alike.

7. Other investigations

a.   Genetic profiling
Research is ever expanding in this novel frontier and will offer the possibility of diagnosis in Orthodontics. It already is promising in screening individuals with high risk of root resorption [102].
Predictive and presymptomatic testing is already creating a lot of media interest and long term implications with financial issues related to health insurance. Most importantly the possibility of genetic discrimination may occur.

Recent work from Hong Kong is uncovering the genetic etiology of Class III malocclusion[103,104]. The genetic profiling is evolving and has to be refined to be reliable. Incomplete penetrance of the genetic code may prove challenging when considering probable malocclusion in the offspring. However, some conditions that have had high genetic predisposition e.g Class III skeletal patterns and anterior open bites (AOB) could provide useful genetic information with the probability of expression. However, caution has to be exercised as malocclusion is multifactorial and with global migrations miscegenation is increasing with leading to genetic dilution so to speak.

b.   Heat sensitive Scans

Moss has shown heat sensitive scans can be utilized to determine the "hot spots" of growth particularly following Orthodontic intervention[79]. This is a simple non-invasive investigation albeit costly.

c.   Image Fusion

The latest of combining various investigations for a holistic evaluation process has been aptly described as "Image fusion" [105] assisting in facial triad analysis (namely soft tissues, skeletal basis and dentition) . The aim of this amalgamation is to recreate a virtual head for treatment planning, long term follow up and documentation. Moreover, it may provide a realistic prediction model in the future.

d.   MRI

MRI (Magnetic Resonance Imaging) although radiation free is a costly investigation, with implication for time and quality of life (unbearably loud at the moment) . MRI usage is limited in Orthodontics needs to be justified and avoided when possible. Previous work [3] has affirmed that lower face height and masseter muscle thickness have inverse correlation. Generally, the increased lower face height individuals will present with classic cephalometric characteristics such as large gonial angles, steep mandibular planes and reduced posterior face height. The masseter muscle cross-section has been measured with MRI's and found to be significantly smaller by nearly 30% in the dolichofacial types [3]. A recent study has compared volume of masseter muscle derived from MRI with vertical facial dimensions and found volume to be more relevant than cross-sectional area [106]. Moreover, the posterior face height was significantly correlated to masseter muscle volume. Although, MRI has shown high resolution for TMJ scans[107] it needs to refine to a small field of view [108] for acceptance as an investigation tool.

e.   Videofluoroscopy

Even though not entirely radiation free Videofluoroscopy is showing promising results in the investigation of Obstructive Sleep Apnoea (OSA)[109]. There is the issue for 2D instead of 3D as found in CBVT scans, perhaps image fusion would add to its limitations.

f.   Videos

Increasingly videos are being recorded to provide a real time assessment on Orthodontic patient smiles, speech and expressions which were not possible with static shots in repose and unnatural smiles. Many of the softwares for diagnosis and data storage are adding video as an important tab to achieve patient motions.

8. Conclusion

Even though the findings from this study are not clinically applicable it has highlighted the importance of finding cost effective and non-invasive methods of investigations for

Orthodontic case assessment. There is now a fast moving trend towards 3D imaging and indeed the lateral cephalometric may phase out in years ahead.

In orthodontics there is a strong association between form and function [110]. This has been studied extensively and proved with strong correlation that exists between the soft tissue and underlying osseous foundation. However, there are not many studies from the soft tissue perspective, particularly, the muscles of mastication including the masseter.

There is definitely scope for US for assessment of masseter muscle in real-time. This has been documented widely in the literature with masseter muscle thickness demonstrating differences in posterior crossbite cases and their relationship to TMDs [5,54,111,112]. But there is concern that some findings can be controversial and others produce non-conclusive results. For example, some studies [113] found the crossbite side had thinner masseter muscle. Obviously, 3D US is superior in function and would be better for oro-facial applications [94] provided it is cost-effective. One can also anticipate the rapid advancement in technology and 4D US may well be the future for diagnosis and treatment evaluation.

Our understanding of malocclusion has come a long way but it is multidimensional, multifaceted and highly variable and needs to be analyzed from several different perspectives. The emphasis is shifting from static, radiation-based investigations which are a snap shot in time to dynamic real-time recordings of soft tissue function. The fourth dimension in combination with conventional diagnostic tools could provide a comprehensive single diagnostic package that would not only assist in complete records diagnosis and evaluation but provide predictions for future expected changes.

9. References

[1] Gibson C: 3D video software the next big thing?, The Adeladian Adelaide, 2009

[2] Raadsheer MC, Van Eijden TM, Van Spronsen PH, et al: A comparison of human masseter muscle thickness measured by ultrasonography and magnetic resonance imaging. Arch Oral Biol 39:1079-84, 1994

[3] van Spronsen PH, Weijs WA, Valk J, et al: A comparison of jaw muscle cross-sections of long-face and normal adults. J Dent Res 71:1279-85, 1992

[4] Hussain AM, Packota G, Major PW, et al: Role of different imaging modalities in assessment of temporomandibular joint erosions and osteophytes: a systematic review. Dentomaxillofac Radiol 37:63-71, 2008

[5] Jank S, Rudisch A, Bodner G, et al: High-resolution ultrasonography of the TMJ: helpful diagnostic approach for patients with TMJ disorders ? J Craniomaxillofac Surg 29:366-71, 2001

[6] Weijs WA, Hillen B: Correlations between the cross-sectional area of the jaw muscles and craniofacial size and shape. Am J Phys Anthropol 70:423-31, 1986

[7] Hajeer MY, Millett DT, Ayoub AF, et al: Applications of 3D imaging in orthodontics: part II. J Orthod 31:154-62, 2004

[8] Kau CH, Richmond S: Three-dimensional analysis of facial morphology surface changes in untreated children from 12 to 14 years of age. Am J Orthod Dentofacial Orthop 134:751-60, 2008

[9] Kau CH, Richmond S, Incrapera A, et al: Three-dimensional surface acquisition systems for the study of facial morphology and their application to maxillofacial surgery. Int J Med Robot 3:97-110, 2007

[10] Broadbent BH: A new x ray technique and its application to orthodontics. Angle Orthod:45-66, 1931

[11] Bjork A, Skieller V: Contrasting mandibular growth and facial development in long face syndrome, juvenile rheumatoid polyarthritis, and mandibulofacial dysostosis. J Craniofac Genet Dev Biol Suppl 1:127-38, 1985

[12] Skieller V, Bjork A, Linde-Hansen T: Prediction of mandibular growth rotation evaluated from a longitudinal implant sample. Am J Orthod 86:359-70, 1984

[13] Vig PS, Spalding PM, Lints RR: Sensitivity and specificity of diagnostic tests for impaired nasal respiration. Am J Orthod Dentofacial Orthop 99:354-60, 1991

[14] Trpkova B, Major P, Prasad N, et al: Cephalometric landmarks identification and reproducibility: a meta analysis. Am J Orthod Dentofacial Orthop 112:165-70, 1997

[15] Mankovich NJ, Samson D, Pratt W, et al: Surgical planning using three-dimensional imaging and computer modeling. Otolaryngol Clin North Am 27:875-89, 1994

[16] Rueda S, Alcaniz M: An approach for the automatic cephalometric landmark detection using mathematical morphology and active appearance models. Med Image Comput Comput Assist Interv Int Conf Med Image Comput Comput Assist Interv 9:159-66, 2006

[17] Naser-Ud-Din S, Sowman PF, Sampson WJ, et al: Masseter length determines muscle spindle reflex excitability during jaw-closing movements. Am J Orthod Dentofacial Orthop 139:e305-13, 2011

[18] Naser-Ud-Din S: Analysis & Correlation Study of Human Masseter Muscle with EMG, US & 3D Imaging Orthodontics. Adelaide, University of Adelaide, 2009, pp 214

[19] Hultborn H: State-dependent modulation of sensory feedback. J Physiol 533:5-13, 2001

[20] Moller E: The chewing apparatus. An electromyographic study of the action of the muscles of mastication and its correlation to facial morphology. Acta Physiol Scand Suppl 280:1-229, 1966

[21] Perry HT, Harris SC: Role of the neuromuscular system in functional activity of the mandible. J Am Dent Assoc 48:665-73, 1954

[22] Ahlgren J: An intercutaneous needle electrode for kinesiologic EMG studies. Acta Odontol Scand 25:15-9, 1967

[23] Huck NL, Abbink JH, Hoogenkamp E, et al: Exteroceptive reflexes in jaw-closing muscle EMG during rhythmic jaw closing and clenching in man. Exp Brain Res 162:230-8, 2005

[24] van der Glas HW, van der Bilt A, Abbink JH, et al: Functional roles of oral reflexes in chewing and biting: phase-, task- and site-dependent reflex sensitivity. Arch Oral Biol 52:365-9, 2007

[25] Goldberg LJ: Masseter muscle excitation induced by stimulation of periodontal and gingival receptors in man. Brain Res 32:369-81, 1971

[26] Fogle LL, Glaros AG: Contributions of facial morphology, age, and gender to EMG activity under biting and resting conditions: a canonical correlation analysis. J Dent Res 74:1496-500, 1995

[27] Ringqvist M: Isometric bite force and its relation to dimensions of the facial skeleton. Acta Odontol Scand 31:35-42, 1973

[28] Ingervall B, Thilander B: Relation between facial morphology and activity of the masticatory muscles. J Oral Rehabil 1:131-47, 1974

[29] Ueda HM, Miyamoto K, Saifuddin M, et al: Masticatory muscle activity in children and adults with different facial types. Am J Orthod Dentofacial Orthop 118:63-8, 2000

[30] Garcia-Morales P, Buschang PH, Throckmorton GS, et al: Maximum bite force, muscle efficiency and mechanical advantage in children with vertical growth patterns. Eur J Orthod 25:265-72, 2003

[31] Morimitsu T, Nokubi T, Nagashima T, et al: [Relationship between orofaciocranial morphologic factors and electromyographic activities of the masticatory muscles]. Nihon Ago Kansetsu Gakkai Zasshi 1:162-71, 1989

[32] Turker KS: Electromyography: some methodological problems and issues. Phys Ther 73:698-710, 1993

[33] Tuxen A, Bakke M, Pinholt EM: Comparative data from young men and women on masseter muscle fibres, function and facial morphology. Arch Oral Biol 44:509-18, 1999

[34] Naser-Ud-Din S, Sowman PF, Dang H, et al: Modulation of masseteric reflexes by simulated mastication. J Dent Res 89:61-5, 2010

[35] Yoshimi H, Sasaguri K, Tamaki K, et al: Identification of the occurrence and pattern of masseter muscle activities during sleep using EMG and accelerometer systems. Head Face Med 5:7, 2009

[36] Turker KS: Reflex control of human jaw muscles. Crit Rev Oral Biol Med 13:85-104, 2002

[37] Kubota K, Masegi T: Muscle spindle supply to the human jaw muscle. J Dent Res 56:901-9, 1977

[38] Yemm R: The orderly recruitment of motor units of the masseter and temporal muscles during voluntary isometric contraction in man. J Physiol 265:163-74, 1977

[39] Eriksson PO, Thornell LE: Relation to extrafusal fibre-type composition in muscle-spindle structure and location in the human masseter muscle. Arch Oral Biol 32:483-91, 1987

[40] Rowlerson A, Mascarello F, Barker D, et al: Muscle-spindle distribution in relation to the fibre-type composition of masseter in mammals. J Anat 161:37-60, 1988

[41] Morimoto T, Inoue T, Masuda Y, et al: Sensory components facilitating jaw-closing muscle activities in the rabbit. Exp Brain Res 76:424-40, 1989

[42] Thexton AJ: Mastication and swallowing: an overview. Br Dent J 173:197-206, 1992

[43] Soboleva U, Laurina L, Slaidina A: The masticatory system--an overview. Stomatologija 7:77-80, 2005

[44] Ueda HM, Ishizuka Y, Miyamoto K, et al: Relationship between masticatory muscle activity and vertical craniofacial morphology. Angle Orthod 68:233-8, 1998

[45] Brinkworth RS, Turker KS: A method for quantifying reflex responses from intra-muscular and surface electromyogram. J Neurosci Methods 122:179-93, 2003

[46] Bakke M, Tuxen A, Vilmann P, et al: Ultrasound image of human masseter muscle related to bite force, electromyography, facial morphology, and occlusal factors. Scand J Dent Res 100:164-71, 1992

[47] Bertram S, Rudisch A, Bodner G, et al: The short-term effect of stabilization-type splints on the local asymmetry of masseter muscle sites. J Oral Rehabil 28:1139-43, 2001

[48] Close PJ, Stokes MJ, L'Estrange PR, et al: Ultrasonography of masseter muscle size in normal young adults. J Oral Rehabil 22:129-34, 1995

[49] Kubo K, Kawata T, Ogawa T, et al: Outer shape changes of human masseter with contraction by ultrasound morphometry. Arch Oral Biol 51:146-53, 2006

[50] Serra MD, Duarte Gaviao MB, dos Santos Uchoa MN: The use of ultrasound in the investigation of the muscles of mastication. Ultrasound Med Biol 34:1875-84, 2008

[51] Naser-Ud-Din S, Sampson WJ, Dreyer CW, et al: Ultrasound measurements of the masseter muscle as predictors of cephalometric indices in orthodontics: a pilot study. Ultrasound Med Biol 36:1412-21, 2010

[52] Prabhu NT, Munshi AK: Measurement of masseter and temporalis muscle thickness using ultrasonographic technique. J Clin Pediatr Dent 19:41-4, 1994

[53] Kiliaridis S, Katsaros C, Karlsson S: Effect of masticatory muscle fatigue on cranio-vertical head posture and rest position of the mandible. Eur J Oral Sci 103:127-32, 1995

[54] Pereira LJ, Gaviao MB, Bonjardim LR, et al: Ultrasound and tomographic evaluation of temporomandibular joints in adolescents with and without signs and symptoms of temporomandibular disorders: a pilot study. Dentomaxillofac Radiol 36:402-8, 2007

[55] Farkas L: Anthropometry of the Head and Face (ed second edition). New York Raven Press Ltd, 1994 pp. 405

[56] Raadsheer MC, Kiliaridis S, Van Eijden TM, et al: Masseter muscle thickness in growing individuals and its relation to facial morphology. Arch Oral Biol 41:323-32, 1996

[57] Kiliaridis S, Kalebo P: Masseter muscle thickness measured by ultrasonography and its relation to facial morphology. J Dent Res 70:1262-5, 1991

[58] Hatch JP, Shinkai RS, Sakai S, et al: Determinants of masticatory performance in dentate adults. Arch Oral Biol 46:641-8, 2001

[59] Emshoff R, Bertram S, Brandlmaier I, et al: Ultrasonographic assessment of local cross-sectional dimensions of masseter muscle sites: a reproducible technique? J Oral Rehabil 29:1059-62, 2002

[60] Satiroglu F, Arun T, Isik F: Comparative data on facial morphology and muscle thickness using ultrasonography. Eur J Orthod 27:562-7, 2005

[61] Burke PH, Beard FH: Stereophotogrammetry of the face. A preliminary investigation into the accuracy of a simplified system evolved for contour mapping by photography. Am J Orthod 53:769-82, 1967

[62] Da Silveira AC, Daw JL, Jr., Kusnoto B, et al: Craniofacial applications of three-dimensional laser surface scanning. J Craniofac Surg 14:449-56, 2003

[63] Ireland AJ, McNamara C, Clover MJ, et al: 3D surface imaging in dentistry - what we are looking at. Br Dent J 205:387-92, 2008

[64] Kopp S, Kuhmstedt P, Notni G, et al: G-scan--mobile multiview 3-D measuring system for the analysis of the face. Int J Comput Dent 6:321-31, 2003

[65] Mah J: 3D imaging in private practice. Am J Orthod Dentofacial Orthop 121:14A, 2002

[66] Sachdeva RC: SureSmile technology in a patient--centered orthodontic practice. J Clin Orthod 35:245-53, 2001

[67] Arridge S, Moss JP, Linney AD, et al: Three dimensional digitization of the face and skull. J Maxillofac Surg 13:136-43, 1985

[68] Moss JP, Linney AD, Grindrod SR, et al: Three-dimensional visualization of the face and skull using computerized tomography and laser scanning techniques. Eur J Orthod 9:247-53, 1987

[69] Aung M, Sobel DF, Gallen CC, et al: Potential contribution of bilateral magnetic source imaging to the evaluation of epilepsy surgery candidates. Neurosurgery 37:1113-20; discussion 1120-1, 1995

[70] Moss JP: The use of three-dimensional imaging in orthodontics. Eur J Orthod 28:416-25, 2006

[71] Quintero JC, Trosien A, Hatcher D, et al: Craniofacial imaging in orthodontics: historical perspective, current status, and future developments. Angle Orthod 69:491-506, 1999

[72] Lee JY, Han Q, Trotman CA: Three-dimensional facial imaging: accuracy and considerations for clinical applications in orthodontics. Angle Orthod 74:587-93, 2004

[73] Nguyen CX, Nissanov J, Ozturk C, et al: Three-dimensional imaging of the craniofacial complex. Clin Orthod Res 3:46-50, 2000

[74] Vallance S: *(personal communications)* Structured Light technique with Mona Lisa Imaging System. Adelaide, 2009

[75] Naser-ud-Din S, Thoirs K, Sampson WJ: Ultrasonography, lateral cephalometry and 3D imaging of the human masseter muscle. Orthod Craniofac Res 14:33-43, 2011

[76] Nute SJ, Moss JP: Three-dimensional facial growth studied by optical surface scanning. J Orthod 27:31-8, 2000

[77] Darwis WE, Messer LB, Thomas CD: Assessing growth and development of the facial profile. Pediatr Dent 25:103-8, 2003

[78] Bourne CO, Kerr WJ, Ayoub AF: Development of a three-dimensional imaging system for analysis of facial change. Clin Orthod Res 4:105-111, 2001

[79] Clark W: Twin Block Functional Therapy Applications in Dentofacial Orthopaedics (ed Second). London, Mosby, 2002

[80] Sharma AA, Lee RT: Prospective clinical trial comparing the effects of conventional Twin-block and mini-block appliances: Part 2. Soft tissue changes. Am J Orthod Dentofacial Orthop 127:473-82, 2005

[81] Ismail SF, Moss JP, Hennessy R: Three-dimensional assessment of the effects of extraction and nonextraction orthodontic treatment on the face. Am J Orthod Dentofacial Orthop 121:244-56, 2002

[82] Ayoub AF, Xiao Y, Khambay B, et al: Towards building a photo-realistic virtual human face for craniomaxillofacial diagnosis and treatment planning. Int J Oral Maxillofac Surg 36:423-8, 2007

[83] Baik HS, Lee HJ, Lee KJ: A proposal for soft tissue landmarks for craniofacial analysis using 3-dimensional laser scan imaging. World J Orthod 7:7-14, 2006

[84] Nechala P, Mahoney J, Farkas LG: Maxillozygional anthropometric landmark: a new morphometric orientation point in the upper face. Ann Plast Surg 41:402-9, 1998

[85] Smith R, Zaitoun H, Coxon T, et al: Defining new dental phenotypes using 3-D image analysis to enhance discrimination and insights into biological processes. Arch Oral Biol, 2008

[86] Brook AH, Pitts NB, Yau F, et al: An image analysis system for the determination of tooth dimensions from study casts: comparison with manual measurements of mesio-distal diameter. J Dent Res 65:428-31, 1986

[87] Kneafsey LC, Cunningham SJ, Petrie A, et al: Prediction of soft-tissue changes after mandibular advancement surgery with an equation developed with multivariable regression. Am J Orthod Dentofacial Orthop 134:657-64, 2008

[88] Kusnoto B, Evans CA: Reliability of a 3D surface laser scanner for orthodontic applications. Am J Orthod Dentofacial Orthop 122:342-8, 2002

89] Papadopoulos MA, Christou PK, Athanasiou AE, et al: Three-dimensional craniofacial reconstruction imaging. Oral Surg Oral Med Oral Pathol Oral Radiol Endod 93:382-93, 2002

90] Fuhrmann RA, Schnappauf A, Diedrich PR: Three-dimensional imaging of craniomaxillofacial structures with a standard personal computer. Dentomaxillofac Radiol 24:260-3, 1995

91] Cevidanes LH, Franco AA, Gerig G, et al: Assessment of mandibular growth and response to orthopedic treatment with 3-dimensional magnetic resonance images. Am J Orthod Dentofacial Orthop 128:16-26, 2005

92] Budai M, Farkas LG, Tompson B, et al: Relation between anthropometric and cephalometric measurements and proportions of the face of healthy young white adult men and women. J Craniofac Surg 14:154-61; discussion 162-3, 2003

93] Khalaf K, Robinson DL, Elcock C, et al: Tooth size in patients with supernumerary teeth and a control group measured by image analysis system. Arch Oral Biol 50:243-8, 2005

94] Benington PC, Gardener JE, Hunt NP: Masseter muscle volume measured using ultrasonography and its relationship with facial morphology. Eur J Orthod 21:659-70, 1999

95] Raadsheer MC, van Eijden TM, van Ginkel FC, et al: Contribution of jaw muscle size and craniofacial morphology to human bite force magnitude. J Dent Res 78:31-42, 1999

96] Hartmann J, Meyer-Marcotty P, Benz M, et al: Reliability of a Method for Computing Facial Symmetry Plane and Degree of Asymmetry Based on 3D-data. J Orofac Orthop 68:477-90, 2007

97] Baumrind S: Taking stock: a critical perspective on contemporary orthodontics. Orthod Craniofac Res 7:150-6, 2004

98] Togashi K, Kitaura H, Yonetsu K, et al: Three-dimensional cephalometry using helical computer tomography: measurement error caused by head inclination. Angle Orthod 72:513-20, 2002

99] Martensson B, Ryden H: The holodent system, a new technique for measurement and storage of dental casts. Am J Orthod Dentofacial Orthop 102:113-9, 1992

100] Terajima M, Endo M, Aoki Y, et al: Four-dimensional analysis of stomatognathic function. Am J Orthod Dentofacial Orthop 134:276-87, 2008

101] Takashima M, Kitai N, Murakami S, et al: Volume and shape of masticatory muscles in patients with hemifacial microsomia. Cleft Palate Craniofac J 40:6-12, 2003

102] Low E, Zoellner H, Kharbanda OP, et al: Expression of mRNA for osteoprotegerin and receptor activator of nuclear factor kappa beta ligand (RANKL) during root resorption induced by the application of heavy orthodontic forces on rat molars. Am J Orthod Dentofacial Orthop 128:497-503, 2005

103] Xue F, Wong R, Rabie AB: Identification of SNP markers on 1p36 and association analysis of EPB41 with mandibular prognathism in a Chinese population. Arch Oral Biol 55:867-72, 2010

104] Xue F, Wong RW, Rabie AB: Genes, genetics, and Class III malocclusion. Orthod Craniofac Res 13:69-74, 2010

[105] Plooij JM, Maal TJ, Haers P, et al: Digital three-dimensional image fusion processes for planning and evaluating orthodontics and orthognathic surgery. A systematic review. Int J Oral Maxillofac Surg 40:341-52, 2011

[106] Boom HP, van Spronsen PH, van Ginkel FC, et al: A comparison of human jaw muscle cross-sectional area and volume in long- and short-face subjects, using MRI. Arch Oral Biol 53:273-81, 2008

[107] Toll DE, Popovic N, Drinkuth N: The use of MRI diagnostics in orthognathic surgery: prevalence of TMJ pathologies in Angle Class I, II, III patients. J Orofac Orthop 71:68-80, 2010

[108] Antonio GE, Griffith JF, Yeung DK: Small-field-of-view MRI of the knee and ankle. AJR Am J Roentgenol 183:24-8, 2004

[109] Johal A, Sheriteh Z, Battagel J, et al: The use of videofluoroscopy in the assessment of the pharyngeal airway in obstructive sleep apnoea. Eur J Orthod 33:212-9, 2011

[110] Naini FB, Moss JP: Three-dimensional assessment of the relative contribution of genetics and environment to various facial parameters with the twin method. Am J Orthod Dentofacial Orthop 126:655-65, 2004

[111] Emshoff R, Brandlmaier I, Bodner G, et al: Condylar erosion and disc displacement: detection with high-resolution ultrasonography. J Oral Maxillofac Surg 61:877-81, 2003

[112] Landes CA, Sterz M: Evaluation of condylar translation by sonography versus axiography in orthognathic surgery patients. J Oral Maxillofac Surg 61:1410-7, 2003

[113] Kiliaridis S, Mahboubi PH, Raadsheer MC, et al: Ultrasonographic thickness of the masseter muscle in growing individuals with unilateral crossbite. Angle Orthod 77:607-11, 2007

10

Phage-Displayed Recombinant Peptides for Non-Invasive Imaging Assessment of Tumor Responsiveness to Ionizing Radiation and Tyrosine Kinase Inhibitors

Hailun Wang, Miaojun Han and Zhaozhong Han
Department of Radiation Oncology and Cancer Biology,
Vanderbilt-Ingrim Cancer Center,
School of Medicine, Vanderbilt University, Nashville, TN,
USA

1. Introduction

Recent studies have resulted in a variety of therapeutic options for cancer. However, tumor patients, even a same patient at different disease stages, respond(s) to a treatment protocol with different efficacy. A concept of personalized medicine has been developed to deliver individually tailored treatment upon the unique responsiveness of each patient. Currently, it is still challenging to predict treatment outcomes due to the genetic complexity and heterogeneity of cancers, which underlies the varied responses to treatment. To efficiently design treatment strategies and monitor the outcomes of therapies for individual patient, tumor responsiveness to a specific treatment regimen needs to be assesses in a time- and cost-efficient manner.

Non-invasive imaging technologies have demonstrated great potentials in diagnosis and treatment management by monitoring individual patient's disease condition and progression. Currently, anatomic and functional imaging modalities have been generally applied to detect, stage, and monitor tumors. Compared to the anatomic imaging that measures tumor size, functional or molecular imaging provides more information on tumor metabolism, biomarker expression, cell death or proliferation and thus is more relevant to the imaging assessment of tumor responsiveness to a treatment regimen, especially when the treatment affects tumors through blocking the tumor progression instead of shrinking the tumor size. Discovery of novel probes that specifically binds to tumor-limited targets with sound biological relevance is a limiting factor to develop such functional imaging modality.

Compared to antibody (~150 kD), peptide is in a much smaller size (1-2 kD) that enables an improved tissue penetration, faster clearance from circulation system, and less immunogenic property that are expected for a imaging probe, especially in the repeated assessment of treatment responsiveness in solid tumors. Advances in phage display-related technologies have facilitated the discovery and development of peptide derivatives as imaging probes for a variety of tumors. By using one example of HVGGSSV peptide that has been discovered and tested for non-invasive imaging assessment of tumor response to ionizing radiation (IR) and receptor tyrosine kinase (RTK) inhibitors in multiple tumor types, this review demonstrates that phage-displayed peptides hold potentials in personalized medicine by facilitating molecular imaging, discovery of diagnostic biomarker or therapeutic target, and tumor-targeted drug delivery.

2. Advancement in radiation therapy of cancer

Cancer is the leading cause of death worldwide, deaths from cancer worldwide are projected to continue rising, with an estimated 12 million deaths in 2030 (World Health Organization). Currently, radiotherapy is one of the most important modalities for the treatment of cancers. Over 60% of cancer patients received radiotherapy as part of their treatments.

2.1 Radiotherapy

Radiotherapy is the medical use of ionizing radiation as part of cancer treatment to control malignant cells. Radiation therapy may be used to treat localized solid tumors, such as cancers of the skin, tongue, larynx, brain, breast, lung, prostate or uterine cervix. It can also be used to treat leukemia and lymphoma. It works by damaging the DNA of cancerous cells through the use of one of two types of high energy radiation, photon or charged particle. This damage is either direct or indirect ionizing the atoms which make up the DNA chain. Indirect damage happens as a result of the ionization of water by high energy radiations, such as X-ray or gamma ray, forming free radicals, notably hydroxyl radicals, which then damage the DNA and form single- or double-stranded DNA breaks. Direct damage to DNA occurs through charged particles such as proton, boron, carbon or neon ions. Due to their relatively large mass, charged particles directly strike DNA and transfer high energy to DNA molecules and usually cause double-stranded DNA breaks. The accumulating damages to cancer cells' DNA cause them to die or proliferate more slowly. To minimize the damage to normal cells, the total dose of radiation therapy is usually fractionated into several smaller doses to allow normal cells time to recover. In clinics, to spare normal tissues from the treatment, shaped radiation beams are aimed from several angles of exposure to intersect at the tumor, providing a much larger absorbed dose in the tumors than in the surrounding tissues. Brachytherapy, in which a radiation source is placed inside or next to the cancer area, is another technique to minimize exposure to healthy tissues during treatment of cancers in the breast, prostate and other organs. It is also common to combine radiotherapy with surgery, chemotherapy, hormone therapy or immunotherapy to maximize treatment efficiency.

2.2 Radiosensitizer

Besides the rapid advances in radiotherapy technologies, the increased understanding of cancer biology and signaling networks behind radiotherapy has led to the development of newer chemotherapy agents that help to increase radiation treatment efficiency. Pathways targeted for radiosensitization include DNA damage repair, cell cycle progression, cell survival and death, angiogenesis, or modulation of tumor microenvironment. For example, hypoxia is one general characteristic associated with fast-growing solid tumors. It stimulates tumor malignant progression and induces HIF-1a. A few studies have found that low

Phage-Displayed Recombinant Peptides for Non-Invasive Imaging Assessment of Tumor
Responsiveness to Ionizing Radiation and Tyrosine Kinase Inhibitors
229

oxygen levels in tumors are associated with a poor response to radiotherapy (Overgaard, 2007). Well-oxygenated cells show an approximately 2-3 fold increases in radiosensitivity compared to hypoxic cells (Dasu and Denekamp, 1998). This discovery results in the development of a family of drugs – oxygen radiosensitizers. From initial attempts to increase oxygen delivery to the tumor by using hyperbaric oxygen in radiotherapy (Mayer et al., 2005), to later use oxygen mimetics/Electron-affinity agents, such as nitroimidazoles (Brown, 1975), or transition metal complexes, such as cisplatin (Liu et al., 1997), oxygen radiosensitizers significantly increase the radiotherapy efficiency. Currently, attention has been given to hypoxic cytotoxins, a group of drugs that selectively or preferably destroys cells in a hypoxic environment. These classes of compounds, such as mitomycin (De Ridder et al., 2008; Moore, 1977), are different from classic radiosensitizer in that they can be converted to cytotoxic agents under low oxygenation states, and they provide valuable adjuncts to radiotherapy. Recently, a wide variety of drugs that influence the DNA damage and repair pathways are being evaluated in conjunction with radiation. It includes topoisomerase inhibitors (e.g. camptothecin, topotecan), the hypoxia-activated anthraquinone AQ4N, and alkylating agents such as temozolomide. Proteins involved in tumor malignant progresses are also drawn attention as attractive targets of radiosensitizers, such as HIF-1a (Palayoor et al., 2008), survivin (Miyazaki et al.), Ras (Cengel and McKenna, 2005), epidermal growth factor receptors and related kinases (Sartor, 2004; Williams et al., 2004). Inhibitors for receptor tyrosine kinases such as vascular endothelial growth factor have been extensively studied and applied to improve the therapeutic efficacy of radiotherapy (Vallerga et al., 2004).

3. Assessment of tumor responses to radiotherapy

Different types of cancers possessed different mutations. Even the same type of cancers, they show different growth characteristics at different locations and in different patients. The heterogeneity of cancers underlies the different responses of cancers to the same treatment. Currently, cancer response is measured by imaging assessment of tumor volumes or by repeated biopsy. The whole processes are time consuming and inefficient. The recent advancement in imaging technologies has revolutionized medical diagnosis and prognosis. From the macroscopic anatomical sites down to a functional assessment of processes within tumors, imaging provide us a method to evaluate tumor response to irradiation treatment in a non-invasive, reliable and repeatable way (Lowery et al., 2011). So far, a few biomarkers have been explored for imaging to predict patients' outcomes after radiation treatment.

3.1 Cell metabolism

Cell metabolism is the earliest biomarker being studied after radiation treatment. Positron emission tomography (PET) has been used to evaluate tumor metabolism. ^{18}F-fuorodeoxyglucose (FDG) is the most common PET tracer for metabolism study. FDG, a glucose analog, is taken up by high-glucose-consuming cells, such as cancer cells. But FDG cannot be further metabolized during glycolysis and it becomes trapped and rapidly accumulates within the cell. As a result, the distribution of ^{18}F-FDG is a good reflection of the location of cancer cells. It is routinely used for the staging of cancer and for the monitoring of therapy (Allal et al., 2004).

3.2 Cell proliferation

The development of proliferation probes for PET imaging has enabled the *in vivo* evaluation of cell proliferation (Shields et al., 1998). Among those probes, nucleoside-based imaging probes (3'deoxy-3'-[18F]-fluorothymidine, FLT) or amino acids based imaging probes are gaining popularity. [18F]-FLT is a pyrimidine analog that, after uptake into the cell, is phosphorylated by thymidine kinase 1. The phosphorylated [18F]-FLT can not leave the cell and result in the intracellular accumulation of radioactivity. Thymidine kinase 1 is a principal enzyme in the salvage pathway of DNA synthesis and exhibits increased activity during the S phase of the cell cycle. [18F]-FLT uptake, therefore, reflects cellular proliferation. Amino acid metabolism is increased in fast proliferating tumor cells. Among the 20 essential amino acids, l-[11C]MET, [18F]fluorotyrosine, l-[11C]leucine, and [18F]fluoro-α-methyl tyrosine have been widely used in the detection of tumors (Laverman et al., 2002). Changes in l-[11C]MET uptake have already been shown to reflect response to radiotherapy treatment in patients suffering from a wide variety of tumors (Team, 2005b).

3.3 Tumor vasculature and hypoxia

Although being characterized as vasculature-rich structures, tumors often develop regions of hypoxia due to the leaky and disorganized tumor blood vessels. Low oxygen environment will promote tumor angiogenesis, metastasis and render tumors resistant to radiation treatment (Tatum et al., 2006). Therefore, the tumor vasculature structure and oxygen level are valuable biomarkers for prognosis after treatment. [18F]-fluoromisonidazole is the most widely used PET tracer for detecting tumor hypoxia. After uptake in cell, it is reduced and binds selectively to macromolecules under hypoxic conditions (Team, 2005a). One recent study indicates that [18F]-fluoromisonidazole uptake is correlated with radiation treatment outcome in Head and neck cancer (Thorwarth et al., 2005). As to the tumor vasculature, several studies have been proposed using two different techniques - quantified power Doppler sonography or Dynamic contrast-enhanced MRI (DCE-MRI). And both showed promising results (Hormigo et al., 2007; Kim et al., 2006; Mangla et al., 2010).

3.4 Apoptosis

Since its recognition as one of the major forms of cell death after radiation, apoptosis is being increasingly studied as a biomarker of cellular radiosensitivity and a prognosis marker for radiotherapy outcome. During the apoptosis process, phosphatidylserine (PS) flips from the inner leaflet of the cell membrane to the exterior of the cells. Annexin V, a cellular protein of the Annexin family, binds to the exposed PS. To date, Annexin V has been fluorinated for PET and radioiodinated for SPECT. Annexin V labeled with 99mTc has demonstrated significant uptake in patients suffering form myocardial infarction (Narula et al., 2001). Studies assessing quantitative ^{99m}Tc-Annexin V uptake in human tumors and their relationship to radiotherapy outcome are underway.

4. Phage-displayed peptides as novel imaging probes for assessing tumor response to treatment

Recently, advances in phage display-related technologies facilitate the use of small peptide derivatives as probe molecules for recognition and targeting tumors. Phage display enables discovery and optimization of affinity probes for the known tumor-specific biomarkers.

Phage-Displayed Recombinant Peptides for Non-Invasive Imaging Assessment of Tumor
Responsiveness to Ionizing Radiation and Tyrosine Kinase Inhibitors
231

Furthermore, this technology makes it possible to *de novo* discover novel imaging probes, and eventually identify novel diagnostic markers or therapeutic targets of cancer. In vivo screening against heterogeneous tumor targets have generated a diverse group of peptides for cancer-targeted delivery of imaging or therapeutic agents.

4.1 Principle of phage display

A phage is a type of viruses that infect bacteria. Typically, phages consist of a protein capsid enclosing genetic materials. Due to its simple structure, phages have been developed into a powerful tool in biological studies. Phage display was originally invented by George P. Smith in 1985 when he demonstrated the display of exogenous peptides on the surface of filamentous phage by fusing the DNA of the peptide on to the capsid gene of filamentous phages (Smith, 1985) (Fig. 1). This technology was further developed and improved to display large proteins such as enzymes and antibodies (Fernandez-Gacio et al., 2003; Han et al., 2004). The connection between genotype and phenotype enables large libraries of peptides or proteins to be screened in a relative fast and economic way. The most common phages used in phage display are M13 filamentous phage and T7 phage (Krumpe et al., 2006; Smith and Petrenko, 1997). The functional moiety on the phage surface can be short peptides, recombinant proteins, engineered antibody fragments or scaffold proteins. Screening can be conducted on the purified organic or inorganic materials, cells, or tissues.

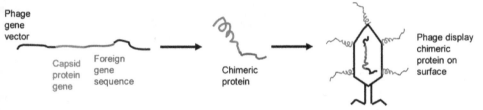

Phage
gene
vector

Capsid protein gene   Foreign gene sequence   Chimeric protein   Phage display chimeric protein on surface

Fig. 1. Schematic illustration of phage display. Foreign gene sequences encoding short peptides, recombinant proteins or large antibody fragments can be fused with capsid protein genes with recombinant DNA technologies. As a result, the recombinant phages express the foreign peptides or proteins on the phage surface for affinity-based selection. The affinity-selected phages can be replicated in bacterial host for further rounds of selection or DNA-sequencing to identify the affinity peptides or proteins expressed on the phage surface.

4.2 Applications of phage display
4.2.1 General applications

The application of phage display technology include determination of binding partners of organic (proteins, polysaccharides, or DNAs) (Gommans et al., 2005) or inorganic materials (Hattori et al., 2010; Whaley et al., 2000). The technique is also used to study enzyme evolution *in vitro* for engineering biocatalysts (Pedersen et al., 1998). Phage display has been widely applied in drug discovery. It can be used for finding new ligands, such as enzyme inhibitors, receptor agonists and antagonists, to target proteins (Hariri et al., 2008; Pasqualini et al., 1995; Perea et al., 2004; Ruoslahti, 1996; Uchino et al., 2005). Invention of antibody phage display revolutionized the drug discovery (Han et al., 2004). Millions of different single chain antibodies on phages are used for isolating highly specific therapeutic antibody leads. One of the most successful examples was adalimumab (Abbott Laboratories), the first fully human antibody targeted to TNF alpha (Spector and Lorenzo, 1975).

4.2.2 *In vivo* phage display and its application in clinical oncology

Because isolating or producing recombinant membrane proteins for use as target molecules in phage library screening is often facing insurmountable obstacles, innovative selection strategies such as panning against whole cells or tissues were devised (Jaboin et al., 2009; Molek et al., 2011; Pasqualini and Ruoslahti, 1996). Due to cells inside the body may express different surface markers and possess different characteristics from cell lines in culture, *in vivo* phage bio-panning was developed to identify more physiologically relevant biomarkers (Fig. 2) (Pasqualini and Ruoslahti, 1996). Since its invention, *in vivo* phage display has been used extensively to screen for novel targets for tumor therapy. Majority of those studies focused on analyzing the structure and molecular diversity of tumor vasculature and selecting tumor stage- and type-specific markers on tumor blood vessels (Arap et al., 2002; Rajotte and Ruoslahti, 1999; Sugahara et al., 2010; Valadon et al., 2006). Recently, the use of this technique was expanded to the field of discovering new biomarkers for evaluation of cancer treatment efficacy. (Han et al., 2008; Passarella et al., 2009).

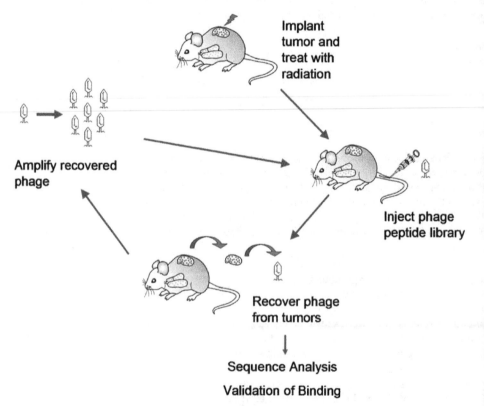

Fig. 2. *In vivo* phage display for screening peptides specifically target to radiation- or drug-treated tumors.

4.3 Peptides as probes for tumor targeted imaging
4.3.1 Advantages of peptide vs. antibody for tumor targeting

Antibodies, especially monoclonal antibodies, have been successfully utilized as cancer-targeting therapeutics and diagnostics due to their high target specificity and affinity. However, due to antibody large size (150 kDa) and limited tissue permeability, non-specific uptake into the reticuloendothelial system, and immunogenicity, most antibody-based therapeutics are of limited efficacy (Lin et al., 2005; Stern and Herrmann, 2005). In contrast to antibodies, peptides are much smaller molecules (1-2 kDa). Peptides have favorable biodistribution profiles compared to antibody, characterized by high uptake in the tumor tissue and rapid clearance from the blood. In addition, peptides have increased capillary permeability, allowing more efficient penetration into tumor tissues. Also peptides are easy to make and safe to use, they will not elicit an immune response (Ladner et al., 2004). With all these advantages, peptides have been increasingly considered as a good tumor targeted imaging probe (Aloj and Morelli, 2004; Okarvi, 2004; Reubi and Maecke, 2008).

4.3.2 Peptide as imaging probe

To date, a large number of peptides derived from natural proteins have already been successfully identified and characterized for tumor targeting and tumor imaging, such as integrin (RGD), somatostatin, gastrin-releasing peptide, cholecystokinin, glucagon-like peptides-1 and neuropeptide-Y (Cai et al., 2008; Hallahan et al., 2003; Korner et al., 2007; Miao and Quinn, 2007; Reubi, 2003; Reubi, 2007). A list of a few tumor homing peptides isolated using phage display technique is shown in Table 1.

Tumor Types	Tumor- targeting peptides
Prostate carcinoma	IAGLATPGWSHWLAL (Newton et al., 2006) ANTPCGPYTHDCPVKR (Deutscher et al., 2009) R/KXXR/K (Sugahara et al., 2009)
Colon carcinoma	CPIEDRPMC (Kelly et al., 2004)
Breast carcinoma	EGEVGLG (Passarella et al., 2009)
Hepatocellular carcinoma	AWYPLPP (Jia et al., 2007) AGKGTAALETTP (Du et al., 2010)
Pancreatic carcinoma	KTLLPTP (Kelly et al., 2008)
Head and Neck Cancer	SPRGDLAVLGHKY (Nothelfer et al., 2009)
Osteosarcoma	ASGALSPSRLDT (Sun et al., 2010)
Fibrosarcoma	SATTHYRLQAAN (Hadjipanayis et al., 2010)
Esophageal Cancer	YSXNXW and PXNXXN (Zhivotosky and Orrenius, 2001)
Bladder Cancer	CSNRDARRC (Ginestier et al., 2007)

Table 1. Phage display-derived tumor-targeting peptides

For use as *in vivo* imaging probes, peptides can be directly or indirectly labeled with a wide range of imaging moieties according to the imaging modality. For instance, near-infrared (NIR) fluorescent dyes or quantum dots have been labeled for optical imaging (Fig. 3), several radionuclides have been employed for positron emission tomography (PET) or single-photon emission computed tomography (SPECT), and paramagnetic agents have been used for magnetic resonance imaging (MRI) (Frangioni, 2003; Reubi and Maecke, 2008). Peptides can also be conjugated to other tumor targeted polymers or nanoparticles and dramatically increase their tumor targeted selectivity and efficiency (Hariri et al., 2010; Lowery et al., 2010; Passarella et al., 2009).

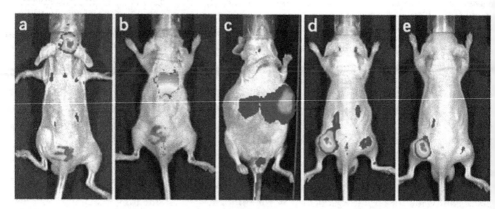

Fig. 3. HVGGSSV peptide labeled with near-infrared (NIR) fluorescent dyes specifically located to radiation-treated tumors. a) brain tumor (D54 human glioblastoma cell), b) lung tumor (H460 cell), c) colon cancer liver metastasis (HT22 cell), d) prostate cancer subcutaneous model (PC-3 cell), and e) breast cancer subcutaneous model (MDA-MB-231 cell). (Adapted from Han et al., 2008).

5. HVGGSSV peptide as one imaging probes to detect tumor response to radiation and tyrosine kinase inhibitor (TKI) *in vivo*

5.1 Discovery of HVGGSSV peptide
In our recent studies, we employed *in vivo* phage display technique and intended to identify peptides that will specifically home to radiation or drug treatment responsive tumors (Han et al., 2008; Passarella et al., 2009). During the studies, we first treated tumors in mice with radiation and tyrosine kinase inhibitors. Then a peptide phage library was injected from the tail vein of tumor bearing mice for tumor binding screening. After several rounds of *in vivo* screening and enrichment of phages isolated from the treated tumors (Fig. 2), one phage clone, encoding HVGGSSV peptides, was identified preferentially target to treatment responsive tumors. The binding preference of those phages were confirmed by fluorescence labeled phage or peptide imaging (Han et al., 2008; Passarella et al., 2009).

5.2 HVGGSSV peptide as imaging/targeting probe for radiation responsive tumors
To explore HVGGSSV peptide's clinical application in noninvasive imaging of tumor response to treatment, fluorescent labeled HVGGSSV peptide were used to target human tumors in several mouse models. Optical imaging studies indicated that the signal intensities of peptide binding within tumors correlate to the overall efficacy of treatment regimens on tumor growth control in multiple tumor models that had been treated with a variety of RTK inhibitors with or without combination of radiation (Han et al., 2008). SPECT/CT provides high spatial resolution and sensitivity in functional imaging. We

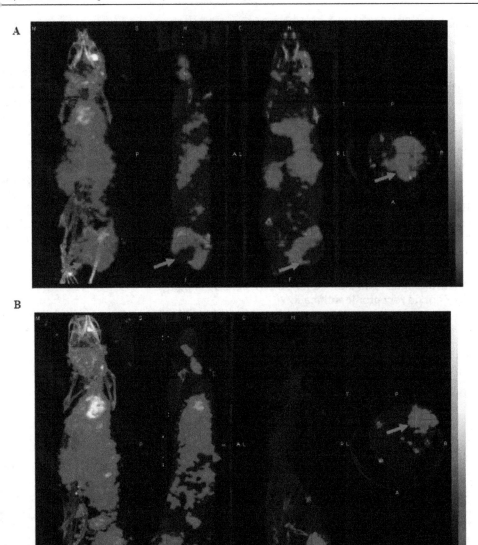

Fig. 4. SPECT/CT imaging of the HVGGSSV peptide within LLC tumors after treatment.
The biotinylated HVGGSSV peptide was complexed with iodine-125-labeled streptavidin.
The implanted tumor was treated once with radiation (5 Gy) alone (**A**), or combination of
Sunitinib (40 mg/Kg) and radiation (5 Gy) (**B**) before intravenous administration of the
imaging probe. Shown are 3D virtual rendering (3D-VR) images (far left) and hybrid
SPECT/CT fusion images in the coronal, sagittal, and transaxial planes (the second to the
fouth from the left, respectively) acquired 4 hours after the imaging probe administration.
The LLC tumors was pointed with arrows, high resolution images enable spatially localizing
the radiation-responding cells within the peripheral and central regions of the tumors.

employed this imaging modality to detect tumor response to radiation by using the HVGGSSV peptide. The mice were treated with radiation alone or combination of radiation and one TKI - Sunitinib (40 mg/Kg,). After the treatment, the HVGGSSV peptide complexed with ^{125}I-labeled streptavidin was selectively targeted to the tumors treated with radiation or radiation combined with Sunitinib. High resolution SPECT/CT images (Fig. 4) also showed that majority of the imaging probes were located in the peripheral area of the tumors that were treated with radiation alone. However, treatment with radiation and Sunitinib extended the imaging probe binding to both the peripheral and central parts of the subcutaneous tumors. This data might reflect the radiosensitization effect of Sunitinib.

The tumor targeting potential of HVGGSSV peptide has been further explored in several drug delivery studies. In these studies, HVGGSSV peptide has been conjugated to different nanoparticles, such as liposome, FePt, and nanoparticle albumin bound (nab) (Hariri et al., 2011; Hariri et al., 2010; Lowery et al., 2011), and selectively targeted those nanoparticles to irradiated tumors. One study also showed >5-fold increase in paclitaxel levels within irradiated tumors in HVGGSSV-nab-paclitaxel-treated groups and significantly increase tumor growth delay as compared with controls (Hariri et al., 2010).

5.3 The biological basis of the HVGGSSV peptide imaging
5.3.1 Peptide receptor identification
To understand the physiology underlines peptides binding, we need to identify the molecular targets of peptides. However, peptides are usually unstable. Their surface charges and structures will change dramatically in different environment. And peptides usually interact with their targets with low binding affinity due to their small sizes. Therefore, traditional affinity purification methods are of little use because of high background of non-specific binding. To date, there are very few identified receptors for peptides in contrast to the great number of discovered cancer targeting peptides (Sugahara et al., 2009). New strategies are needed for identifying peptide's receptors.

5.3.2 TIP-1 as a molecular target of HVGGSSV peptide
In our recent studies of one peptide (HVGGSSV), we utilized a phage cDNA library screening to search for peptide's receptors. Because several rounds of phage display screening can significantly enrich the low-affinity or low-abundance proteins, we successfully identified a PDZ protein - TIP-1 as the target of HVGGSSV peptide (Wang et al., 2010). Through the PDZ domain, TIP-1 binds to the classic C-terminal PDZ motif within the HVGGSSV peptide. One TIP-1-specific antibody that inhibited the in vitro interaction between TIP-1 and the HVGGSSV peptide attenuated the peptide's accumulation within irradiated tumors. Imaging with TIP-1-specific antibody recapitulated the pattern of peptide imaging in tumor-bearing mice. Mutation in the classic PDZ binding motif of the HVGGSSV peptide destroyed the specific binding within irradiated tumors. These results also demonstrated the potentials of screening phage-displayed cDNA library in discovery of molecular targets of the peptides with a simple structure and low affinity.

5.3.3 The biological relevance of TIP-1 relocation onto tumor cell surface to the radiation response of tumor cells
With a TIP-1 specific antibody, it was further identified that radiation induced translocation of the basically intracellular TIP-1 protein onto the cell surface in a dose-dependent manner.

The treatment-induced TIP-1 expression on the cell surface is detectable in the first few hours after the treatment and before the onset of treatment associated apoptosis or cell death. Majority of the cells expressing TIP-1 on the cell surface are the live albeit such cells are less potent in proliferation and more susceptible to subsequent radiation treatment (Wang et al., 2010). The increased susceptibility to the subsequent irradiation might explain why the peptide binding is predictive in assessing tumor overall responsiveness in the early stage of treatment course. The treatment-inducible TIP-1 translocation before the onset of cell apoptosis or death further suggests potentials of the HVGGSSV peptide in non-invasive imaging assessment of tumor response to radiation and tyrosine kinase inhibitors.

6. Perspectives

The development of imaging technologies revolutionizes medial diagnosis and clinical management. Functional molecular imaging becomes one critical part of personalized medicine. Peptide, with its small sizes and versatile structures, is increasingly recognized as a promising imaging probe to predict the outcomes of radiotherapy and other medical treatments. Although some disadvantages associated with peptides, such as its degradation inside human body and its low affinity with its targets, with chemical modification to improve its stability and association with nanoparticles to increase its binding affinity, peptides will play a major role in the future molecular imaging.

7. References

Allal, A. S., Slosman, D. O., Kebdani, T., Allaoua, M., Lehmann, W., and Dulguerov, P. (2004). Prediction of outcome in head-and-neck cancer patients using the standardized uptake value of 2-[18F]fluoro-2-deoxy-D-glucose. International journal of radiation oncology, biology, physics 59, 1295-1300.

Aloj, L., and Morelli, G. (2004). Design, synthesis and preclinical evaluation of radiolabeled peptides for diagnosis and therapy. Current pharmaceutical design 10, 3009-3031.

Arap, W., Kolonin, M. G., Trepel, M., Lahdenranta, J., Cardo-Vila, M., Giordano, R. J., Mintz, P. J., Ardelt, P. U., Yao, V. J., Vidal, C. I., et al. (2002). Steps toward mapping the human vasculature by phage display. Nat Med 8, 121-127.

Brown, J. M. (1975). Selective radiosensitization of the hypoxic cells of mouse tumors with the nitroimidazoles metronidazole and Ro 7-0582. Radiation research 64, 633-647.

Cai, W., Niu, G., and Chen, X. (2008). Imaging of integrins as biomarkers for tumor angiogenesis. Current pharmaceutical design 14, 2943-2973.

Cengel, K. A., and McKenna, W. G. (2005). Molecular targets for altering radiosensitivity: lessons from Ras as a pre-clinical and clinical model. Critical reviews in oncology/hematology 55, 103-116.

Dasu, A., and Denekamp, J. (1998). New insights into factors influencing the clinically relevant oxygen enhancement ratio. Radiother Oncol 46, 269-277.

De Ridder, M., Van Esch, G., Engels, B., Verovski, V., and Storme, G. (2008). Hypoxic tumor cell radiosensitization: role of the iNOS/NO pathway. Bulletin du cancer 95, 282-291.

Deutscher, S. L., Figueroa, S. D., and Kumar, S. R. (2009). Tumor targeting and SPECT imaging properties of an (111)In-labeled galectin-3 binding peptide in prostate carcinoma. Nuclear medicine and biology 36, 137-146.

Du, B., Han, H., Wang, Z., Kuang, L., Wang, L., Yu, L., Wu, M., Zhou, Z., and Qian, M. (2010). targeted drug delivery to hepatocarcinoma in vivo by phage-displayed specific binding peptide. Mol Cancer Res 8, 135-144.

Fernandez-Gacio, A., Uguen, M., and Fastrez, J. (2003). Phage display as a tool for the directed evolution of enzymes. Trends in biotechnology 21, 408-414.

Frangioni, J. V. (2003). In vivo near-infrared fluorescence imaging. Current opinion in chemical biology 7, 626-634.

Ginestier, C., Hur, M. H., Charafe-Jauffret, E., Monville, F., Dutcher, J., Brown, M., Jacquemier, J., Viens, P., Kleer, C. G., Liu, S., et al. (2007). ALDH1 is a marker of normal and malignant human mammary stem cells and a predictor of poor clinical outcome. Cell Stem Cell 1, 555-567.

Gommans, W. M., Haisma, H. J., and Rots, M. G. (2005). Engineering zinc finger protein transcription factors: the therapeutic relevance of switching endogenous gene expression on or off at command. Journal of molecular biology 354, 507-519.

Hadjipanayis, C. G., Machaidze, R., Kaluzova, M., Wang, L., Schuette, A. J., Chen, H., Wu, X., and Mao, H. (2010). EGFRvIII antibody-conjugated iron oxide nanoparticles for magnetic resonance imaging-guided convection-enhanced delivery and targeted therapy of glioblastoma. Cancer Res 70, 6303-6312.

Hallahan, D., Geng, L., Qu, S., Scarfone, C., Giorgio, T., Donnelly, E., Gao, X., and Clanton, J. (2003). Integrin-mediated targeting of drug delivery to irradiated tumor blood vessels. Cancer cell 3, 63-74.

Han, Z., Fu, A., Wang, H., Diaz, R., Geng, L., Onishko, H., and Hallahan, D. E. (2008). Noninvasive assessment of cancer response to therapy. Nature medicine 14, 343-349.

Han, Z., Karatan, E., Scholle, M. D., McCafferty, J., and Kay, B. K. (2004). Accelerated screening of phage-display output with alkaline phosphatase fusions. Comb Chem High Throughput Screen 7, 55-62.

Hariri, G., Wellons, M. S., Morris, W. H., 3rd, Lukehart, C. M., and Hallahan, D. E. (2011). Multifunctional FePt nanoparticles for radiation-guided targeting and imaging of cancer. Annals of biomedical engineering 39, 946-952.

Hariri, G., Yan, H., Wang, H., Han, Z., and Hallahan, D. E. (2010). Nanoparticle Albumin Bound Paclitaxel Retargeted to Radiation Inducible TIP-1 in Cancer. Clin Cancer Res.

Hariri, G., Zhang, Y., Fu, A., Han, Z., Brechbiel, M., Tantawy, M. N., Peterson, T. E., Mernaugh, R., and Hallahan, D. (2008). Radiation-guided P-selectin antibody targeted to lung cancer. Ann Biomed Eng 36, 821-830.

Hattori, T., Umetsu, M., Nakanishi, T., Togashi, T., Yokoo, N., Abe, H., Ohara, S., Adschiri, T., and Kumagai, I. (2010). High affinity anti-inorganic material antibody generation by integrating graft and evolution technologies: potential of antibodies as biointerface molecules. J Biol Chem 285, 7784-7793.

Hormigo, A., Gutin, P. H., and Rafii, S. (2007). Tracking normalization of brain tumor vasculature by magnetic imaging and proangiogenic biomarkers. Cancer cell 11, 6-8.

Jaboin, J. J., Han, Z., and Hallahan, D. E. (2009). Using in vivo biopanning for the development of radiation-guided drug delivery systems. Methods in molecular biology (Clifton, NJ 542, 285-300.

Jia, W. D., Sun, H. C., Zhang, J. B., Xu, Y., Qian, Y. B., Pang, J. Z., Wang, L., Qin, L. X., Liu, Y.
K., and Tang, Z. Y. (2007). A novel peptide that selectively binds highly metastatic
hepatocellular carcinoma cell surface is related to invasion and metastasis. Cancer
letters 247, 234-242.

Kelly, K., Alencar, H., Funovics, M., Mahmood, U., and Weissleder, R. (2004). Detection of
invasive colon cancer using a novel, targeted, library-derived fluorescent peptide.
Cancer research 64, 6247-6251.

Kelly, K. A., Bardeesy, N., Anbazhagan, R., Gurumurthy, S., Berger, J., Alencar, H., Depinho,
R. A., Mahmood, U., and Weissleder, R. (2008). Targeted nanoparticles for imaging
incipient pancreatic ductal adenocarcinoma. PLoS medicine 5, e85.

Kim, D. W., Huamani, J., Niermann, K. J., Lee, H., Geng, L., Leavitt, L. L., Baheza, R. A.,
Jones, C. C., Tumkur, S., Yankeelov, T. E., et al. (2006). Noninvasive assessment of
tumor vasculature response to radiation-mediated, vasculature-targeted therapy
using quantified power Doppler sonography: implications for improvement of
therapy schedules. J Ultrasound Med 25, 1507-1517.

Korner, M., Stockli, M., Waser, B., and Reubi, J. C. (2007). GLP-1 receptor expression in
human tumors and human normal tissues: potential for in vivo targeting. J Nucl
Med 48, 736-743.

Krumpe, L. R., Atkinson, A. J., Smythers, G. W., Kandel, A., Schumacher, K. M., McMahon,
J. B., Makowski, L., and Mori, T. (2006). T7 lytic phage-displayed peptide libraries
exhibit less sequence bias than M13 filamentous phage-displayed peptide libraries.
Proteomics 6, 4210-4222.

Ladner, R. C., Sato, A. K., Gorzelany, J., and de Souza, M. (2004). Phage display-derived
peptides as therapeutic alternatives to antibodies. Drug discovery today 9, 525-529.

Laverman, P., Boerman, O. C., Corstens, F. H., and Oyen, W. J. (2002). Fluorinated amino
acids for tumour imaging with positron emission tomography. European journal of
nuclear medicine and molecular imaging 29, 681-690.

Lin, M. Z., Teitell, M. A., and Schiller, G. J. (2005). The evolution of antibodies into versatile
tumor-targeting agents. Clin Cancer Res 11, 129-138.

Liu, T. Z., Lin, T. F., Chiu, D. T., Tsai, K. J., and Stern, A. (1997). Palladium or platinum
exacerbates hydroxyl radical mediated DNA damage. Free radical biology &
medicine 23, 155-161.

Lowery, A., Onishko, H., Hallahan, D. E., and Han, Z. (2010). Tumor-targeted delivery of
liposome-encapsulated doxorubicin by use of a peptide that selectively binds to
irradiated tumors. J Control Release.

Lowery, A., Onishko, H., Hallahan, D. E., and Han, Z. (2011). Tumor-targeted delivery of
liposome-encapsulated doxorubicin by use of a peptide that selectively binds to
irradiated tumors. J Control Release 150, 117-124.

Mangla, R., Singh, G., Ziegelitz, D., Milano, M. T., Korones, D. N., Zhong, J., and Ekholm, S.
E. (2010). Changes in relative cerebral blood volume 1 month after radiation-
temozolomide therapy can help predict overall survival in patients with
glioblastoma. Radiology 256, 575-584.

Mayer, R., Hamilton-Farrell, M. R., van der Kleij, A. J., Schmutz, J., Granstrom, G., Sicko, Z.,
Melamed, Y., Carl, U. M., Hartmann, K. A., Jansen, E. C., et al. (2005). Hyperbaric
oxygen and radiotherapy. Strahlenther Onkol 181, 113-123.

Miao, Y., and Quinn, T. P. (2007). Alpha-melanocyte stimulating hormone peptide-targeted melanoma imaging. Front Biosci 12, 4514-4524.

Miyazaki, A., Kobayashi, J., Torigoe, T., Hirohashi, Y., Yamamoto, T., Yamaguchi, A., Asanuma, H., Takahashi, A., Michifuri, Y., Nakamori, K., et al. Phase I clinical trial of survivin-derived peptide vaccine therapy for patients with advanced or recurrent oral cancer. Cancer science 102, 324-329.

Molek, P., Strukelj, B., and Bratkovic, T. (2011). Peptide phage display as a tool for drug discovery: targeting membrane receptors. Molecules (Basel, Switzerland) 16, 857-887.

Moore, H. W. (1977). Bioactivation as a model for drug design bioreductive alkylation. Science (New York, NY 197, 527-532.

Narula, J., Acio, E. R., Narula, N., Samuels, L. E., Fyfe, B., Wood, D., Fitzpatrick, J. M., Raghunath, P. N., Tomaszewski, J. E., Kelly, C., et al. (2001). Annexin-V imaging for noninvasive detection of cardiac allograft rejection. Nat Med 7, 1347-1352.

Newton, J. R., Kelly, K. A., Mahmood, U., Weissleder, R., and Deutscher, S. L. (2006). In vivo selection of phage for the optical imaging of PC-3 human prostate carcinoma in mice. Neoplasia (New York, NY 8, 772-780.

Nothelfer, E. M., Zitzmann-Kolbe, S., Garcia-Boy, R., Kramer, S., Herold-Mende, C., Altmann, A., Eisenhut, M., Mier, W., and Haberkorn, U. (2009). Identification and characterization of a peptide with affinity to head and neck cancer. J Nucl Med 50, 426-434.

Okarvi, S. M. (2004). Peptide-based radiopharmaceuticals: future tools for diagnostic imaging of cancers and other diseases. Medicinal research reviews 24, 357-397.

Overgaard, J. (2007). Hypoxic radiosensitization: adored and ignored. J Clin Oncol 25, 4066-4074.

Palayoor, S. T., Mitchell, J. B., Cerna, D., Degraff, W., John-Aryankalayil, M., and Coleman, C. N. (2008). PX-478, an inhibitor of hypoxia-inducible factor-1alpha, enhances radiosensitivity of prostate carcinoma cells. International journal of cancer 123, 2430-2437.

Pasqualini, R., Koivunen, E., and Ruoslahti, E. (1995). A peptide isolated from phage display libraries is a structural and functional mimic of an RGD-binding site on integrins. The Journal of cell biology 130, 1189-1196.

Pasqualini, R., and Ruoslahti, E. (1996). Organ targeting in vivo using phage display peptide libraries. Nature 380, 364-366.

Passarella, R. J., Zhou, L., Phillips, J. G., Wu, H., Hallahan, D. E., and Diaz, R. (2009). Recombinant peptides as biomarkers for tumor response to molecular targeted therapy. Clin Cancer Res 15, 6421-6429.

Pedersen, H., Holder, S., Sutherlin, D. P., Schwitter, U., King, D. S., and Schultz, P. G. (1998). A method for directed evolution and functional cloning of enzymes. Proceedings of the National Academy of Sciences of the United States of America 95, 10523-10528.

Perea, S. E., Reyes, O., Puchades, Y., Mendoza, O., Vispo, N. S., Torrens, I., Santos, A., Silva, R., Acevedo, B., Lopez, E., et al. (2004). Antitumor effect of a novel proapoptotic peptide that impairs the phosphorylation by the protein kinase 2 (casein kinase 2). Cancer research 64, 7127-7129.

Rajotte, D., and Ruoslahti, E. (1999). Membrane dipeptidase is the receptor for a lung-targeting peptide identified by in vivo phage display. J Biol Chem 274, 11593-11598.

Reubi, J. C. (2003). Peptide receptors as molecular targets for cancer diagnosis and therapy. Endocrine reviews 24, 389-427.

Reubi, J. C. (2007). Targeting CCK receptors in human cancers. Current topics in medicinal chemistry 7, 1239-1242.

Reubi, J. C., and Maecke, H. R. (2008). Peptide-based probes for cancer imaging. J Nucl Med 49, 1735-1738.

Ruoslahti, E. (1996). RGD and other recognition sequences for integrins. Annual review of cell and developmental biology 12, 697-715.

Sartor, C. I. (2004). Mechanisms of disease: Radiosensitization by epidermal growth factor receptor inhibitors. Nature clinical practice 1, 80-87.

Shields, A. F., Grierson, J. R., Dohmen, B. M., Machulla, H. J., Stayanoff, J. C., Lawhorn-Crews, J. M., Obradovich, J. E., Muzik, O., and Mangner, T. J. (1998). Imaging proliferation in vivo with [F-18]FLT and positron emission tomography. Nature medicine 4, 1334-1336.

Smith, G. P. (1985). Filamentous fusion phage: novel expression vectors that display cloned antigens on the virion surface. Science 228, 1315-1317.

Smith, G. P., and Petrenko, V. A. (1997). Phage Display. Chemical reviews 97, 391-410.

Spector, R., and Lorenzo, A. V. (1975). Myo-inositol transport in the central nervous system. Am J Physiol 228, 1510-1518.

Stern, M., and Herrmann, R. (2005). Overview of monoclonal antibodies in cancer therapy: present and promise. Critical reviews in oncology/hematology 54, 11-29.

Sugahara, K. N., Teesalu, T., Karmali, P. P., Kotamraju, V. R., Agemy, L., Girard, O. M., Hanahan, D., Mattrey, R. F., and Ruoslahti, E. (2009). Tissue-penetrating delivery of compounds and nanoparticles into tumors. Cancer Cell 16, 510-520.

Sugahara, K. N., Teesalu, T., Karmali, P. P., Kotamraju, V. R., Agemy, L., Greenwald, D. R., and Ruoslahti, E. (2010). Coadministration of a tumor-penetrating peptide enhances the efficacy of cancer drugs. Science 328, 1031-1035.

Sun, X., Niu, G., Yan, Y., Yang, M., Chen, K., Ma, Y., Chan, N., Shen, B., and Chen, X. (2010). Phage display-derived peptides for osteosarcoma imaging. Clin Cancer Res 16, 4268-4277.

Tatum, J. L., Kelloff, G. J., Gillies, R. J., Arbeit, J. M., Brown, J. M., Chao, K. S., Chapman, J. D., Eckelman, W. C., Fyles, A. W., Giaccia, A. J., et al. (2006). Hypoxia: importance in tumor biology, noninvasive measurement by imaging, and value of its measurement in the management of cancer therapy. International journal of radiation biology 82, 699-757.

Team, T. M. R. (2005a). [18F]Fluoromisonidazole. Molecular Imaging and Contrast Agent Database (MICAD) [Internet], http://www.ncbi.nlm.nih.gov/books/NBK23099/.

Team, T. M. R. (2005b). 1-[methyl-11C]Methionine. Molecular Imaging and Contrast Agent Database (MICAD) [Internet], http://www.ncbi.nlm.nih.gov/books/NBK23696/.

Thorwarth, D., Eschmann, S. M., Scheiderbauer, J., Paulsen, F., and Alber, M. (2005). Kinetic analysis of dynamic 18F-fluoromisonidazole PET correlates with radiation treatment outcome in head-and-neck cancer. BMC cancer 5, 152.

Uchino, H., Matsumura, Y., Negishi, T., Koizumi, F., Hayashi, T., Honda, T., Nishiyama, N., Kataoka, K., Naito, S., and Kakizoe, T. (2005). Cisplatin-incorporating polymeric micelles (NC-6004) can reduce nephrotoxicity and neurotoxicity of cisplatin in rats. Br J Cancer 93, 678-687.

Valadon, P., Garnett, J. D., Testa, J. E., Bauerle, M., Oh, P., and Schnitzer, J. E. (2006). Screening phage display libraries for organ-specific vascular immunotargeting in vivo. Proc Natl Acad Sci U S A *103*, 407-412.

Vallerga, A. K., Zarling, D. A., and Kinsella, T. J. (2004). New radiosensitizing regimens, drugs, prodrugs, and candidates. Clin Adv Hematol Oncol *2*, 793-805.

Wang, H., Yan, H., Fu, A., Han, M., Hallahan, D., and Han, Z. (2010). TIP-1 translocation onto the cell plasma membrane is a molecular biomarker of tumor response to ionizing radiation. PloS one *5*, e12051.

Whaley, S. R., English, D. S., Hu, E. L., Barbara, P. F., and Belcher, A. M. (2000). Selection of peptides with semiconductor binding specificity for directed nanocrystal assembly. Nature *405*, 665-668.

Williams, K. J., Telfer, B. A., Brave, S., Kendrew, J., Whittaker, L., Stratford, I. J., and Wedge, S. R. (2004). ZD6474, a potent inhibitor of vascular endothelial growth factor signaling, combined with radiotherapy: schedule-dependent enhancement of antitumor activity. Clin Cancer Res *10*, 8587-8593.

Zhivotosky, B., and Orrenius, S. (2001). Assessment of apoptosis and necrosis by DNA fragmentation and morphological criteria. Curr Protoc Cell Biol *Chapter 18*, Unit 18 13.

11

Biosafety in the Use of Radiation: Biological Effects Comparison Between Laser Radiation, Intense Pulsed Light and Infrared and Ultraviolet Lamps in an Experimental Model in Chicken Embryos

Rodolfo Esteban Avila, Maria Elena Samar, Gustavo Juri,
Juan Carlos Ferrero and Hugo Juri
Department of Cell Biology, Histology and Embryology,
Faculty of Medical Sciences, National University of Cordoba,
Argentina

1. Introduction

Physics is the science that studies the interaction between matter and energy. This discipline studies the general properties of bodies, the forces that modify the transfer of energy and the interaction between particles. Physics now has many branches, and one of them is the Biomedical Physics.

Biomedical Physics in Medicine applies the principles and methods of physics. This will generate new knowledge and progress towards new horizons in the management of certain diseases. Thus the study of radiation led to progress in the field of diagnosis and therapy of some diseases.

Radiation it's called all energy that propagates as a wave through space. Radiation can be classified as ionizing radiation (cosmic rays, gamma rays, x-rays) and non-ionizing radiation. The concept of non-ionizing radiation includes ultraviolet (UV), infrared (IR) and others lasers.

Non-ionizing radiations are those that radiation interaction with matter does not generate ions due to its energy content is relatively low.

Non-ionizing radiation (laser, ultraviolet and infrared) are commonly used in medicine for diagnosis and therapy. However, they have a deleterious effect on organogenesis (organ formation).

1.1 Ultraviolet radiation (UV)

UV radiation is part of the natural light. According to its wavelength is recognized groups A (400 - 320 nm), B (320 - 290 nm) and C (290 - 200 nm).

UV radiation sources are natural (the sun) and artificial (hospitals, industries, cosmetics, etc.).

UVC radiation does not reach the surface because it is retained by the layer of ozone in the stratosphere.

The natural radiation that reaches us is UVA and UVB. The UVC is the most dangerous to health because of its higher energy.

As for the benefits of using ultraviolet (UV), is recognized for use in phototherapy of patients with psoriasis, vitiligo and other skin diseases. In addition, ultraviolet rays have a bactericidal action, allowing its use in pressure ulcers.

However, involvement of UV radiation in different pathologies has led to the approach of concepts such as photo-aging, accelerated aging process due to modification by UV radiation, DNA (deoxyribonucleic acid) and lipids membrane and induction of programmed cell death (apoptosis) of epithelial cells and the activation of enzymes degrading the collagen in the skin.

While studying the effects of ultraviolet radiation on the apoptotic induction and angiogenesis the results are mixed for use in the therapy of skin diseases including cancer and photocarcinogenesis, understood as the mechanism of DNA mutation due to the alteration of repair when damaged by UV radiation.

Ultrviolet radiation is involved in the alterations of protein synthesis, the immunosuppressive properties and Its relationship to skin cancer and change in the synthesis of melanin.

Ultraviolet radiation B are related to the induction and progression of cutaneous melanoma in mice, so its use helps to know the specific immune processes committed to these cancers, and thus to develop new treatments for melanoma. This type of radiation has also been used experimentally to elucidate the mechanisms of cellular radioresistance in ovarian cell lines.

1.2 Infrared radiation (IR)

Other natural rays are infrared (IR) or infrared radiation (the prefix infra means below, correspond to an emission of energy in the form of electromagnetic waves in the spectrum located immediately after the red zone). These rays have wavelengths between 800 and 0.25 nm and cause heating of the exposed regions, being the less penetrating wavelength formerly called them "calorific rays."

As for the infrared rays have a wavelength between 800 and 0.25 nm, cause heating of the exposed areas and those are more penetrating radiation of shorter wavelength. Special lamps are used commercially to produce 14,000 to infrared.

Its heating effect allows blood vessels to dilate and increase blood flow.

They are also used in conjunction with hyperthermia and photodynamic therapy to treat tumor in the ablation of organs, and therapy of processes related to inflammation of nerves, muscles and joints.

It has been postulated that IR produce a higher activity of phagocytosis and metabolic reactivity in infectious and has been used in phototherapy of patients with certain skin tumors.

However, the intracellular level, has shown that infrared radiation acting microtubule disrupting the centrosome (organoids) in mammalian cells in vitro.

1.3 Laser radiation

The laser light (artificial) could be defined as an amplification process that culminates in the production of a light. In turn, electricity is the band of the electromagnetic spectrum including UV radiation, visible light and IR radiation. The laser light is not normally found in nature and is of extraordinary intensity.

The best known property is the emission of beams of highly monochromatic, coherent and directional light.

There are different laser for medical use, classified as high-power-like carbon dioxide (CO2), for surgical use, and low power-such as Helium-Neon (He-Ne) for clinical use.

The term laser means in English Light Amplification by Stimulated Emission of Radiation.

The laser is a device or devices that produce a very special light (visible or invisible depending on their wavelength), created by man and acts like solid matter.

Einstein laid the theoretical basis for his work on the Quantum Theory of Radiation in 1917, by which the energy (light) can be converted into mass and mass into energy.

As light is reflected, the laser is absorbed, burning, and changes its direction through different lenses. As solid mass you can cut, melt, burn, and transmit.

The use of laser energy is not without several problems as the high price of the apparatus and the need for prior learning to use. In addition there is one type of laser that fits all applications, but each type has it sitting indications, many of which are unique to it.

From the medical point of view the use of laser can be used to remove tumors, seal blood vessels to reduce blood loss, sealed lymph vessels to reduce swelling and decrease the spread of tumor cells and nerve endings sealed to reduce postoperative pain.

Aesthetic laser therapy treatments can be used to remove warts and moles and to remove tattoos. It is also used to treat stretch marks, cellulite, sagging skin, acne sequels.It is based on the fundamental physiological properties of the laser, anti-inflammatory, spasmolytic and antiphlogistic effects and bio-stimulants. (Andreu & Valiente Zaldivar 1996).

Lasers and biomedical use:

- Solid lasers: the most used is Neodymium-Yag laser. In ophthalmology it is used to coagulate tissue. It is also used to treat hyperpigmentation of the skin
- Gas lasers: they are the most widely used therapeutic and amongst these the helium-neon laser, used in beauty treatments, therapies reductive, etc. (red), the argon laser used in dermatology (bluish green) and the laser CO2 used in surgery as a scalpel.

The Helium-Neon laser is used to treat various conditions "satisfactory" mainly osteomioarticular (rehabilitation), skin disorders and wound healing.

Among the lasers used in medical practice are the 308nm excimer. This type of laser emits 85% of ultraviolet (UV) B and 15% of UV radiation type A.

Contraindications: The laser should not be used in patients affected by neoplasia (cancer). Nor can it be used in the presence of acute infections and in patients treated with photosensitizing drugs or remedies. Not recommended laser surgery in people with pacemakers or pregnant women.

In a previous study, we irradiated chick embryos and new born chickens with He-Ne laser, infrared and ultraviolet radiation, finding post irradiation histopathologic changes. (Samar et al., 1993, Avila et al., 1994 , Samar et al., 1995)

1.4 Intense Pulsed Light (IPL)

In addition to the laser, since 1995 it also has been available a device that emits at 308 nm, called intense pulsed light (IPL) which is basically a XeCl lamp that has proved to be a useful tool for the treatment of the changes of the skin. The IPL provides high energy pulsed excitation consisting of 85% of ultraviolet radiations UV B and 15% of radiations UV A. This technology is also known as "Photoderm". Although the active medium of this lamp is also XeCl, the emission is polychromatic but also non-coherent; therefore is not a laser. However, the use in medicine and the risks associated with it are comparable to the medical lasers of the high energy (class 3b and 4) and therefore their use should be subject to the same safety guidelines. (Spencer & Hadi, 2004). (Town, et al. 2007).

The IPL is used in a similar way to the excimer laser in the treatment of various pathologies of the skin: psoriasis, vitiligo, etc. Clinically it is also used to stimulate the regeneration of the cartilage in degenerative processes. It has been postulated that its action is based on the activation of the cell division, collagenous and elastic fibers formation, regeneration of blood vessels, cicatrisation of bone tissue and reepithelization of damaged tissue. (Schoenewolf et al. 2011)

Nevertheless, there are some discrepancies in results obtained with animal experimentation on the medical use of not-ionizing radiations. (Chan et al. 2007)

However, the biological effect of the 308 nm IPL has not been investigated using the chick embryos as sensor.

1.5 Objective

Different studies led us to establish preliminary results concerning the action of He-Ne laser, ultraviolet, infrared, LIP on the different tissues of chick embryos and in salivary glands and tongue of newly hatched chicks.

We also carry out the observation of effects that these types of radiation have on the chorioallantoic membrane of chick embryos, particularly on the formation of new blood vessels.

Moreover, the exposure time and the amount of non-ionizing radiation doses used in animal experiments and in medical practice are not yet fully known. Thus, our experimental design established in embryonated eggs and newly hatched chicks will contribute to the quantitative analysis (exposure and dose) of radiation mentioned above.

The purpose of this chapter describes the biological effects comparison between laser radiation, intense pulsed light and infrared and ultraviolet lamps in an experimental model in chicken embryos.

The chick embryo is a good model to evaluate direct effects of ionizing radiation because of easy handling and availability ITS.

Describe the most important results with different sources that emit non-ionizing radiation.

2. Materials and methods

2.1 Experimental model of chicken embryo

We used the chick embryo model as a mechanism for measuring biological effects of radiation on tissues, and this is an interesting method to be easily replicable.

Also to compare the results obtained by applying the radiation we have established in previous studies, the sequence of morphological changes, biochemical and histochemical occurring during differentiation and growth of the tongue, stomach, and ovaries both mesonephros "in ovo" as "in vitro" from 7 to 21 days post-birth immediately.

The problem groups were irradiated for 5 minutes through a window opened in the egg shell, and the eggs were maintained aseptically for 24hs in an incubator.We used the following radiations: intense pulsed light (excimer laser wavelength 308 nm) He-Ne laser (power 5mV, wavelength 632.8 nm), UV germicidal lamp and IR lamp OSRAM infraphil.

The control group were irradiated and only opened a window in the eggshell. All controls and aseptically problems remained in an incubator at 37 ° C for 24 hours post-irradiation.

The samples taken was fixed in Bouin and were processed by routine histological technique and stained with: hematoxylin and eosin; conventional Histochemistry: PAS for the demonstration of glycoproteins, Alcian Blue at pH 2.5 and 1.0 for the demonstration of glycosaminoglycans sulfated and sulfated; Toluidine blue at pH 3.8 to demonstrate basophilic and metachromatic substances, alcohol-resistant.

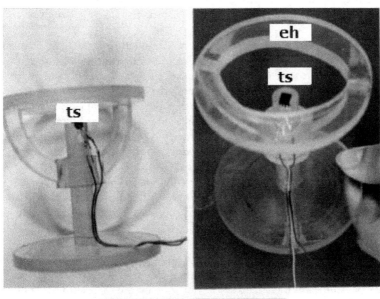

Sample (egg) with thermic sensor (ts). Egg holder (eh). Window in the egg shell (we).

Fig. 1. The system adaptable to the experimental requirements.

2.2 Infrared radiations

In 1996 we published design of a thermic detection system applied to chick embryonated eggs irradiated with infrared rays (Avila et al. 1996).

Infrared radiations are widely used in medical therapeutics. It has been argued that doses and periods of time employed in experimental animals are higher than those used in clinic. Thus, we considered of interest to analyse aspects of dosimetry and thermic effects of infrared rays with current methods in medical practice, using in ovo chick embryo as a model of easy control.

To this end we designed a system to measure temperatures and their acquisitions and software for its handling. The system consists of: a) thermic points: thermocouples or termistores adaptable to the experimental requirements and calibrated with a greater precision within a range of ten degrees around the incubation temperature; b) acquirer circuit of thermic data (hardware): it generates a time base that varies with the thermic sensor. Software: the PC XT or AT detects changes in the time base by means of a programs[1] in a Turbo Pascal; c) storage and analysis of data allows, through a menu the scale selection, time of program data to be acquired, storage and recovery of the diskette information and graphic impression; d) chick's embryonated eggs.

This system allows to measure temperature distribution in small physical spaces with little disturbance of the system to be measured in irradiated bodies, to analyse variations of the temperatures in time and to secure a greater confidence and automatism to obtain the required data. Figure 1.

2.3 Experimental Intense Pulsed Light (IPL) system with and without different filters (Avila et al. 2009)

To carry out the studies of the effects of the radiation on the different tissues in embryos of chicken we designed the experimental plan that is shown in the Figure 2.

The system consists of an Intense Pulsed Light (IPL) radiation source of 308 nm from DEKA model: Excilite (XeCl Excimer Light) wave length of 308 nm, capable to emit a radiation density of 4,5 J/cm2. The emission spectrum of this lamp was characterized in this work and is shown in the Figure 2.

Besides we used a filter set:

Acetate: Colorless, Orange, Green, Blue and Yellow.

Cellophane: Colorless, Orange, Green, Blue, Yellow and Red

A spectrometer (E1) Ocean Optics model HR4000 was used to characterize the emission of the IPL and the filters used in this work. The measurements of energy were carried out with an energy meter (E2) Scientech model 756 and a sensor (E3) model 362 with complete range (200nm to 1m). With this system we measured the total emitted energy per surface unit, mJ/cm2, on each sample with the different filters used and also the variation of the emission spectrum of the IPL after passing through the filter.

Optical characteristics of the light emitted by the XeCl lamp were determined, Intense Pulsed Light (IPL), using the experimental plan shown in Figure 2. The typical emission spectrum of this lamp was obtained by means of this system We determined that 85% of UVB radiation emitted by the IPL corresponds to 307.94 nm and only 15% to UVA centered in 367,5 nm. This implies that the effects seen on the tissue samples were due mainly by only one emission with very similar characteristics to the effects that could be caused by a laser radiation of this wave length.

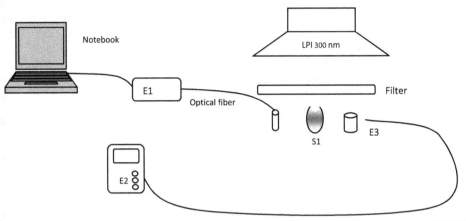

E1: Spectrometer Ocean Optics HR4000, E2: Energy meter, E3: Energy sensor, S1: sample (egg), LPI: IPL
source (308 nm). (Avila et al, 2009)

Fig. 2. Scheme of the experimental setup

In Table 1 we shown the light intensity transmitted in the different units by the IPL with the
different filters used of acrylic (Ac) and cellophane (Cel) with the exposure time that are
normally given for skin treatment. The selection criteria used for the filters to work with
were those which showed a transmittance grater than 30%. This transmittance percentage
guarantees the possibility to observe the effects of radiation on tissues.

Filter	Exposure time /s	Transmitted Intensity (mW)	Transmitted Fluence (mJ/cm²)	Filter absorbance	Filter transmittance (%)
None	20	18.1	147.50	0.00	100
F1: Green acetate	20	0.46	3.74	1.59	2.5
F2: Blue acetate	20	0.00	0.00	2.00	0.0
F3: Clear acetate	20	0.00	0.00	2.00	0.0
F4: Yellow acetate	20	0.00	0.00	2.00	0.0
F5: Orange acetate	20	6.70	54.60	0.43	37.0
F6: Red Cellophane	10	3.29	13.40	1.04	9.1
F7: Clear Cellophane	10	18.0	73.34	0.30	49.7
F8:Yellow Cellophane	10	2.2	8.96	1.21	6.07
F9: Blue Cellophane	10	0.48	1.95	1.87	1.32
F10: Orange Cellophane	10	9.2	37.50	0.60	25.0
F11:Green Cellophane	20	5.25	42.8	0.53	30.0

Table 1. Intensity of the light from the XeCl lamp (IPL) transmitted through the different
filters, measured at 30 cm from the lamp. (Avila et al. 2009).

Once we characterized the IPL emission and the filter absorption we proceeded to radiate the chicken embryos by previously removing their white membrane (below the shell of the egg). We first carried out a set of measurements without using any filters and later using the filters that were previously selected.

Before starting the radiation process in the different samples we determined the absorption spectrum of the albumen of the egg and of the white membrane. From these results we determined that the white membrane should be removed before the radiation of each sample since it has a very large absorption coefficient to the study wave length and would not permit to determine the effects of the radiation in the embryo tissues.

The acetate filters (transparent and orange) and cellophane filters (transparent orange and green) allow to the passage of the radiations producing biological effects in the organs of the chicken embryo, for this reason it is not recommended to be used as a filter.

The acetate filters (green, blue and yellow) and cellophane filters (red, yellow and blue) do not allow to the passage of radiation, without showing any biological effects in the chicken embryo.

3. Results

3.1 He-Ne laser irradiation
3.1.1 Chick embryo mesonephros
A study on the effects of He-Ne laser irradiation on glomeruli and renal tubules of the chick embryo mesonephros at 7 days of in ovo development was made. (Avila et al, 1992)

To this purpose, He-Ne laser irradiation (potency: 5 mW, wavelength: 632.8 nm) was beamed for 5 minutes through a window opened in the egg shell, and the eggs were maintained aseptically for 24 h in an incubator.

Mesonephros were dissected out and processed for hematoxylin/eosin, periodic acid-Schiff, and Alcian blue at pH 2.5 and 1.0, toluidine Blue at pH 3.8.

In controls, the glomeruli were formed by coiled capillaries with a homogeneous basement membrane, the proximal tubules presented a high cubic epithelium with acidophilic cytoplasm and a developed brush border. Distal and collecting segments were lined by cubic epithelium. An alcianophilic and PAS-positive reaction stood out at the membrane coats of the proximal tubules and tubular and glomerular basement membranes. No glomerular alterations were observed during the experiment. However, there was a marked enlargement of the tubular interstitium, with edema and lymphohistiocytic inflammatory infiltration. Some tubular cells desquamated to the lumen. Other cells presented a raveled apical surface. Some nuclei were dispersed out, and mixing with chromatin, formed diminutive granules. Pyknotic nuclei were seen occasionally. Epithelial necrosis and cytoplasmic debris in the lumen were also noted. For mucins, some zones showed brush border coats of the proximal tubules as discontinuous.

3.1.2 Chick embryo ovary
We published in 1992 the structural changes induced by He-Ne laser on the chick embryo ovary. (Avila et al. 1992)

The morpho-histochemical alterations that occur in the chicken ovary at 7 days of incubation after irradiation with He-Ne laser of a potency of 5 mw and at a wavelength of 632.8 nm were studied.

The embryos were irradiated for 5 minutes through a window opened in the eggshell and aseptically maintained in incubator for 24 hours. The gonads were dissected out and processed for the following techniques: H/E, PAS, Alcian blue, and Toluidine blue. Controls: The ovaries were formed by a germinative or superficial epithelium, with germ and epithelial cells, and by primary sex cords compressed between them, although separated by a reduced stroma. The cords contained germ cells The surface coat of the germinative epithelium presented a thin layer of PAS positive, alcianophilic at pH 2.5 and orthochromatic material. Basement membranes and intercord extracellular substance were also PAS positive. Problems: Disorganization of the tissue structure was well manifest in irradiated gonads, accompanied by negativization of the histochemical reactions. A lymphocytic infiltration was also found. No structural alterations were observed in germ or epithelial cells.

It is concluded that laser radiations would act producing decrease of the mucosubstances associated to the plasma membrane and basement membrane.

They would also provoke the appearance of an inflammatory mononuclear infiltration.

3.1.3 Newborn chicken

3.1.3.1 Structural and cytochemical modifications in the lingual glands of the newborn chicken irradiated with He-Ne laser (Avila et al. 1997)

Despite the increasing and successful use of laser in Medicine and Odontology, the possible iatrogenic and otherwise deleterious side effects of this radiation remain mostly unknown. In previous studies, it was shown that both the embryonic and the post-hatched chicken constitute reliable experimental models for this type of studies.

Hence, the purpose of the present work was to analyze the structural and cytochemical alterations of the lingual glands of the newborn chicken irradiated with low energy He-Ne laser.

This laser produced regressive structural changes of the glands towards the embryonic stage as well as hyperplasia of the reserve glandular basal cells. Furthermore, a decrease in the glycoprotein content and a rise in the sulfated glycosaminoglycans were also found.

These results corroborate the pathogenic effects of the He-Ne- laser on the experimental model employed and, at the same time, emphasize the importance of considering, regarding clinical applications, possible previous neoplastic alterations as well as adverse reactions which might appear once laser therapy has been installed.

3.2 Intense Pulsed Light (IPL)
3.2.1 Tongue of the chicken embryo

3.2.1.1 Biological effects on the lingual cartilage of the chicken embryo

In 2009 we described the biological effects produced by intense pulsed light (Xe-Cl) on the cartilage of the tongue chick embryo using various filters. (Avila et al. 2009)

The Laser used correctly in the medical practice offers clear advantages compared with traditional therapies. The improvement and even the elimination of many significant skin lesions can be achieved with reduced risks to patients. However, it is important to keep security measures and understand the possible effects on an experimental model.

The chick embryo is a good model to evaluate the direct effects of non-ionizing radiation for its easy handling and availability. The purpose of this communication is to show our histological findings in organs of the chick embryo with and without protective barrier to be subjected to radiation excimer.

We used the following emitter: intense pulsed light (excimer Xe-Cl laser of 308 nm wavelength). It was irradiated embryos through an open window on eggshells. Aseptically the eggs were kept for 24 hours in an incubator. The protective barriers were used with and without colored glass, latex, cellophane, paper, polycarbonate of different colors and thicknesses.

The tissue changes observed are consistent with possible side effects of these fotothermical radiations we warned about possible side effects when they are applied indiscriminately. We believe it is important to explore different means to safeguard the safety of operators and patients. Figure 3.

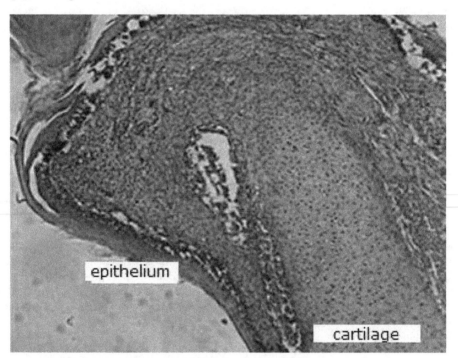

Fig. 3. Tongue of chicken 15 days of incubation in "ovo" irradiated with Intense Pulsed Light (Xe-Cl) 308 nm wavelength. Stain: hematoxylin and eosin. 5 x . The most striking findings consisted degenerative changes at the level of cartilage necrosis following groups glandular mucosa nature, accompanied by leukocyte infiltration, vascular wall thickening at the expense of the tunica media, perivascular edema.

3.2.2 Chick embryo wing

3.2.2.1 Histopathologic findings in epithelium and stroma of the chick embryo wing irradiated with intense pulsed light of 308 nm

Intense Pulsed Light (IPL) is used in medical practice enabling improved significantly even elimination of many skin lesions of patients. However, it is important to keep safety measures and know the possible effects on an experimental model. The chick embryo is a good model to evaluate the direct effects of ionizing radiation is its easy handling and availability.

The purpose of this communication is to show our histopathological findings in chick embryo wings with and without barrier protection when subjected to intense pulsed light radiation.

We used intense pulsed light of 308 nm wavelength Excilite model DEKA brand, Luz Excimer (XeCl) Wavelength 308 nm. Embryos were irradiated through an open window in the eggshell without barriers (controls) or barriers (problems). The eggs were maintained aseptically for 24 hours in an incubator. The barrier used was transparent cellophane without color and a green.

The most outstanding results obtained without a barrier, with clear cellophane and green were epithelial haematic fibrinous exudate and epithelial hyperplasia. In the stroma are varying degrees of vasocongestion, erythrocyte extravasation, and focal hemorrhage and edema, mononuclear infiltrate. It is concluded that tissue changes observed are consistent with possible changes produced by these collateral photothermal radiation, which warns of possible adverse effects when these are applied indiscriminately. Figure 4.

We believe it is important to study the means to ensure the safety of operators and patients.

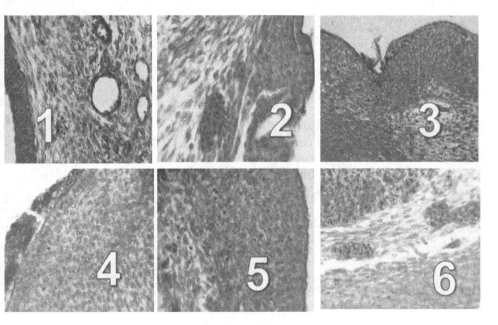

1. Control. 10 X H / E. 2. Problem:epithelial hyperplasia. 40 X. H / E. 3. Problem: Extravasation erythrocytes. 40 X H / E. 4. Problem: Fibrinous exudate 40 X. H / E. 5. Problem: epithelial hyperplasia, edema and mononuclear infiltrate (3), which migrate into the epithelium sectors 40 XH / E. 6. Issue: Stromal with vessels congestion and epithelial hyperplasia. 10 X. H / E

Fig. 4. Epithelium and stroma of the chick embryo wing 15 days irradiated with (problem) or without (control) intense pulsed light of 308 nm.

3.2.3 Heart of the chick embryo

3.2.3.1 Biological effects on the heart of the chick embryo

The 308 nm excimer laser is a new application in cardiology for the treatment of congenital heart malformations, vascular and ischemic cardiomyopathy. (Spencer & Hadi, S. M. 2004). This type of laser emits 85% of ultraviolet (UV) type B and 15% of UV radiation type A. However, the literature does not describe the changes that occur in myocardial cells and surrounding embryonic tissue when exposed to this type of laser.

Moreover, the intense pulsed light therapy with high-energy UVB spectrum is used in a manner similar to the excimer laser in different conditions. Intense Pulsed Light (IPL), is based on generating a polychromatic light source, high intensity incoherent.

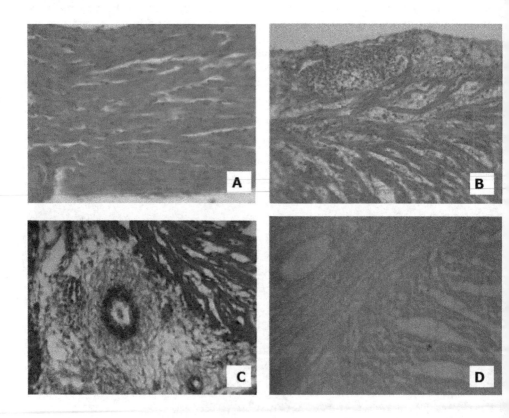

A. Control: cardiomyocytes. Coloration hematoxylin / eosin. B. Problem: disorganized tissue with interstitial edema, vasocongestion, exudate. Pericardium with leukocyte exudate. Coloration hematoxylin / eosin. C. Problem: Interstitium with blood vessels. Cardiomyocytes. Masson's Trichrome Staining. D. Problem: Interstitium and cardiomyocytes. Reduction of glycoproteins. PAS Staining

Fig. 5. Heart of the chick embryo 15 days days irradiated with (problem) or without (control) intense pulsed light of 308 nm.

4. Discussion and conclusion

The chick embryo is a good model to evaluate direct effects of ionizing radiation is its easy handling and availability.

We irradiated chick embryos and newly hatched chicks with He-Neon laser and infrared radiation (IR) and ultraviolet (UV) radiation found histopathological changes post. On the other hand, in the databases searched found no experimental studies on the biological effects of radiation from pulsed light of 308 nm (XeCl Excimer Light) in chick embryo organs. So also did experiments with intense pulsed light corroborating deleterious effects on organs of chicken embryos.

On the other hand, in the databases searched found no experimental studies on the biological effects of radiation from pulsed light of 308 nm (XeCl Excimer Light) in chick embryo organs. So also did experiments with intense pulsed light corroborating deleterious effects on organs of chicken embryos.

Authors studied the Influence of UV-B radiation on embryonic development of chickens Hampshire breed and considered that it is possible to state that short-lasting UV radiation appealing can have positive influence on organisms, which can be used in medicine for preventive end treating purposes. (Veterany et al. 2004)

Some authors (Schroeder et al. 2008) imply that IRA-radiation is capable of altering gene expression which brings forward a pro-aging phenotype of the skin. Apart from minimising exposure to natural IRA and responsible use of artificial IRA sources, the questions arises how a protection against the detrimental effects of IRA can be achieved.

The XeCl (IPL) lamp used in these experiments present a very monochromatic spectrum of emission of 307,94 nm and its effects could be very similar to those corresponding to a XeCl laser. The IPL is normally used for skin treatment for which the spectroscopic characterization carried out lets us determine the effects of this wave length on the tissues. (Nahavandi et al., 2008) The filters used basically acted only as intensity attenuators and they let us determine which optic material is more convenient for the protection of the personnel that uses this type of IPL.

It is because of this that we investigated the histopathological changes produced on an organ of the oral cavity of the chicken, the tongue and especially those centered in the cartilage, being similar to those described in prior publications produced in different organs of the embryo of chicken radiated with ultraviolet radiation, infrared and HeNe Lasers. (Samar et al., 1993, Avila et al., 1994 , Samar et al., 1995)

Clinically it is used as cartilage stimulus for its regeneration in the degenerative processes. It's been postulated that its action is based on the cell multiplication, the formation of collagen and elastic fibers, the vessel regeneration, the scar formation of bony tissues and the re-epithelization of the damaged tissue. (Baumanna et al., 2006 and Andreu & Zaldivar 1996).

Others authors. (Baumanna et al., 2006) studied the influence of laser radiation in human osteoartritic condrocites using different wave lengths (laser diode of 690nm and laser Nd: YAG of 1064 nm), power densities and exposition times and they observed that uaing a specific set of parameters (2 W; 16W/cm2; 60 s; 120 J) they observed an increase in the synthesis of the matrix (material of cartilage originating from 36 patients). They also describe that, using a very high power density, but a constant energy density, a reduction of the rate was produced (28%) of synthesis of the matrix.

By using only the filters with a transmittance grater than 30%, we could guarantee the possibility to observe the biological effects in tissue radiation.

Many countries have published safety guidelines for the use of lasers and most of them are harmonized with the international standards of the International Electrotecnic Commission (IEC). The maximum limit permitted of the exposure with the (MPE) laser used in international safety standards of the IEC are based on guidelines of the International Commission in the protection against the not-ionizing radiation (ICNIRP). (Moseley 1994; Sliney 2006; Parker,2007).

The CIS 825-1 norm (1993) is set for the manufacturers; However, it also offers some limitations oriented on the safety for users. All commercial lasers must exhibit the risk classification indicated.

We repeat that the use of IPL in spite of having similar biological effects than UV Lasers, do not have an equal standard safety classification.

The study of different types of radiation in our experimental model allowed to obtain information regarding the behavior of cells and tissues. We believe it is important to study the various means to ensure the safety of patients and operators. Also allow those responsible for advising on the lamps to regulate medical and aesthetic use.

We conclude that these tissue changes are compatible with photothermal side effects of this radiation that warn us about possible adverse effects of dose and time used experimentally.

5. References

Andreu, M. I. & Valiente Zaldivar, C. (1996). Efectos biológicos de la radiación láser de baja potencia en la reparación hística. *Rev Cubana Estomatol*, Vol. 33, No. 2, pp. 60-63. 9.

Avila, R. E., Ferraris, R. & Samar, M. E. (2001). Efectos biológicos inmediatos de las radiaciones ultravioletas en el mesonefros del embrión de pollo. *Rev Med Cordoba*, No. 89, pp. 9-11.

Avila, R. E., Plivelic, T., Samar, M. E. & Benavidez, E. (1996). Diseño de un sistema de detectores térmicos aplicado en huevos embrionados de pollo, irradiados con rayos infrarrojos. *Rev Fac Cienc Med Univ Nac Cordoba*, No. 54, pp. 13-18.

Avila, R. E., Samar, M. E., De Fabro, S. P., Leguina, M. & Juri, H. O. (1994). Histologic changes produced by non ionizing radiation in the chick embryo. *Rev. Fac. Cien. Med. Univ. Nac. Cordoba*, Vol 52 No.1, pp. 27-30.

Avila, R. E., Samar, M. E., de Fabro, S. P. & Plivelic, T. S. (1997). Structural and cytochemical modifications in the lingual glands of the newborn chicken irradiated with He-Ne laser. *Acta Odontol Latinoam*, Vol. 10, No. 2, pp. 81-88.

Avila, R. E., Samar, M. E., Fabro, S. P. de Juri, H. & Leguina, M. (1994) Alteraciones histológicas por radiaciones no ionizantes en el embrión de pollo. *Rev FacCienc Med Univ Nac Cordoba*, Vol. 52, No. 1, pp. 27-30.

Avila, R. E., Samar, M. E., Juri, H., Fabro, S. P. de, Centurión, C. & Sánchez Mazzaferro, R. (1992). Structural changes induced by He-Ne laser on the chick embryo ovary. *Rev Fac Cienc Med Univ NacCordoba*, No. 50, pp. 7-10.

Avila, R. E., Samar, M.E., Juri, H. & Fabro, S. P. de. (1992). Effects of He Ne laser
irradiation on chick embryo mesonephros. *J Clin Laser Med Surg*, No. 10, pp.
287-290.

Avila, R. E.; Samar, M. E.; Juri, H.; Femopase, G. A.; Hidalgo, M.; Ferrero, J. C.; Rinaldi, C.;
Fonseca, I. & Juri, G. (2009). Biological Effects Produced by Intense Pulsed Light
(Xe-Cl) on the Cartilage of the Tongue Chick Embryo Using Various Filters. *Int. J.
Morphol.*, 27(4):1003-1008.

Baumanna, M., Bjorn, J., Rohdec E., Bindigd, U., Mullerc, G. & Schellera, E. (2006). Influence
of wavelength, power density and exposure time of laser radiation on chondrocyte
cultures – An in-vitro investigation. *Med Laser Appl*, No. 21, pp. 191–198.

Chan, H. H., Yang, C. H., Leung, J. C., Wei, W. I. & Lai, K. N. (2007). An animal study of
the effects on p16 and PCNA expression of repeated treatment with high-
energy laser and intense pulsed light exposure. *Lasers Surg. Med.*, Vol. 39, No. 1,
pp. 8-13.

Moseley, H. (1994). Ultraviolet and laser radiation safety. *Phys. Med. Biol.*, No. 39, pp. 1765-
1799.

Nahavandi, H., Neumann, R., Holzer, G. & Knobler, R. (2008). Evaluation of safety and
efficacy of variable pulsed light in the treatment of unwanted hair in 77 volunteers.
J. Eur. Acad. Dermatol. Venereol., Vol. 22, No. 3, pp. 311-315.

Parker, S. (2007). Laser regulation and safety in general dental practice. *Br Dent J*. Vol. 12,
No. 202(9), pp. 523-532.

Samar, M. E., Avila, R. E., de Ferrais, M. E., Ferraris, R. V. & de Fabro, S. P. (1993).
Embryogeny of human labial glands: a structural, ultrastructural and cytochemical
study. *Acta Odontol Latinoam*, Vol. 7, No. 2, pp. 23-32.

Samar, M. E., Avila, R. E., Juri, H., Centurión, C. & Fabro, S. P. de. (1993). Histologycal
changes produced by He-Ne laser on different tissues from chick embryo. *J Clin
Laser Med Surg*, No. 11, pp. 87-89.

Samar, M. E., Avila, R. E., Juri, H., Plivelic, T. & Fabro, S. P. de. (1995). Histopathological
alterations induced by He-Ne laser in the salivary glands of the posthatched
chicken. *J Clin Laser Med Surg*, Vol. 13, No. 4, pp. 267-272.

Samar, M. E.; Avila, R. E. & Juri, H. (1993). Histological changes produced by He-Ne
laser on different tissues from chick embryo. *J. Clin. Laser Med. Surg.*, No. 11, pp.
87-89.

Schroeder, P., Haendeler, J. & Krutmann. (2008). The role of near infrared radiation in
photoaging of the skinJ. *Exp Gerontol*, Vol. 43, No. 7, pp. 629-632.

Schoenewolf, N. L., Barysch, M. J. & Dummer, R. (2011). Intense pulsed light. *Curr Probl
Dermatol*. No. 42, pp. 166-172.

Sliney D. H. (2006) Risks of occupational exposure to optical radiation. *Med Lav, Vol. 97, No.
2, pp. 215-20.

Spencer, J. M. & Hadi, S. M. (2004). The excimer lasers. *J. Drugs Dermatol.*, No. 5, pp.
522-525.

Town, G., Ash, C., Eadie, E. & Moseley, H. (2007). Measuring key parameters of intense
pulsed light (IPL) devices. *J. Cosmet. Laser Ther.*, Vol. 9, No. 3, pp. 148-160.

Veterány, L., Hluchý, S. & Veterányová, A. (2004). The Influence of Ultra-Violet Radiation on Chicken Hatching. *J Environ Sci Health A Tox Hazard Subst Environ Eng*. Vol. 39, No. 9, pp. 2333-2339.

Permissions

The contributors of this book come from diverse backgrounds, making this book a truly international effort. This book will bring forth new frontiers with its revolutionizing research information and detailed analysis of the nascent developments around the world.

We would like to thank Dr. Mitsuru Nenoi, for lending his expertise to make the book truly unique. He has played a crucial role in the development of this book. Without his invaluable contribution this book wouldn't have been possible. He has made vital efforts to compile up to date information on the varied aspects of this subject to make this book a valuable addition to the collection of many professionals and students.

This book was conceptualized with the vision of imparting up-to-date information and advanced data in this field. To ensure the same, a matchless editorial board was set up. Every individual on the board went through rigorous rounds of assessment to prove their worth. After which they invested a large part of their time researching and compiling the most relevant data for our readers. Conferences and sessions were held from time to time between the editorial board and the contributing authors to present the data in the most comprehensible form. The editorial team has worked tirelessly to provide valuable and valid information to help people across the globe.

Every chapter published in this book has been scrutinized by our experts. Their significance has been extensively debated. The topics covered herein carry significant findings which will fuel the growth of the discipline. They may even be implemented as practical applications or may be referred to as a beginning point for another development. Chapters in this book were first published by InTech; hereby published with permission under the Creative Commons Attribution License or equivalent.

The editorial board has been involved in producing this book since its inception. They have spent rigorous hours researching and exploring the diverse topics which have resulted in the successful publishing of this book. They have passed on their knowledge of decades through this book. To expedite this challenging task, the publisher supported the team at every step. A small team of assistant editors was also appointed to further simplify the editing procedure and attain best results for the readers.

Our editorial team has been hand-picked from every corner of the world. Their multi-ethnicity adds dynamic inputs to the discussions which result in innovative outcomes. These outcomes are then further discussed with the researchers and contributors who give their valuable feedback and opinion regarding the same. The feedback is then collaborated with the researches and they are edited in a comprehensive manner to aid the understanding of the subject.

Apart from the editorial board, the designing team has also invested a significant amount of their time in understanding the subject and creating the most relevant covers. They scrutinized every image to scout for the most suitable representation of the subject and create an appropriate cover for the book.

The publishing team has been involved in this book since its early stages. They were actively engaged in every process, be it collecting the data, connecting with the contributors or procuring relevant information. The team has been an ardent support to the editorial, designing and production team. Their endless efforts to recruit the best for this project, has resulted in the accomplishment of this book. They are a veteran in the field of academics and their pool of knowledge is as vast as their experience in printing. Their expertise and guidance has proved useful at every step. Their uncompromising quality standards have made this book an exceptional effort. Their encouragement from time to time has been an inspiration for everyone.

The publisher and the editorial board hope that this book will prove to be a valuable piece of knowledge for researchers, students, practitioners and scholars across the globe.

List of Contributors

Otto G. Raabe
University of California Davis, USA

Varut Vardhanabhuti and Carl A. Roobottom
Derriford Hospital and Peninsula College of Medicine and Dentistry, Plymouth, United Kingdom

L. A. Darchuk, A. Worobiec and R. Van Grieken
University of Antwerp, Antwerp, Belgium

L. V. Zaverbna
National University of Medicine, Lviv, Ukraine

Michela Zacchino and Fabrizio Calliada
Fondazione IRCCS Policlinico "San Matteo", Radiology Department, Pavia - Piazzale Golgi, Pavia, Italy

Yehuwdah E. Chad-Umoren
Department of Physics, University of Port Harcourt, Choba, Port Harcourt, Nigeria

Nicolaas A. P. Franken
Academic Medical Centre, University of Amsterdam, Laboratory for Experimental Oncology and Radiobiology (LEXOR), Centre for Experimental Molecular Medicine, Department of Radiation Oncology, The Netherlands

Suzanne Hovingh, Arlene Oei, Paul Cobussen, Chris van Bree, Hans Rodermond, Lukas Stalpers, Petra Kok, Gerrit W. Barendsen and Johannes Crezee
Academic Medical Centre, University of Amsterdam, Laboratory for Experimental Oncology and Radiobiology (LEXOR), Centre for Experimental Molecular Medicine, Department of Radiation Oncology, The Netherlands

Gerrit W. Barendsen
Department of Cell Biology and Histology, Amsterdam, The Netherlands

Judith W. J. Bergs
Institute of Molecular Biology and Tumor Research, Philipps-University Marburg, Emil-Mannkopff-Str. 2, Marburg, Germany

Maria Konopacka and Jacek Rogoliński
Center for Translational Research and Molecular Biology of Cancer, Poland

Krzysztof Ślosarek
Department of Radiotherapy and Brachytherapy Treatment Planning, Maria Sklodowska-Curie Memorial Cancer Center and Institute of Oncology, Gliwice, Poland

M. A. El-Missiry and A. I. Othman
Department of Zoology, Faculty of Sciences, Mansoura University, Mansoura, Egypt

M. A. Alabdan
Department of Zoology, Faculty of Science, Princess Nora Bint AbdulRahman University, Riyadh, Kingdom of Saudi Arabia

Shazia Naser-Ud-Din
School of Dentistry, University of Queensland, Brisbane, Australia

Hailun Wang, Miaojun Han and Zhaozhong Han
Department of Radiation Oncology and Cancer Biology, Vanderbilt-Ingrim Cancer Center, School of Medicine, Vanderbilt University, Nashville, TN, USA

Rodolfo Esteban Avila, Maria Elena Samar, Gustavo Juri, Juan Carlos Ferrero and Hugo Juri
Department of Cell Biology, Histology and Embryology, Faculty of Medical Sciences, National University of Cordoba, Argentina